Quantum Field Theory Demystified

Demystified Series

Accounting Demystified
Advanced Calculus Demystified
Advanced Physics Demystified
Advanced Statistics Demystified
Algebra Demystified
Alternative Energy Demystified
Anatomy Demystified
asp.net 2.0 Demystified
Astronomy Demystified
Audio Demystified
Biology Demystified
Biotechnology Demystified
Business Calculus Demystified
Business Math Demystified
Business Statistics Demystified
C++ Demystified
Calculus Demystified
Chemistry Demystified
Circuit Analysis Demystified
College Algebra Demystified
Corporate Finance Demystified
Data Structures Demystified
Databases Demystified
Diabetes Demystified
Differential Equations Demystified
Digital Electronics Demystified
Earth Science Demystified
Electricity Demystified
Electronics Demystified
Engineering Statistics Demystified
Environmental Science Demystified
Everyday Math Demystified
Fertility Demystified
Financial Planning Demystified
Forensics Demystified
French Demystified
Genetics Demystified
Geometry Demystified
German Demystified
Global Warming and Climate Change Demystified
Hedge Funds Demystified
Home Networking Demystified
Investing Demystified
Italian Demystified
Java Demystified
JavaScript Demystified
Lean Six Sigma Demystified

Linear Algebra Demystified
Macroeconomics Demystified
Management Accounting Demystified
Math Proofs Demystified
Math Word Problems Demystified
MATLAB® Demystified
Medical Billing and Coding Demystified
Medical Terminology Demystified
Meteorology Demystified
Microbiology Demystified
Microeconomics Demystified
Nanotechnology Demystified
Nurse Management Demystified
OOP Demystified
Options Demystified
Organic Chemistry Demystified
Personal Computing Demystified
Pharmacology Demystified
Physics Demystified
Physiology Demystified
Pre-Algebra Demystified
Precalculus Demystified
Probability Demystified
Project Management Demystified
Psychology Demystified
Quality Management Demystified
Quantum Field Theory Demystified
Quantum Mechanics Demystified
Real Estate Math Demystified
Relativity Demystified
Robotics Demystified
Sales Management Demystified
Signals and Systems Demystified
Six Sigma Demystified
Spanish Demystified
SQL Demystified
Statics and Dynamics Demystified
Statistics Demystified
Technical Analysis Demystified
Technical Math Demystified
Trigonometry Demystified
UML Demystified
Visual Basic 2005 Demystified
Visual C# 2005 Demystified
Vitamins and Minerals Demystified
XML Demystified

Quantum Field
Theory Demystified

David McMahon

New York   Chicago   San Francisco   Lisbon   London
Madrid   Mexico City   Milan   New Delhi   San Juan
Seoul   Singapore   Sydney   Toronto

5 6 7 8 9 0   QFR/QFR   0 1 4 3

ISBN 978-0-07-154382-8
MHID 0-07-154382-1

Sponsoring Editor
Judy Bass

Production Supervisor
Pamela A. Pelton

Editing Supervisor
Stephen M. Smith

Project Manager
Sam RC (International Typesetting
and Composition)

Copy Editor
Priyanka Sinha (International
Typesetting and Composition)

Proofreader
Megha RC (International Typesetting
and Composition)

Indexer
Broccoli Information Management

Art Director, Cover
Jeff Weeks

Composition
International Typesetting and
Composition

Printed and bound by Quebecor/Fairfield.

McGraw-Hill books are available at special quantity discounts to use as premiums and sales promotions, or for use in corporate training programs. To contact a special sales representative, please visit the Contact Us page at www.mhprofessional.com.

ABOUT THE AUTHOR

David McMahon works as a researcher at Sandia National Laboratories. He has advanced degrees in physics and applied mathematics, and is the author of *Quantum Mechanics Demystified, Relativity Demystified, MATLAB® Demystified*, and several other successful books.

CONTENTS AT A GLANCE

CHAPTER 1	Particle Physics and Special Relativity	1
CHAPTER 2	Lagrangian Field Theory	23
CHAPTER 3	An Introduction to Group Theory	49
CHAPTER 4	Discrete Symmetries and Quantum Numbers	71
CHAPTER 5	The Dirac Equation	85
CHAPTER 6	Scalar Fields	109
CHAPTER 7	The Feynman Rules	139
CHAPTER 8	Quantum Electrodynamics	163
CHAPTER 9	Spontaneous Symmetry Breaking and the Higgs Mechanism	187
CHAPTER 10	Electroweak Theory	209
CHAPTER 11	Path Integrals	233
CHAPTER 12	Supersymmetry	245
	Final Exam	263
	Solutions to Quizzes and Final Exam	281
	References	289
	Index	291

CONTENTS

Preface   xv

CHAPTER 1   **Particle Physics and Special Relativity**   **1**

Special Relativity   5

A Quick Overview of Particle Physics   12

Elementary Particles   14

The Higgs Mechanism   18

Grand Unification   18

Supersymmetry   19

String Theory   19

Summary   20

Quiz   20

CHAPTER 2   **Lagrangian Field Theory**   **23**

Basic Lagrangian Mechanics   23

The Action and the Equations of Motion   26

Canonical Momentum and the Hamiltonian   29

Lagrangian Field Theory   30

Symmetries and Conservation Laws   35

Conserved Currents   38

The Electromagnetic Field   39

Gauge Transformations   43

Summary   47

Quiz   47

CHAPTER 3	**An Introduction to Group Theory**	**49**
	Representation of the Group	50
	Group Parameters	52
	Lie Groups	52
	The Rotation Group	54
	Representing Rotations	55
	SO(N)	58
	Unitary Groups	62
	Casimir Operators	67
	Summary	68
	Quiz	68
CHAPTER 4	**Discrete Symmetries and Quantum Numbers**	**71**
	Additive and Multiplicative Quantum Numbers	71
	Parity	72
	Charge Conjugation	76
	CP Violation	78
	The CPT Theorem	80
	Summary	82
	Quiz	83
CHAPTER 5	**The Dirac Equation**	**85**
	The Classical Dirac Field	85
	Adding Quantum Theory	87
	The Form of the Dirac Matrices	89
	Some Tedious Properties of the Dirac Matrices	91
	Adjoint Spinors and Transformation Properties	94
	Slash Notation	95
	Solutions of the Dirac Equation	95
	Free Space Solutions	99
	Boosts, Rotations, and Helicity	103
	Weyl Spinors	104
	Summary	107
	Quiz	108

CHAPTER 6	**Scalar Fields**	**109**
	Arriving at the Klein-Gordon Equation	110
	Reinterpreting the Field	117
	Field Quantization of Scalar Fields	117
	States in Quantum Field Theory	127
	Positive and Negative Frequency Decomposition	128
	Number Operators	128
	Normalization of the States	130
	Bose-Einstein Statistics	131
	Normal and Time-Ordered Products	134
	The Complex Scalar Field	135
	Summary	137
	Quiz	137
CHAPTER 7	**The Feynman Rules**	**139**
	The Interaction Picture	141
	Perturbation Theory	143
	Basics of the Feynman Rules	146
	Calculating Amplitudes	151
	Steps to Construct an Amplitude	153
	Rates of Decay and Lifetimes	160
	Summary	160
	Quiz	160
CHAPTER 8	**Quantum Electrodynamics**	**163**
	Reviewing Classical Electrodynamics Again	165
	The Quantized Electromagnetic Field	168
	Gauge Invariance and QED	170
	Feynman Rules for QED	173
	Summary	185
	Quiz	185
CHAPTER 9	**Spontaneous Symmetry Breaking and the Higgs Mechanism**	**187**
	Symmetry Breaking in Field Theory	189

Mass Terms in the Lagrangian   192
Aside on Units   195
Spontaneous Symmetry Breaking and Mass   196
Lagrangians with Multiple Particles   199
The Higgs Mechanism   202
Summary   207
Quiz   207

CHAPTER 10   Electroweak Theory   209
Right- and Left-Handed Spinors   210
A Massless Dirac Lagrangian   211
Leptonic Fields of the Electroweak Interactions   212
Charges of the Electroweak Interaction   213
Unitary Transformations and the Gauge Fields
    of the Theory   215
Weak Mixing or Weinberg Angle   219
Symmetry Breaking   220
Giving Mass to the Lepton Fields   222
Gauge Masses   224
Summary   231
Quiz   231

CHAPTER 11   Path Integrals   233
Gaussian Integrals   233
Basic Path Integrals   238
Summary   242
Quiz   243

CHAPTER 12   Supersymmetry   245
Basic Overview of Supersymmetry   246
Supercharge   247
Supersymmetric Quantum Mechanics   249
The Simplified Wess-Zumino Model   253
A Simple SUSY Lagrangian   254
Summary   260
Quiz   260

Contents

Final Exam   263

Solutions to Quizzes and Final Exam   281

References   289

Index   291

PREFACE

Quantum field theory is the union of Einstein's special relativity and quantum mechanics. It forms the foundation of what scientists call the standard model, which is a theoretical framework that describes all known particles and interactions with the exception of gravity. There is no time like the present to learn it—the Large Hadron Collider (LHC) being constructed in Europe will test the final pieces of the standard model (the Higgs mechanism) and look for physics beyond the standard model. In addition quantum field theory forms the theoretical underpinnings of string theory, currently the best candidate for unifying all known particles and forces into a single theoretical framework.

Quantum field theory is also one of the most difficult subjects in science. This book aims to open the door to quantum field theory to as many interested people as possible by providing a simplified presentation of the subject. This book is useful as a supplement in the classroom or as a tool for self-study, but be forewarned that the book includes the math that comes along with the subject.

By design, this book is not thorough or complete, and it might even be considered by some "experts" to be shallow or filled with tedious calculations. But this book is not written for the experts or for brilliant graduate students at the top of the class, it is written for those who find the subject difficult or impossible. Certain aspects of quantum field theory have been selected to introduce new people to the subject, or to help refresh those who have been away from physics.

After completing this book, you will find that studying other quantum field theory books will be easier. You can master quantum field theory by tackling the reference list in the back of this book, which includes a list of textbooks used in the development of this one. Frankly, while all of those books are very good and make fine references, most of them are hard to read. In fact many quantum field theory books are impossible to read. My recommendation is to work through this book first, and then tackle *Quantum Field Theory in a Nutshell* by Anthony Zee. Different than all other books on the subject, it's very readable and is packed with great

physical insight. After you've gone through that book, if you are looking for mastery or deep understanding you will be well equipped to tackle the other books on the list.

Unfortunately, learning quantum field theory entails some background in physics and math. The bottom line is, I assume you have it. The background I am expecting includes quantum mechanics, some basic special relativity, some exposure to electromagnetics and Maxwell's equations, calculus, linear algebra, and differential equations. If you lack this background do some studying in these subjects and then give this book a try.

Now let's forge ahead and start learning quantum field theory.

David McMahon

Quantum Field
Theory Demystified

CHAPTER 1

Particle Physics and Special Relativity

Quantum field theory is a theoretical framework that combines *quantum mechanics* and *special relativity*. Generally speaking, quantum mechanics is a theory that describes the behavior of small systems, such as atoms and individual electrons. Special relativity is the study of high energy physics, that is, the motion of particles and systems at velocities near the speed of light (but without gravity). What follows is an introductory discussion to give you a flavor of what quantum field theory is like. We will explore each concept in more detail in the following chapters.

There are three key ideas we want to recall from quantum mechanics, the first being that physical observables are *mathematical operators* in the theory.

For instance, the Hamiltonian (i.e., the energy) of a *simple harmonic oscillator* is the operator

$$\hat{H} = \hbar\omega\left(\hat{a}^{\dagger}\hat{a} + \frac{1}{2}\right)$$

where $\hat{a}^{\dagger}$, $\hat{a}$ are the creation and annihilation operators, and $\hbar$ is Planck's constant.

The second key idea you should remember from quantum mechanics is the uncertainty principle. The uncertainty relation between the position operator $\hat{x}$ and the momentum operator $\hat{p}$ is

$$\Delta\hat{x}\,\Delta\hat{p} \geq \frac{\hbar}{2} \tag{1.1}$$

There is also an uncertainty relation between energy and time.

$$\Delta E\,\Delta t \geq \frac{\hbar}{2} \tag{1.2}$$

When considering the uncertainty relation between energy and time, it's important to remember that time is only a parameter in nonrelativistic quantum mechanics, not an operator.

The final key idea to recall from quantum mechanics is the commutation relations. In particular,

$$[\hat{x}, \hat{p}] = \hat{x}\hat{p} - \hat{p}\hat{x} = i\hbar$$

Now let's turn to special relativity. We can jump right to Einstein's famous equation that every lay person knows something about, in order to see how special relativity is going to impact quantum theory. This is the equation that relates energy to mass.

$$E = mc^2 \tag{1.3}$$

What should you take away from this equation? The thing to notice is that if there is enough energy—that is, enough energy proportional to a given particle's mass as described by Eq. (1.3)—then we can "create" the particle. Due to conservation laws, we actually need twice the particle's mass, so that we can create a particle and its antiparticle. So in high energy processes,

- Particle number is not fixed.
- The types of particles present are not fixed.

These two facts are in direct conflict with nonrelativistic quantum mechanics. In nonrelativistic quantum mechanics, we describe the dynamics of a system with the

Schrödinger equation, which for a particle moving in one dimension with a potential V is

$$-\frac{\hbar^2}{2m}\frac{\partial^2\psi}{\partial x^2} + V\psi = i\hbar\frac{\partial\psi}{\partial t} \tag{1.4}$$

We can extend this formalism to treat the case when several particles are present. However, the number and types of particles are absolutely fixed. The Schrödinger equation cannot in any shape or form handle changing particle number or new types of particles appearing and disappearing as relativity allows.

In fact, there is no wave equation of the type we are used to from nonrelativistic quantum mechanics that is truly compatible with both relativity and quantum theory. Early attempts to merge quantum mechanics and special relativity focused on generating a relativistic version of the Schrödinger equation. In fact, Schrödinger himself derived a relativistic equation prior to coming up with the wave equation he is now famous for. The equation he derived, which was later discovered independently by Klein and Gordon (and is now known as the Klein-Gordon equation) is

$$\frac{1}{c^2}\frac{\partial^2\varphi}{\partial t^2} - \frac{\partial^2\varphi}{\partial x^2} = \frac{m^2c^2}{\hbar^2}\varphi$$

We will have more to say about this equation in future chapters. Schrödinger discarded it because it gave the wrong fine structure for the hydrogen atom. It is also plagued by an unwanted feature—it appears to give negative probabilities, something that obviously contradicts the spirit of quantum mechanics. This equation also has a funny feature—it allows negative energy states.

The next attempt at a relativistic quantum mechanics was made by Dirac. His famous equation is

$$i\hbar\frac{\partial\psi}{\partial t} = -i\hbar c\vec{\alpha}\cdot\vec{\nabla}\psi + \beta mc^2\psi$$

Here, $\vec{\alpha}$ and β are actually matrices. This equation, which we will examine in detail in later chapters, resolves some of the problems of the Klein-Gordon equation but also allows for negative energy states.

As we will emphasize later, part of the problem with these relativistic wave equations is in their interpretation. We move forward into a quantum theory of fields by changing how we look at things. In particular, in order to be truly compatible with special relativity we need to discard the notion that φ and ψ in the

Klein-Gordon and Dirac equations, respectively describe single particle states. In their place, we propose the following new ideas:

- The wave functions φ and ψ are not wave functions at all, instead they are *fields*.
- The fields are operators that can create new particles and destroy particles.

Since we have promoted the fields to the status of operators, they must satisfy commutation relations. We will see later that we make a transition of the type

$$[\hat{x}, \hat{p}] \rightarrow \left[\hat{\varphi}(x,t), \hat{\pi}(y,t)\right]$$

Here, $\hat{\pi}(y,t)$ is another field that plays the role of momentum in quantum field theory. Since we are transitioning to the continuum, the commutation relation will be of the form

$$\left[\hat{\varphi}(x,t), \hat{\pi}(y,t)\right] = i\hbar\delta(x-y)$$

where x and y are two points in space. This type of relation holds within it the notion of causality so important in special relativity—if two fields are spatially separated they cannot affect one another.

With fields promoted to operators, you might wonder what happens to the ordinary operators of quantum mechanics. There is one important change you should make sure to keep in mind. In quantum mechanics, position $\hat{x}$ is an operator while time t is just a parameter. In relativity, since time and position are on a similar footing, we might expect that in relativistic quantum mechanics we would also put time and space on a similar footing. This could mean promoting time to an operator $\hat{t}$. This is not what is done in ordinary quantum field theory, where we take the opposite direction—and demote position to a parameter x. So in quantum field theory,

- Fields φ and ψ are operators.
- They are parameterized by spacetime points (x, t).
- Position x and time t are just numbers that fix a point in spacetime—they are not operators.
- Momentum continues to play a role as an operator.

In quantum field theory, we frequently use tools from classical mechanics to deal with fields. Specifically, we often use the Lagrangian

$$L = T - V \tag{1.5}$$

The Lagrangian is important because symmetries (such as rotations) leave the form of the Lagrangian invariant. The classical path taken by a particle is the one which minimizes the action.

$$S = \int L \, dt \qquad (1.6)$$

We will see how these methods are applied to fields in Chap. 2.

Special Relativity

The arena in which quantum field theory operates is the high energy domain of special relativity. Therefore, brushing up on some basic concepts in special relativity and familiarizing ourselves with some notation is important to gain some understanding of quantum field theory.

Special relativity is based on two simple postulates. Simply stated, these are:

- The laws of physics are the same for all inertial observers.
- The speed of light c is a constant.

An *inertial frame of reference* is one for which Newton's first law holds. In special relativity, we characterize spacetime by an *event*, which is something that happens at a particular time t and some spatial location (x, y, z). Also notice that the speed of light c can serve in a role as a conversion factor, transforming time into space and vice versa. Space and time therefore form a unified framework and we denote coordinates by (ct, x, y, z).

One consequence of the second postulate is the *invariance of the interval*. In special relativity, we measure distance in space and time together. Imagine a flash of light emitted at the origin at $t = 0$. At some later time t the spherical wavefront of the light can be described by

$$c^2 t^2 = x^2 + y^2 + z^2$$
$$\Rightarrow \quad c^2 t^2 - x^2 - y^2 - z^2 = 0$$

Since the speed of light is invariant, this equation must also hold for another observer, who is measuring coordinates with respect to a frame we denote by (ct', x', y', z'). That is,

$$c^2 t'^2 - x'^2 - y'^2 - z'^2 = 0$$

It follows that

$$c^2t^2 - x^2 - y^2 - z^2 = c^2t'^2 - x'^2 - y'^2 - z'^2$$

Now, in ordinary space, the differential distance from the origin to some point (x, y, z) is given by

$$dr^2 = dx^2 + dy^2 + dz^2$$

We define an analogous concept in spacetime, called the *interval*. This is denoted by ds^2 and is written as

$$ds^2 = c^2dt^2 - dx^2 - dy^2 - dz^2 \qquad (1.7)$$

From Eq. (1.7) it follows that the interval is invariant. Consider two observers in two different inertial frames. Although they measure different spatial coordinates (x, y, z) and (x', y', z') and different time coordinates t and t' to label events, the interval for each observer is the same, that is,

$$ds^2 = c^2dt^2 - dx^2 - dy^2 - dz^2 = c^2dt'^2 - dx'^2 - dy'^2 - dz'^2 = ds'^2$$

This is a consequence of the fact that the speed of light is the same for all inertial observers.

It is convenient to introduce an object known as the *metric*. The metric can be used to write down the coefficients of the differentials in the interval, which in this case are just +/−1. The metric of special relativity ("flat space") is given by

$$\eta^{\mu\nu} = \begin{pmatrix} 1 & 0 & 0 & 0 \\ 0 & -1 & 0 & 0 \\ 0 & 0 & -1 & 0 \\ 0 & 0 & 0 & -1 \end{pmatrix} \qquad (1.8)$$

The metric has an inverse, which in this case turns out to be the same matrix. We denote the inverse with lowered indices as

$$\eta_{\mu\nu} = \begin{pmatrix} 1 & 0 & 0 & 0 \\ 0 & -1 & 0 & 0 \\ 0 & 0 & -1 & 0 \\ 0 & 0 & 0 & -1 \end{pmatrix}$$

The symbol $\eta_{\mu\nu}$ is reserved for the metric of special relativity. More generally, the metric is denoted by $g_{\mu\nu}$. This is the convention that we will follow in this book. We have

$$g_{\mu\nu}g^{\nu\rho} = \delta_\mu^\rho \qquad (1.9)$$

where δ_μ^ρ is the *Kronecker delta function* defined by

$$\delta_\mu^\rho = \begin{cases} 1 \text{ if } \mu = \rho \\ 0 \text{ if } \mu \neq \rho \end{cases}$$

Hence Eq. (1.9) is just a statement that

$$gg^{-1} = I$$

where I is the identity matrix.

In relativity, it is convenient to label coordinates by a number called an *index*. We take $ct = x^0$ and $(x, y, z) \rightarrow (x^1, x^2, x^3)$. Then an event in spacetime is labeled by the coordinates of a *contravariant* vector.

$$x^\mu = (x^0, x^1, x^2, x^3) \qquad (1.10)$$

Contravariant refers to the way the vector transforms under a Lorentz transformation, but just remember that a contravariant vector has raised indices. A *covariant* vector has lowered indices as

$$x_\mu = (x_0, x_1, x_2, x_3)$$

An index can be raised or lowered using the metric. Specifically,

$$x_\alpha = g_{\alpha\beta}x^\beta \qquad x^\alpha = g^{\alpha\beta}x_\beta \qquad (1.11)$$

Looking at the metric, you can see that the components of a covariant vector are related to the components of a contravariant vector by a change in sign as

$$x_0 = x^0 \qquad x_1 = -x^1 \qquad x_2 = -x^2 \qquad x_3 = -x^3$$

We use the *Einstein summation convention* to represent sums. When an index is repeated in an expression once in a lowered position and once in a raised position, this indicates a sum, that is,

$$s_\alpha s^\alpha \equiv \sum_{\alpha=0}^{3} s_\alpha s^\alpha = s_0 s^0 + s_1 s^1 + s_2 s^2 + s_3 s^3$$

So for example, the index lowering expression in Eq. (1.11) is really shorthand for

$$x_\alpha = g_{\alpha\beta}x^\beta = g_{\alpha 0}x^0 + g_{\alpha 1}x^1 + g_{\alpha 2}x^2 + g_{\alpha 3}x^3$$

Greek letters such as α, β, μ, and ν are taken to range over all spacetime indices, that is, $\mu = 0, 1, 2,$ and 3. If we want to reference spatial indices only, a Latin letter such as $i, j,$ and k is used. That is, $i = 1, 2,$ and 3.

LORENTZ TRANSFORMATIONS

A *Lorentz transformation* Λ allows us to transform between different inertial reference frames. For simplicity, consider an inertial reference frame x'^μ moving along the x axis with respect to another inertial reference frame x^μ with speed $v < c$. If we define

$$\beta = \frac{v}{c} \qquad \gamma = \frac{1}{\sqrt{1-\beta^2}} \qquad (1.12)$$

Then the Lorentz transformation that connects the two frames is given by

$$\Lambda^\mu{}_\nu = \begin{pmatrix} \gamma & -\beta\gamma/c & 0 & 0 \\ -\beta\gamma/c & \gamma & 0 & 0 \\ 0 & 0 & 1 & 0 \\ 0 & 0 & 0 & 1 \end{pmatrix} \qquad (1.13)$$

Specifically,

$$x'^0 = \gamma\left(x^0 - \frac{\beta}{c}x^1\right)$$

$$x'^1 = \gamma\left(x^1 - \frac{\beta}{c}x^0\right) \qquad (1.14)$$

$$x'^2 = x^2$$

$$x'^3 = x^3$$

We can write a compact expression for a Lorentz transformation relating two sets of coordinates as

$$x'^\mu = \Lambda^\mu{}_\nu x^\nu \qquad (1.15)$$

The *rapidity* ϕ is defined as

$$\tanh \phi = \beta = \frac{v}{c} \tag{1.16}$$

Using the rapidity, we can view a Lorentz transformation as a kind of rotation (mathematically speaking) that rotates time and spatial coordinates into each other, that is,

$$x'^0 = -x^1 \sinh \phi + x^0 \cosh \phi$$
$$x'^1 = x^1 \cosh \phi - x^0 \sinh \phi$$

Changing velocity to move from one inertial frame to another is done by a Lorentz transformation and we refer to this as a *boost*.

We can extend the shorthand index notation used for coordinates to derivatives. This is done with the following definition:

$$\frac{\partial}{\partial t} \rightarrow \frac{\partial}{\partial x^0} = \partial_0 \qquad \frac{\partial}{\partial x} \rightarrow \frac{\partial}{\partial x^1} = \partial_1$$
$$\frac{\partial}{\partial y} \rightarrow \frac{\partial}{\partial x^2} = \partial_2 \qquad \frac{\partial}{\partial z} \rightarrow \frac{\partial}{\partial x^3} = \partial_3$$

We can raise an index on these expressions so that

$$\partial^\mu = g^{\mu\nu} \partial_\nu$$
$$\partial^0 = \partial_0 \qquad \partial^i = -\partial_i$$

In special relativity many physical vectors have spatial and time components. We call such objects *4-vectors* and denote them with italic font (sometimes with an index) reserving the use of an arrow for the spatial part of the vector. An arbitrary 4-vector A^μ has components

$$A^\mu = (A^0, A^1, A^2, A^3)$$
$$A_\mu = (A_0, -A_1, -A_2, -A_3)$$
$$A^\mu = (A^0, \vec{A})$$
$$A_\mu = (A_0, -\vec{A})$$

We denote the ordinary vector part of a 4-vector as a 3-vector. So the 3-vector part of A^μ is $\bar{A}$. The magnitude of a vector is computed using a generalized dot product, like

$$A \cdot A = A^\mu A_\mu = A^0 A_0 - A^1 A_1 - A^2 A_2 - A^3 A_3$$
$$= g^{\mu\nu} A_\mu A_\nu$$

This magnitude is a *scalar*, which is invariant under Lorentz transformations. When a quantity is invariant under Lorentz transformations, all inertial observers agree on its value which we call the *scalar product*. A consequence of the fact that the scalar product is invariant, meaning that $x'^\mu x'_\mu = x^\mu x_\mu$, is

$$\Lambda^{\alpha\beta} \Lambda_{\alpha\mu} = \delta^\beta_\mu \tag{1.17}$$

Now let's consider derivatives using relativistic notation. The derivative of a field is written as

$$\frac{\partial\varphi}{\partial x^\mu} = \partial_\mu \varphi \tag{1.18}$$

The index is lowered because as written, the derivative is a covariant 4-vector. The components of the vector are

$$\left(\frac{\partial\varphi}{\partial x^0}, \frac{\partial\varphi}{\partial x^1}, \frac{\partial\varphi}{\partial x^2}, \frac{\partial\varphi}{\partial x^3} \right)$$

We also have

$$\frac{\partial\varphi}{\partial x_\mu} = \partial^\mu \varphi$$

which is a contravariant 4-vector. Like any 4-vector, we can compute a scalar product, which is the four-dimensional generalization of the Laplacian called the *D'Alembertian operator* which using ordinary notation is $\frac{1}{c^2}\frac{\partial^2}{\partial t^2} - \nabla^2 \equiv \Box$. Using the relativistic notation for derivatives together with the generalized dot product we have

$$\partial^\mu \partial_\mu = \frac{1}{c^2}\frac{\partial^2}{\partial t^2} - \nabla^2 \equiv \Box \tag{1.19}$$

One 4-vector that is of particular importance is the *energy-momentum 4-vector* which unifies the energy and momentum into a single object. This is given by

$$p_\mu = (E, -\vec{p}) = (E, -p_1, -p_2, -p_3)$$
$$\Rightarrow p^\mu = (E, \vec{p}) = (E, p_1, p_2, p_3) \qquad (1.20)$$

The magnitude of the energy-momentum 4-vector gives us the Einstein relation connecting energy, momentum, and mass.

$$E^2 = \vec{p}^2 c^2 + m^2 c^4 \qquad (1.21)$$

We can always choose a Lorentz transformation to boost to a frame in which the 3-momentum of the particle is zero $\vec{p} = 0$ giving Einstein's famous relation between energy and rest mass, like

$$E = mc^2$$

Another important 4-vector is the current 4-vector J. The time component of this vector is the charge density ρ while the 3-vector part of J is the current density $\vec{J}$. That is,

$$J^\mu = (\rho, J_x, J_y, J_z) \qquad (1.22)$$

The current 4-vector is conserved, in the sense that

$$\partial_\mu J^\mu = 0 \qquad (1.23)$$

which is nothing other than the familiar relation for conservation of charge as shown here.

$$\partial_\mu J^\mu = \partial_t \rho + \partial_x J^x + \partial_y J^y + \partial_z J^z = 0$$
$$\Rightarrow \frac{\partial \rho}{\partial t} + \nabla \cdot \vec{J} = 0$$

A Quick Overview of Particle Physics

The main application of quantum field theory is to the study of particle physics. This is because quantum field theory describes the fundamental particles and their interactions using what scientists call the *standard model*. In this framework, the standard model is believed to describe all physical phenomena with the exception of gravity. There are three fundamental interactions or forces described in the standard model:

- The electromagnetic interaction
- The weak interaction
- The strong interaction

Each force is mediated by a force-carrying particle called a *gauge boson*. Being a boson, a force-carrying particle has integral spin. The gauge bosons for the electromagnetic, weak, and strong forces are all spin-1 particles. If gravity is quantized, the force-carrying particle (called the *graviton*) is a spin-2 particle.

Forces in nature are believed to result from the exchange of the gauge bosons. For each interaction, there is a field, and the gauge bosons are the quanta of that field. The number of gauge bosons that exist for a particular field is given by the number of *generators* of the field. For a particular field, the generators come from the unitary group used to describe the symmetries of the field (this will become clearer later in the book).

THE ELECTROMAGNETIC FORCE

The symmetry group of the electromagnetic field is a unitary transformation, called $U(1)$. Since there is a single generator, the force is mediated by a single particle, which is known to be massless. The electromagnetic force is due to the exchange of photons, which we denote by γ. The photon is spin-1 and has two polarization states. If a particle is massless and spin-1, it can only have two polarization states. Photons do not carry charge.

THE WEAK FORCE

The gauge group of the weak force is $SU(2)$ which has three generators. The three physical gauge bosons that mediate the weak force are W^+, W^-, and Z. As we will see, these particles are superpositions of the generators of the gauge group. The gauge bosons for the weak force are massive.

- W^+ has a mass of 80 GeV/c^2 and carries +1 electric charge.
- W^- has a mass of 80 GeV/c^2 and carries −1 electric charge.
- Z has a mass of 91 GeV/c^2 and is electrically neutral.

The massive gauge bosons of the weak interaction are spin-1 and can have three polarization states.

THE STRONG FORCE

The gauge group of the strong force is $SU(3)$ which has eight generators. The gauge bosons corresponding to these generators are called *gluons*. Gluons mediate interactions between quarks (see below) and are therefore responsible for binding neutrons and protons together in the nucleus. A gluon is a massless spin-1 particle, and like the photon, has two polarization states. Gluons carry the charge of the strong force, called *color*. Since gluons also carry color charge they can interact among themselves, something that is not possible with photons since photons carry no charge. The theory that describes the strong force is called *quantum chromodynamics*.

THE RANGE OF A FORCE

The *range of a force* is dictated primarily by the mass of the gauge boson that mediates this force. We can estimate the range of a force using simple arguments based on the uncertainty principle. The amount of energy required for the exchange of a force mediating particle is found using Einstein's relation for rest mass as

$$\Delta E \approx mc^2$$

Now we use the uncertainty principle to determine how long the particle can exist as shown here.

$$\Delta t \approx \frac{\hbar}{\Delta E} = \frac{\hbar}{mc^2}$$

The special theory of relativity tells us that nothing travels faster than the speed of light c. So, we can use the speed of light to set an upper bound on the velocity of the force-carrying particle, and estimate the range it travels in a time Δt, that is,

$$\text{Velocity} = \frac{\text{distance}}{\text{time}}$$

$$\Rightarrow \Delta x = c\Delta t = \frac{c\hbar}{mc^2} = \frac{\hbar}{mc}$$

This is the range of the force. From this relation, you can see that if $m \to 0$, $\Delta x \to \infty$. So the range of the electromagnetic force is infinite. The range of the weak force, however, is highly constrained because the gauge bosons of the weak force have large masses. Plugging in the mass of the W as 80 GeV/c^2 you can verify that the range is

$$\Delta x \approx 10^{-3} \text{ fm}$$

This explains why the weak force is only felt over nuclear distances. This argument does not apply to gluons, which are massless, because the strong force is more complicated and involves a concept known as *confinement*. As stated above, the charge of the strong force is called *color charge*, and gluons carry color. Color charge has a strange property in that it exerts a constant force that binds color-carrying particles together. This can be visualized using the analogy of a rubber band. The stronger you pull on the rubber band, the tighter it feels. If you don't pull on it at all, it hangs loose. The strong force acts like a rubber band. At very short distances, it is relaxed and the particles behave as free particles. As the distance between them increases, the force gets them back in stronger pulling. This limits the range of the strong force, which is believed to be on the order of 10^{-15} m, the dimension of a nuclear particle. As a result of confinement, gluons are involved in mediating interactions between quarks, but are only indirectly responsible for the binding of neutrons and protons, which is accomplished through secondary particles called *mesons*.

Elementary Particles

The elementary particles of quantum field theory are treated as mathematical point-like objects that have no internal structure. The particles that make up matter all carry spin-1/2 and can be divided into two groups, *leptons* and *quarks*. Each group comes in three "families" or "generations." All elementary particles experience the gravitational force.

LEPTONS

Leptons interact via the electromagnetic and weak interaction, but do not participate in the strong interaction. Since they do not carry color charge, they do not participate in the strong interaction. They can carry electric charge e, which we denote as -1

(the charge of the electron), or they can be electrically neutral. The leptons include the following particles:

- The electron e carries charge -1 and has a mass of 0.511 MeV/c^2.
- The muon μ^- carries charge -1 and has a mass of 106 MeV/c^2.
- The tau τ^- carries charge -1 and has a mass of 1777 MeV/c^2.

Each type of lepton described above defines one of the three families that make up the leptons. In short, the muon and tau are just heavy copies of the electron. Physicists are not sure why there are three families of particles. The muon and tau are unstable and decay into electrons and neutrinos.

Corresponding to each particle above, there is a neutrino. It was thought for a long time that neutrinos were massless, but recent evidence indicates this is not the case, although experiment puts small bounds on their masses. Like the electron, muon, and tau, the three types of neutrinos come with masses that increase with each family. They are electrically neutral and are denoted by

- Electron neutrino ν_e
- Muon neutrino ν_μ
- Tau neutrino ν_τ

Since they are electrically neutral, the neutrinos do not participate in the electro-magnetic interaction. Since they are leptons, they do not participate in the strong interaction. They interact only via the weak force.

To each lepton there corresponds an antilepton. The antiparticles corresponding to the electron, muon, and tau all carry charge of $+1$, but they have the same masses. They are denoted as follows:

- The positron e^+ carries charge $+1$ and has a mass of 0.511 MeV/c^2.
- The antimuon μ^+ carries charge $+1$ and has a mass of 106 MeV/c^2.
- The antitau τ^+ carries charge $+1$ and has a mass of 1777 MeV/c^2.

In particle physics, we often indicate an antiparticle (a particle with the same properties but opposite charge) with an overbar; so if p is a given particle, we can indicate its corresponding antiparticle by $\bar{p}$. We will see later that charge is not the only quantum number of interest; a lepton also carries a quantum number called *lepton number*. It is $+1$ for a particle and -1 for the corresponding antiparticle. The antineutrinos $\bar{\nu}_e, \bar{\nu}_\mu$, and $\bar{\nu}_\tau$, like their corresponding particles, are also electrically neutral, but while the neutrinos ν_e, ν_μ, and ν_τ all have lepton number $+1$, the antineutrinos $\bar{\nu}_e, \bar{\nu}_\mu$, and $\bar{\nu}_\tau$ have lepton number -1.

In particle interactions, lepton number is always conserved. Particles that are not leptons are assigned a lepton number 0. Lepton number explains why there are antineutrinos, because they are neutral like ordinary neutrinos. Consider the beta decay of a neutron as shown here.

$$n \rightarrow p + e + \bar{\nu}_e$$

A neutron and proton are not leptons, hence they carry lepton number 0. The lepton number must balance on each side of the reaction. On the left we have total lepton number 0. On the right we have

$$0 + n_e + n_{\bar{\nu}_e}$$

Since the electron is a lepton, $n_e = 1$. This tells us that the neutrino emitted in this decay must be an antineutrino, and the lepton number is $n_{\bar{\nu}_e} = -1$ allowing lepton number to be conserved in the reaction.

QUARKS

Quarks are fundamental particles that make up the neutron and proton. They carry electrical charge and hence participate in the electromagnetic interaction. They also participate in the weak and strong interactions. *Color charge*, which is the charge of the strong interaction, can come in *red*, *blue*, or *green*. These color designations are just labels, so they should not be taken literally. There is also "anticolor" charge, antired, antiblue, and antigreen. Color charge can only be arranged such that the total color of a particle combination is *white*. There are three ways to get white color charge:

- Put three quarks together, one red, one blue, and one green.
- Put three quarks together, one antired, one antiblue, and one antigreen.
- Put two quarks together, one colored and one anticolored, for example a red quark and an antired quark.

The charge carried by a quark is −1/3 or +2/3 (in units of electric charge *e*). There are six types or "flavors" of quarks:

- Up quark u with charge +2/3
- Down quark d with charge −1/3
- Strange quark s with charge +2/3

- Charmed quark c with charge −1/3
- Top quark t with charge +2/3
- Bottom quark b with charge −1/3

Like the leptons, the quarks come in three families. One member of a family has charge +2/3 and the other has charge −1/3. The families are (u,d), (s,c), and (t,b). With each family, the mass increases. For example, the mass of the up quark is only

$$m_u \approx 4 \text{ MeV/c}^2$$

while the mass of the top quark is a hefty

$$m_t \approx 172 \text{ GeV/c}^2$$

which is as heavy as a single gold nucleus. Like the leptons, there are antiparticles corresponding to each quark.

Bound states of quarks are called *hadrons*. Bound states of observed quarks consist of two or three quarks only. A *baryon* is a hadron with three quarks or three antiquarks. Two famous baryons are

- The proton, which is the three-quark state uud
- The neutron, which is the three-quark state udd

Bound states consisting of a quark and antiquark are called *mesons*. These include:

The pion $\pi^0 = u\bar{u}$ or $d\bar{d}$

The charged pion $\pi^+ = u\bar{d}$ or $\pi^- = \bar{u}d$

SUMMARY OF PARTICLE GENERATIONS OR FAMILIES

The elementary particles come in three generations:

- The first generation includes the electron, electron neutrino, the up quark, the down quark, and the corresponding antiparticles.
- The second generation includes the muon, muon neutrino, strange quark, and charmed quark, along with the corresponding antiparticles.
- The third generation includes the tau, the tau neutrino, the top quark, and the bottom quark, along with the corresponding antiparticles.

The Higgs Mechanism

As the standard model of particle physics is formulated, the masses of all the particles are 0. An extra field called the *Higgs field* has to be inserted by hand to give the particles mass. The quantum of the Higgs field is a spin-0 particle called the *Higgs boson*. The Higgs boson is electrically neutral.

The Higgs field, if it exists, is believed to fill all of empty space throughout the entire universe. Elementary particles acquire their mass through their interaction with the Higgs field. Mathematically we introduce mass into a theory by adding interaction terms into the Lagrangian that couple the field of the particle in question to the Higgs field. Normally, the lowest energy state of a field would have an expectation value of zero. By symmetry breaking, we introduce a nonzero lowest energy state of the field. This procedure leads to the acquisition of mass by the particles in the theory.

Qualitatively, you might think of the Higgs field by imagining the differences between being on land and being completely submerged in water. On dry land, you can move your arm up and down without any trouble. Under water, moving your arm up and down is harder because the water is resisting your movement. We can imagine the movement of elementary particles being resisted by the Higgs field, with each particle interacting with the Higgs field at a different strength. If the coupling between the Higgs field and the particle is strong, then the mass of the particle is large. If it is weak, then the particle has a smaller mass. A particle like the photon with zero rest mass doesn't interact with the Higgs field at all. If the Higgs field didn't exist at all, then all particles would be massless. It is not certain what the mass of the Higgs boson is, but current estimates place an upper limit of $\approx 140 \text{ GeV}/c^2$. When the Large Hadron Collider begins operation in 2008, it should be able to detect the Higgs, if it exists.

Grand Unification

The standard model, as we have described above, consists of the electromagnetic interaction, the weak force, and quantum chromodynamics. Theorists would like to unify these into a single force or interaction. Many problems remain in theoretical physics, and in the past, many problems have been solved via some kind of unification. In many cases two seemingly different phenomena are actually two sides of the same coin. The quintessential example of this type of reasoning is the discovery by Faraday, Maxwell, and others that light, electricity, and magnetism are all the same physical phenomena that we now group together under electromagnetism.

Electromagnetism and the weak force have been unified into a single theoretical framework called *electroweak theory*. A *grand unified theory* or *GUT* is an attempt to bring quantum chromodynamics (and hence the strong force) into this unified framework.

If such a theory is valid, then there is a *grand unification energy* at which the electromagnetic, weak, and strong forces become unified into a single force. There is some support for this idea since the electromagnetic and weak force are known to become unified at high energies (but at lower energies than where unification with the strong force is imagined to occur).

Supersymmetry

There exists yet another unification scheme beyond that tackled by the GUTs. In particle physics, there are two basic types of particles. These include the spin-1/2 matter particles (fermions) and the spin-1 force-carrying particles (bosons). In elementary quantum mechanics, you no doubt learned that bosons and fermions obey different statistics. While the Pauli exclusion principle prevents two fermions from inhabiting the same state, there is no such limitation for bosons.

One might wonder why there are these two types of particles. In supersymmetry, an attempt is made to apply the reasoning of Maxwell and propose that a symmetry exists between bosons and fermions. For each fermion, supersymmetry proposes that there is a boson with the same mass, and vice versa. The partners of the known particles are called *superpartners*. Unfortunately, at this time there is no evidence that this is the case. The fact that the superpartners do not have the same mass indicates either that the symmetry of the theory is broken, in which case the masses of the superpartners are much larger than expected, or that the theory is not correct at all and supersymmetry does not exist.

String Theory

The ultimate step forward for quantum field theory is a unified theory known as *string theory*. This theory was originally proposed as a theory of the strong interaction, but it fell out of favor when quantum chromodynamics was developed. The basic idea of string theory is that the fundamental objects in the universe are not pointlike elementary particles, but are instead objects spread out in one dimension called strings. Excitations of the string give the different particles we see in the universe.

String theory is popular because it appears to be a completely unified theory. Quantum field theory unifies quantum mechanics and special relativity, and as a result is able to describe interactions involving three of the four known forces.

Gravity, the fourth force, is left out. Currently gravity is best described by Einstein's general theory of relativity, a classical theory that does not take quantum mechanics into account.

Efforts to bring quantum theory into the gravitational realm or vice versa have met with some difficulty. One reason is that interactions at a point cause the theory to "blow up"—in other words you get calculations with infinite results. By proposing that the fundamental objects of the theory are strings rather than point particles, interactions are spread out and the divergences associated with gravitational interactions disappear. In addition, a spin-2 state of the string naturally arises in string theory. It is known that the quantum of the gravitational field, if it exists, will be a massless spin-2 particle. Since this arises naturally in string theory, many people believe it is a strong candidate for a unified theory of all interactions.

Summary

Quantum field theory is a theoretical framework that unifies nonrelativistic quantum mechanics with special relativity. One consequence of this unification is that the types and number of particles can change in an interaction. As a result, the theory cannot be implemented using a single particle wave equation. The fundamental objects of the theory are quantum fields that act as operators, able to create or destroy particles.

Quiz

1. A quantum field
   (a) Is a field with quanta that are operators
   (b) Is a field parameterized by the position operator
   (c) Commutes with the Hamiltonian
   (d) Is an operator that can create or destroy particles

2. The particle generations
   (a) Are in some sense duplicates of each other, with each generation having increasing mass
   (b) Occur in pairs of three particles each
   (c) Have varying electrical charge but the same mass
   (d) Consists of three leptons and three quarks each

3. In relativistic situations
    (a) Particle number and type is not fixed
    (b) Particle number is fixed, but particle types are not
    (c) Particle number can vary, but new particle types cannot appear
    (d) Particle number and types are fixed

4. In quantum field theory
    (a) Time is promoted to an operator
    (b) Time and momentum satisfy a commutation relation
    (c) Position is demoted from being an operator
    (d) Position and momentum continue to satisfy the canonical commutation relation

5. Leptons experience
    (a) The strong force, but not the weak force
    (b) The weak force and electromagnetism
    (c) The weak force only
    (d) The weak force and the strong force

6. The number of force-carrying particles is
    (a) Equivalent to the number of generators for the fields gauge group
    (b) Random
    (c) Proportional to the number of fundamental matter particles involved in the interaction
    (d) Proportional to the number of generators minus one

7. The gauge group of the strong force is:
    (a) $SU(2)$
    (b) $U(1)$
    (c) $SU(3)$
    (d) $SU(1)$

8. Antineutrinos
    (a) Have charge −1 and lepton number 0
    (b) Have lepton number +1 and charge 0
    (c) Have lepton number −1 and charge 0
    (d) Are identical to neutrinos, since they carry no charge

9. The lightest family of elementary particles is
   (a) The electron, muon, and neutrino
   (b) The electron, up quark, and down quark
   (c) The electron, electron neutrino, up quark, and down quark
   (d) The electron and its antiparticle

10. The Higgs field
    (a) Couples the W and Z bosons to each other
    (b) Is a zero mass field
    (c) Has zero mass and charge +1
    (d) Gives mass to elementary particles

CHAPTER 2

Lagrangian Field Theory

We begin our study of quantum field theory by building up some fundamental mathematical tools that can be applied to any fundamental physical theory. The main quantity of interest in this chapter is a function known as the *Lagrangian*, which is constructed by taking the difference of the kinetic and potential energies. In classical mechanics the Lagrangian is an equivalent method to Newtonian mechanics that can be used to derive the equations of motion. When Lagrangian methods are applied to fields, we can use the same techniques to derive the field equations.

Basic Lagrangian Mechanics

For now, we work in one spatial dimension x and consider the motion of a single particle. Let T be the kinetic energy of the particle moving in a potential V. The Lagrangian L is defined as

$$L = T - V \qquad (2.1)$$

The Lagrangian is a foundational concept which captures all the dynamics of the system and allows us to determine many useful properties such as averages and dynamic behavior.

Given L we can find the equations of motion from the *Euler-Lagrange equations*. For a single particle moving in one dimension, these are given by the single equation.

$$\frac{d}{dt}\frac{\partial L}{\partial \dot{x}} - \frac{\partial L}{\partial x} = 0 \qquad (2.2)$$

where we have used a dot to denote differentiation with respect to time, that is,

$$\dot{x} = \frac{dx}{dt}$$

EXAMPLE 2.1

Consider a particle of mass m with kinetic energy $T = \frac{1}{2}m\dot{x}^2$ moving in one dimension in a potential $V(x)$. Use the Euler-Lagrange equations to find the equation of motion.

SOLUTION

We follow five basic steps. (1) First we write down the Lagrangian L, then we calculate the derivatives we need for the Euler-Lagrange equation. (2) The first is the derivative of L with respect to $\dot{x}$, $\frac{\partial L}{\partial \dot{x}}$. (3) We take the time derivative of this quantity, $\frac{d}{dt}(\frac{\partial L}{\partial \dot{x}})$. (4) Next is the derivative of L with respect to x, $\frac{\partial L}{\partial x}$. (5) We are finally able to form the equation $\frac{d}{dt}(\frac{\partial L}{\partial \dot{x}}) = \frac{\partial L}{\partial x}$.

Begin by writing the Lagrangian [step (1)] which is

$$L = \frac{1}{2}m\dot{x}^2 - V(x)$$

Next is the derivative of L with respect to $\dot{x}$. When doing calculations involving the Lagrangian, we treat $\dot{x}$ just as though it were an independent variable. For example, $\frac{\partial}{\partial \dot{x}}(\dot{x})^2 = 2\dot{x}$ and $\frac{\partial}{\partial \dot{x}}V(x) = 0$. Applying this [step (2)] we get

$$\frac{\partial L}{\partial \dot{x}} = \frac{\partial}{\partial \dot{x}}\left[\frac{1}{2}m\dot{x}^2 - V(x)\right] = m\dot{x}$$

which is a momentum term: mass times velocity. Now we take the time derivative of this momentum [step (3)] which is

$$\frac{d}{dt}\left(\frac{\partial L}{\partial \dot{x}}\right) = \frac{d}{dt}(m\dot{x}) = \dot{m}\dot{x} + m\ddot{x} = m\ddot{x}$$

and we have mass times acceleration. The remaining derivative is now a simple calculation [step (4)], that is,

$$\frac{\partial L}{\partial x} = \frac{\partial}{\partial x}\left[\frac{1}{2}m\dot{x}^2 - V(x)\right] = -\frac{\partial V}{\partial x}$$

Next [step (5)] is writing the equation which describes the dynamical behavior of the system, like

$$\frac{d}{dt}\left(\frac{\partial L}{\partial \dot{x}}\right) - \frac{\partial L}{\partial x} = 0 \Rightarrow \frac{d}{dt}\left(\frac{\partial L}{\partial \dot{x}}\right) = \frac{\partial L}{\partial x} \Rightarrow m\ddot{x} = -\frac{\partial V}{\partial x}$$

This elegant result is quite familiar. From classical mechanics, we recall that if a force F is conservative then

$$\vec{F} = -\nabla V$$

In one dimension this becomes

$$F = -\frac{\partial V}{\partial x}$$

Therefore, $\frac{\partial L}{\partial x} = -\frac{\partial V}{\partial x} = F$. Using our calculation we have

$$m\ddot{x} = -\frac{\partial V}{\partial x} = F$$

Hence we arrive at Newton's second law,

$$F = m\ddot{x} = ma$$

where the acceleration a is given by $a = d^2x/dt^2 = \ddot{x}$.

EXAMPLE 2.2
Consider a particle of mass m undergoing simple harmonic motion. The force on the particle is given by Hooke's law $F(x) = -kx$. Determine the equation of motion by using the Euler-Lagrange equation.

SOLUTION
Once again, the kinetic energy is given by

$$T = \frac{1}{2}m\dot{x}^2$$

We integrate the force $F(x) = -kx$ to compute the potential and find that

$$V = \frac{1}{2}kx^2$$

Using Eq. (2.1) the Lagrangian is [step (1)]

$$L = T - V = \frac{1}{2}m\dot{x}^2 - \frac{1}{2}kx^2$$

[step (2)] $\frac{\partial}{\partial \dot{x}}(L) = \frac{\partial}{\partial \dot{x}}(\frac{1}{2}m\dot{x}^2 - \frac{1}{2}kx^2) = m\dot{x}$. As above, [step (3)] the time derivative of the momentum is a force $\frac{d}{dt}(m\dot{x}) = m\ddot{x}$ and the last derivative [step (4)] is

$$\frac{\partial L}{\partial x} = \frac{\partial}{\partial x}\left[\frac{1}{2}m\dot{x}^2 - \frac{1}{2}kx^2\right] = -kx$$

To obtain the equation of motion [step (5)] for the particle we use Eq. (2.2). We have

$$m\ddot{x} = -kx$$

This is the familiar equation of a simple harmonic oscillator, that is,

$$\frac{d^2x}{dt^2} + \omega_0^2 x = 0$$

where the natural frequency $\omega_0^2 = k/m$.

The Action and the Equations of Motion

If we integrate the Lagrangian with respect to time, we obtain a new quantity called the *action* which we denote by S

$$S = \int L\, dt \qquad (2.3)$$

The action is *functional* because it takes a *function* as argument and returns a *number*. Particles always follow a path of least action. By *varying (minimzing the variance of)* the action, we can determine the path actually followed by a particle. Consider two fixed points $x(t_1)$ and $x(t_2)$. There are an infinite number of paths connecting these points. This means that there are an infinite number of paths for the particle to follow between these two points. The actual path that the particle follows is the path of least action. The path of least action represents a minimum. To find this path we minimize the variance of the action. We do this by describing the unknown action as minimal term and a variation.

$$S \rightarrow S + \delta S$$

then the path followed by the particle is the one for which

$$\delta S = 0 \qquad (2.4)$$

This is the path with zero variation. This is the path of least action.

Computing the variation δS leads to the equations of motion for the system. To see how this works, we start with a simple example, returning to the derivation of Newton's second law. We consider a small change in coordinates given by

$$x \rightarrow x + \varepsilon$$

where ε is small. This variation is constrained to keep the end points fixed, that is,

$$\varepsilon(t_1) = \varepsilon(t_2) = 0 \qquad (2.5)$$

Using a Taylor expansion, the potential can be approximated as

$$V(x + \varepsilon) \approx V(x) + \varepsilon \frac{dV}{dx}$$

The action then becomes

$$S = \int_{t_1}^{t_2} L\,dt = \int_{t_1}^{t_2} \frac{1}{2}m\dot{x}^2 - V(x)\,dt = \int_{t_1}^{t_2} \frac{1}{2}m(\dot{x} + \dot{\varepsilon})^2 - \left(V(x) - \varepsilon\frac{dV}{dx}\right)dt$$

Now we expand out the first term, the kinetic energy term, giving

$$(\dot{x} + \dot{\varepsilon})^2 = \dot{x}^2 + 2\dot{x}\dot{\varepsilon} + \dot{\varepsilon}^2 \approx \dot{x}^2 + 2\dot{x}\dot{\varepsilon}$$

We dropped the $\dot{\varepsilon}^2$ term, since by assumption ε is small, so squaring it gives a term we can neglect. That is, we only keep the leading order terms. So we have

$$S = \int_{t_1}^{t_2} \frac{1}{2}m(\dot{x}^2 + 2\dot{x}\dot{\varepsilon}) - \left(V(x) + \varepsilon\frac{dV}{dx}\right)dt$$

This expression can be written in a more useful fashion with some manipulation. The idea is to isolate the terms containing ε. We can do this by applying integration by parts to the $2\dot{x}\dot{\varepsilon}$ term, transferring the time derivative from ε to $\dot{x}$. First let's recall the formula for integration by parts

$$\int_{t_1}^{t_2} f(t)\frac{dg}{dt}\,dt = f(t)g(t)\Big|_{t_1}^{t_2} - \int_{t_1}^{t_2} g(t)\frac{df}{dt}\,dt \qquad (2.6)$$

In our case, $f(t) = \dot{x}$ and $\frac{dg}{dt} = \dot{\varepsilon}$. Recalling Eq. (2.5), the fact that the variation vanishes at the end points of the interval, the boundary term vanishes in our case. Hence,

$$\int_{t_1}^{t_2} \frac{1}{2}m(\dot{x}+\dot{\varepsilon})^2\,dt = \frac{1}{2}m\int_{t_1}^{t_2}(\dot{x}+\dot{\varepsilon})^2\,dt \approx \frac{1}{2}m\int_{t_1}^{t_2}(\dot{x}^2 + 2\dot{x}\dot{\varepsilon})\,dt$$

$$\int_{t_1}^{t_2}\frac{1}{2}m(\dot{x}^2 + 2\dot{x}\dot{\varepsilon})\,dt = \int_{t_1}^{t_2}\frac{1}{2}m\dot{x}^2\,dt - \int_{t_1}^{t_2} m\ddot{x}\varepsilon\,dt$$

We can now collect the ε terms

$$S = \int_{t_1}^{t_2}\frac{1}{2}m\dot{x}^2\,dt - \int_{t_1}^{t_2} m\ddot{x}\varepsilon\,dt - \int_{t_1}^{t_2}\left(V(x) + \varepsilon\frac{dV}{dx}\right)dt$$

$$= \int_{t_1}^{t_2}\frac{1}{2}m\dot{x}^2 - V(x)\,dt + \int_{t_1}^{t_2}\left(-m\ddot{x} - \frac{dV}{dx}\right)\varepsilon\,dt$$

$$= S + \delta S$$

The requirement that $\delta S = 0$ can be satisfied only if the integral of the second term is 0. Since the end points are arbitrary, the integrand must be identically 0, that is,

$$m\ddot{x} + \frac{dV}{dx} = 0 \Rightarrow m\ddot{x} = -\frac{dV}{dx}$$

Now let's consider a more general situation where there are N generalized coordinates $q_i(t)$ where $i = 1,\ldots,N$. Considering a Lagrangian expressed in terms of these coordinates and their first derivatives only, we have an action of the form

$$S = \int_{t_1}^{t_2} L(q,\dot{q})\,dt$$

In this case the system evolves from some initial point $q_1 = q(t_1)$ to some final point $q_2 = q(t_2)$. To find the trajectory followed by the system, we apply the principle of least action and solve $\delta S = 0$. Once again, the end points of the trajectory are fixed so that

$$\delta q(t_1) = \delta q(t_2) = 0$$

Also note that

$$\delta \dot{q}(t) = \frac{d}{dt}(\delta q) \tag{2.7}$$

Then,

$$\delta S = \delta \int_{t_1}^{t_2} L(q, \dot{q}) \, dt$$

$$= \int_{t_1}^{t_2} \sum_i \left[\frac{\partial L}{\partial q_i} \delta q_i + \frac{\partial L}{\partial \dot{q}_i} \delta \dot{q}_i \right] dt$$

$$= \int_{t_1}^{t_2} \sum_i \left[\frac{\partial L}{\partial q_i} \delta q_i + \frac{\partial L}{\partial \dot{q}_i} \frac{d}{dt}(\delta q_i) \right] dt$$

To move from the second to the third line, we applied Eq. (2.7). Now we use integration by parts on the second term, giving

$$\delta S = \int_{t_1}^{t_2} \sum_i \left[\frac{\partial L}{\partial q_i} \delta q_i - \frac{d}{dt}\left(\frac{\partial L}{\partial \dot{q}_i} \right) \delta q_i \right] dt$$

For this expression to vanish, since δq_i is arbitrary each coordinate index must satisfy

$$\frac{\partial L}{\partial q_i} - \frac{d}{dt}\left(\frac{\partial L}{\partial \dot{q}_i} \right) = 0 \tag{2.8}$$

which are of course the Euler-Lagrange equations and $i = 1, \ldots, N$. Therefore, we see that the principle of least action gives rise to the Euler-Lagrange equations. Hence the Lagrangian satisfies the Euler-Lagrange equation independently for each coordinate.

Canonical Momentum and the Hamiltonian

In Examples 2.1 and 2.2 we saw that the derivative of the kinetic energy with respect to velocity was the momentum: a completely classical result, that is,

$$\frac{\partial}{\partial \dot{x}}(L) = \frac{\partial}{\partial \dot{x}}(T - V) = \frac{\partial}{\partial \dot{x}}\left(\frac{1}{2}m\dot{x}^2 \right) = m\dot{x}$$

This result can be made more general and more useful. The *canonical* momentum is defined as

$$p_i = \frac{\partial L}{\partial \dot{q}_i} \tag{2.9}$$

This allows us to define a Hamiltonian function, which is given in terms of the Lagrangian and the canonical momentum as

$$H(p,q) = \sum_i p_i \dot{q}_i - L \tag{2.10}$$

Lagrangian Field Theory

Now that we have reviewed the basics of the Lagrangian formalism, we are prepared to generalize these techniques and apply them to fields, that is, functions on spacetime $\varphi(x,t)$ which we write more compactly as $\varphi(x)$. In the continuous case, we actually work with the Lagrangian density.

$$L = T - V = \int \mathcal{L} \, d^3x \tag{2.11}$$

The action S is then the time integral of this expression.

$$S = \int dt \, L = \int \mathcal{L} \, d^4x \tag{2.12}$$

Typically, the Lagrangians encountered in quantum field theory depend only on the fields and their first derivatives.

$$\mathcal{L} = \mathcal{L}(\varphi, \partial_\mu \varphi)$$
$$L \to L(\varphi, \partial_\mu \varphi) \tag{2.13}$$

Moreover we are interested in fields that are *local*, meaning that at a given spacetime point x, the Lagrangian density depends on the fields and their first derivatives evaluated at that point.

Now we apply the principle of least action to Eq. (2.12). We vary the action with respect to the field $\varphi(x)$ and with respect to the first derivative of the field $\partial_\mu \varphi(x)$ as follows:

$$0 = \delta S$$
$$= \delta \int d^4x \, \mathcal{L}$$
$$= \int d^4x \left\{ \frac{\partial \mathcal{L}}{\partial \varphi} \delta \varphi + \frac{\partial \mathcal{L}}{\partial [\partial_\mu \varphi]} \delta(\partial_\mu \varphi) \right\}$$

Now we use the fact that

$$\delta(\partial_\mu \varphi) = \partial_\mu(\delta\varphi)$$

and apply integration by parts to the second term in this expression. The boundary terms vanish because the end points are fixed and the second term becomes

$$\int d^4x \frac{\partial \mathcal{L}}{\partial[\partial_\mu \varphi]} \delta(\partial_\mu \varphi) = \int d^4x \frac{\partial \mathcal{L}}{\partial[\partial_\mu \varphi]} \partial_\mu(\delta\varphi)$$

$$= -\int d^4x \partial_\mu \left(\frac{\partial \mathcal{L}}{\partial[\partial_\mu \varphi]} \right) \delta\varphi$$

All together, the variation of the action is then

$$0 = \delta S = \int d^4x \left\{ \frac{\partial \mathcal{L}}{\partial \varphi} - \partial_\mu \left(\frac{\partial \mathcal{L}}{\partial[\partial_\mu \varphi]} \right) \right\} \delta\varphi$$

What does it mean when an integral is 0? There are two cases: Either there is cancellation because the integrand takes positive and negative values or the integrand is 0 over the entire domain of integration. In this case the domain of integration can vary, so we can't rely on cancellation and we know the integrand must be 0. That is, the term inside the braces vanishes. This gives us the Euler-Lagrange equations for a field φ.

$$\frac{\partial \mathcal{L}}{\partial \varphi} - \partial_\mu \left(\frac{\partial \mathcal{L}}{\partial[\partial_\mu \varphi]} \right) = 0 \tag{2.14}$$

The Einstein summation convention is in force, so there is an implied sum on the second term. That is,

$$\partial_\mu \left(\frac{\partial \mathcal{L}}{\partial[\partial_\mu \varphi]} \right) = \partial_t \left(\frac{\partial \mathcal{L}}{\partial[\partial_t \varphi]} \right) - \partial_x \left(\frac{\partial \mathcal{L}}{\partial[\partial_x \varphi]} \right) - \partial_y \left(\frac{\partial \mathcal{L}}{\partial[\partial_y \varphi]} \right) - \partial_z \left(\frac{\partial \mathcal{L}}{\partial[\partial_z \varphi]} \right)$$

The canonical momentum density for the field is given by

$$\pi(x) = \frac{\partial \mathcal{L}}{\partial \dot{\varphi}} \tag{2.15}$$

just as in the classical case. The Hamiltonian density is then

$$\mathcal{H} = \pi(x)\dot{\varphi}(x) - \mathcal{L} \tag{2.16}$$

To obtain the Hamiltonian, we integrate this density over space

$$H = \int \mathcal{H}\, d^3x \tag{2.17}$$

EXAMPLE 2.3

Find the equation of motion and the Hamiltonian corresponding to the Lagrangian

$$\mathcal{L} = \frac{1}{2}\left\{(\partial_\mu\varphi)^2 - m^2\varphi^2\right\}$$

SOLUTION

We can find the field equations with a straightforward application of Eq. (2.14). Let's follow the five-step process from the previous examples. Since we are given the Lagrangian, step (1) is done. The remaining steps require some insight.

The trick in doing the problems is that we treated $\partial_\mu\varphi$ as a variable. If you need a crutch to get used to thinking this way, as an analogy think of φ as x and $\partial_\mu\varphi$ as y. Then $\frac{\partial}{\partial\varphi}[\frac{1}{2}(\partial_\mu\varphi)^2]$ is like calculating $\frac{\partial}{\partial x}[\frac{1}{2}(y)^2] = 0$.

Now we are ready for step (2), compute $\frac{\partial}{\partial(\partial_\mu\varphi)}(L)$.

Begin by expanding the first term in the Lagrangian:

$$(\partial_\mu\varphi)^2 = (\partial_\mu\varphi)(\partial^\mu\varphi) = (\partial_\mu\varphi)g^{\mu\nu}(\partial_\nu\varphi)$$

Hence, we have

$$\frac{\partial\mathcal{L}}{\partial[\partial_\mu\varphi]} = \frac{\partial}{\partial[\partial_\mu\varphi]}\left(\frac{1}{2}\left\{(\partial_\mu\varphi)^2 - m^2\varphi^2\right\}\right)$$

$$= \frac{1}{2}\frac{\partial}{\partial[\partial_\mu\varphi]}(\partial_\mu\varphi)g^{\mu\nu}(\partial_\nu\varphi) - \frac{1}{2}\frac{\partial}{\partial[\partial_\mu\varphi]}(m^2\varphi^2)$$

Using our observation that we simply act like $\partial_\mu \varphi$ is a variable when computing these derivatives, it is clear that

$$\frac{\partial}{\partial[\partial_\mu \varphi]}(m^2 \varphi^2) = 0$$

This means we are left with

$$\frac{\partial \mathcal{L}}{\partial[\partial_\mu \varphi]} = \frac{1}{2}\frac{\partial}{\partial[\partial_\mu \varphi]}(\partial_\mu \varphi)g^{\mu\nu}(\partial_\nu \varphi)$$

But

$$\frac{\partial}{\partial[\partial_\mu \varphi]}(\partial_\mu \varphi)g^{\mu\nu}(\partial_\nu \varphi) = g^{\mu\nu}(\partial_\nu \varphi) + g^{\mu\nu}(\partial_\mu \varphi) = 2\partial^\mu \varphi$$

Meaning that

$$\frac{\partial \mathcal{L}}{\partial[\partial_\mu \varphi]} = \partial^\mu \varphi$$

Step (3) entails taking the derivative of this result.

$$\partial_\mu \left(\frac{\partial L}{\partial[\partial_\mu \varphi]} \right) = \partial_\mu (\partial^\mu \varphi)$$

Next, step (4) is the easiest.

$$\frac{\partial \mathcal{L}}{\partial \varphi} = \frac{\partial}{\partial \varphi}\left[\frac{1}{2}\left\{ (\partial_\mu \varphi)^2 - m^2 \varphi^2 \right\} \right]$$

$$= \frac{\partial}{\partial \varphi}\left[\frac{1}{2}(\partial_\mu \varphi)^2 \right] - \frac{\partial}{\partial \varphi}\frac{1}{2}(m^2 \varphi^2)$$

$$= -\frac{\partial}{\partial \varphi}\frac{1}{2}(m^2 \varphi^2) = -m^2 \varphi$$

Finally step (5) is to form the equations of motion.

$$0 = \frac{\partial \mathcal{L}}{\partial \varphi} - \partial_\mu \left(\frac{\partial \mathcal{L}}{\partial [\partial_\mu \varphi]} \right)$$

$$= -m^2 \varphi - \partial_\mu (\partial^\mu \varphi)$$

Using $\partial_\mu \partial^\mu = \frac{\partial^2}{\partial t^2} - \nabla^2$, the field equations corresponding to the Lagrangian given in this problem can be written as

$$\frac{\partial^2 \phi}{\partial t^2} - \nabla^2 \varphi + m^2 \varphi = 0$$

EXAMPLE 2.4

Derive the *sine-Gordon equation* $\frac{\partial^2 \varphi}{\partial t^2} - \frac{\partial^2 \varphi}{\partial x^2} + \sin \varphi = 0$ from the Lagrangian

$$\mathcal{L} = \frac{1}{2} \left\{ (\partial_t \varphi)^2 - (\partial_x \varphi)^2 \right\} + \cos \varphi$$

SOLUTION

First we calculate

$$\frac{\partial \mathcal{L}}{\partial \varphi} = \frac{\partial}{\partial \varphi} \left[\frac{1}{2} \left\{ (\partial_t \varphi)^2 - (\partial_x \varphi)^2 \right\} + \cos \varphi \right]$$

$$= \frac{\partial}{\partial \varphi} \left[\frac{1}{2} \left\{ (\partial_t \varphi)^2 - (\partial_x \varphi)^2 \right\} \right] + \frac{\partial}{\partial \varphi} \cos \varphi$$

$$= \frac{\partial}{\partial \varphi} (\cos \varphi) = -\sin \varphi$$

The Lagrangian only has one spatial coordinate, so

$$\partial_\mu \left(\frac{\partial \mathcal{L}}{\partial [\partial_\mu \varphi]} \right) = \partial_t \left(\frac{\partial \mathcal{L}}{\partial [\partial_t \varphi]} \right) - \partial_x \left(\frac{\partial \mathcal{L}}{\partial [\partial_x \varphi]} \right)$$

Now tackling the time and space derivatives separately leads us to

$$\frac{\partial \mathcal{L}}{\partial [\partial_t \varphi]} = \frac{\partial}{\partial [\partial_t \varphi]} \left[\frac{1}{2} \left\{ (\partial_t \varphi)^2 - (\partial_x \varphi)^2 \right\} + \cos \varphi \right] = \partial_t \varphi$$

$$\frac{\partial \mathcal{L}}{\partial [\partial_x \varphi]} = \frac{\partial}{\partial [\partial_x \varphi]} \left[\frac{1}{2} \left\{ (\partial_t \varphi)^2 - (\partial_x \varphi)^2 \right\} + \cos \varphi \right] = \partial_x \varphi$$

and so

$$\partial_\mu \left(\frac{\partial \mathcal{L}}{\partial [\partial_\mu \varphi]} \right) = \partial_t \left(\frac{\partial \mathcal{L}}{\partial [\partial_t \varphi]} \right) - \partial_x \left(\frac{\partial \mathcal{L}}{\partial [\partial_x \varphi]} \right) = \partial_t (\partial_t \varphi) - \partial_x (\partial_x \varphi)$$

$$= \frac{\partial^2 \varphi}{\partial t^2} - \frac{\partial^2 \varphi}{\partial x^2} = \Box \varphi$$

Therefore, the equation of motion in this case is

$$\Box \varphi + \sin \varphi = (\Box + \sin) \varphi = 0$$

This equation is called the *sine-Gordon equation*, due to its resemblance to the Klein-Gordon equation.

Symmetries and Conservation Laws

A *symmetry* is a change in perspective that leaves the equations of motion invariant. For example, the change could be a translation in space, a change in time or a rotation. These are external symmetries, that is, they depend upon changes in spacetime. There also exist *internal symmetries*, changes in the fields that do not involve changes with respect to spacetime at all.

A famous theorem in classical mechanics which is very important is known as *Noether's theorem.* This theorem allows us to relate symmetries to conserved quantities like charge, energy, and momentum. Mathematically, a symmetry is some kind of variation to the fields or the Lagrangian that leaves the equations of motion invariant. We will see how to make such a change and then deduce a conserved quantity.

Two of the most fundamental results of physics, conservation of energy and momentum, are due to a symmetry that results from a small displacement in spacetime. That is we let the spacetime coordinates vary according to

$$x^\mu \to x^\mu + a^\mu \tag{2.18}$$

where a^μ is a small and arbitrary parameter describing a displacement in spacetime. Expanding in Taylor, the field changes according to

$$\varphi(x) \to \varphi(x+a) = \varphi(x) + a^\mu \partial_\mu \varphi \tag{2.19}$$

Under a small variation (perturbation), the field can be described as

$$\varphi \to \varphi + \delta\varphi$$

Quantum Field Theory Demystified

This means we can write the variation in the field explicitly as

$$\delta\varphi = a^{\mu}\partial_{\mu}\varphi \qquad (2.20)$$

Now let's reconsider the variation of the Lagrangian. We have, in the case of a Lagrangian depending only on the field and its first derivatives,

$$\delta\mathcal{L} = \frac{\partial\mathcal{L}}{\partial\varphi}\delta\varphi + \frac{\partial\mathcal{L}}{\partial(\partial_{\mu}\varphi)}\delta(\partial_{\mu}\varphi)$$

From the Euler-Lagrange equations [Eq. (2.14)] we know that

$$\frac{\partial\mathcal{L}}{\partial\varphi} = \partial_{\mu}\left(\frac{\partial\mathcal{L}}{\partial[\partial_{\mu}\varphi]}\right)$$

Hence the variation in the Lagrangian can be written as

$$\delta\mathcal{L} = \partial_{\mu}\left(\frac{\partial\mathcal{L}}{\partial[\partial_{\mu}\varphi]}\right)\delta\varphi + \frac{\partial\mathcal{L}}{\partial(\partial_{\mu}\varphi)}\delta(\partial_{\mu}\varphi)$$

$$= \partial_{\mu}\left(\frac{\partial\mathcal{L}}{\partial[\partial_{\mu}\varphi]}\right)\delta\varphi + \frac{\partial\mathcal{L}}{\partial(\partial_{\mu}\varphi)}\partial_{\mu}(\delta\varphi)$$

Now we have an expression that can be written as a total derivative. Remember the product rule from ordinary calculus that $(fg)' = f'g + g'f$. We take

$$f = \frac{\partial\mathcal{L}}{\partial[\partial_{\mu}\varphi]} \qquad g = \delta\varphi$$

Allowing us to write

$$\delta\mathcal{L} = \partial_{\mu}\left(\frac{\partial\mathcal{L}}{\partial[\partial_{\mu}\varphi]}\delta\varphi\right) = (fg)'$$

Now we can apply Eq. (2.20). Note that the same index can only be used twice in a single expression, so we need to change the label used in Eq. (2.20) to another dummy index, say $\delta\varphi = a^{\mu}\partial_{\mu}\varphi = a^{\nu}\partial_{\nu}\varphi$. The quantities involved are just ordinary scalars, so we can also move them around and write $\delta\varphi = \partial_{\nu}\varphi a^{\nu}$. So, we arrive at the following expression

$$\delta\mathcal{L} = \partial_{\mu}\left(\frac{\partial\mathcal{L}}{\partial[\partial_{\mu}\varphi]}\partial_{\nu}\varphi\right)a^{\nu}$$

Equivalently, we can also write the variation of the Lagrangian as

$$\delta\mathcal{L} = \partial_\mu(\mathcal{L})a^\mu = \delta_\nu^\mu\partial_\mu(\mathcal{L})a^\nu$$

That is, we consider how it varies directly with respect to the displacement Eq. (2.18). Equating these two results gives

$$\delta\mathcal{L} = \delta_\nu^\mu\partial_\mu(\mathcal{L})a^\nu = \partial_\mu\left(\frac{\partial\mathcal{L}}{\partial[\partial_\mu\varphi]}\partial_\nu\varphi\right)a^\nu$$

Moving both terms to the same side of the equation gives

$$\partial_\mu\left(\frac{\partial\mathcal{L}}{\partial[\partial_\mu\varphi]}\partial_\nu\varphi - \delta_\nu^\mu\mathcal{L}\right)a^\nu = 0$$

Now, recall that a^ν is arbitrary. So in order for this expression to vanish, the derivative must be zero. That is,

$$\partial_\mu\left(\frac{\partial\mathcal{L}}{\partial[\partial_\mu\varphi]}\partial_\nu\varphi - \delta_\nu^\mu\mathcal{L}\right) = 0$$

This expression is so important that we give this quantity its own name. It turns out this is the *energy-momentum tensor*. We write this as

$$T_\nu^\mu = \frac{\partial\mathcal{L}}{\partial[\partial_\mu\varphi]}\partial_\nu\varphi - \delta_\nu^\mu\mathcal{L} \qquad (2.21)$$

Hence the conservation relation expressed by zero total divergence is

$$\partial_\mu T_\nu^\mu = 0 \qquad (2.22)$$

Notice that

$$T_0^0 = \frac{\partial\mathcal{L}}{\partial\dot{\varphi}}\dot{\varphi} - \mathcal{L} = \mathcal{H} \qquad (2.23)$$

That is, T_0^0 is nothing other than the Hamiltonian density: it's the energy density and the equation $\partial_0 T_0^0 = 0$ reflects *conservation of energy*. The components of momentum density are given by T_i^0 where i runs over the spatial indices. The components of momentum of the field are found by integrating each of these terms over space, that is,

$$P_i = \int d^3x\, T_i^0 \qquad (2.24)$$

Conserved Currents

Now let's go through the process used in the last section again to see how Noether's theorem can be applied to derive a conserved current and an associated conserved charge. We let the field vary by a small amount

$$\varphi \to \varphi + \delta\varphi \qquad (2.25)$$

We then start from the premise that under this variation, the Lagrangian does not change. The variation in the Lagrangian due to Eq. (2.25) will be of the form

$$\mathcal{L} \to \mathcal{L} + \delta\mathcal{L} \qquad (2.26)$$

So what we mean is that

$$\delta\mathcal{L} = 0 \qquad (2.27)$$

Now following the usual procedure the variation of the Lagrangian due to a variation in the field will be

$$\delta\mathcal{L} = \frac{\partial\mathcal{L}}{\partial\varphi}\delta\varphi + \frac{\partial\mathcal{L}}{\partial(\partial_\mu\varphi)}\partial_\mu(\delta\varphi)$$

Once again, from the Euler-Lagrange equations [Eq. (2.14)] we can write

$$\frac{\partial\mathcal{L}}{\partial\varphi} = \partial_\mu\left(\frac{\partial\mathcal{L}}{\partial[\partial_\mu\varphi]}\right)$$

Therefore, we have

$$\delta\mathcal{L} = \partial_\mu\left(\frac{\partial\mathcal{L}}{\partial[\partial_\mu\varphi]}\right)\delta\varphi + \frac{\partial\mathcal{L}}{\partial[\partial_\mu\varphi]}\partial_\mu(\delta\varphi) = \partial_\mu\left(\frac{\partial\mathcal{L}}{\partial[\partial_\mu\varphi]}\delta\varphi\right)$$

Since we are operating under the premise that the variation in the field does not change the Lagrangian [Eq. (2.27)], this leads us to the result

$$\partial_\mu\left(\frac{\partial\mathcal{L}}{\partial[\partial_\mu\varphi]}\delta\varphi\right) = 0 \qquad (2.28)$$

We call the quantity in the parentheses a *conserved current*. In analogy with electrodynamics we denote it with the letter J and write

$$J^\mu = \frac{\partial \mathcal{L}}{\partial [\partial_\mu \varphi]} \delta\varphi \qquad (2.29)$$

The conservation law Eq. (2.28) then can be written as

$$\partial_\mu J^\mu = 0 \qquad (2.30)$$

This is the central result of Noether's theorem:

- For every continuous symmetry of the Lagrangian—that is, a variation in the field that leaves form of the Lagrangian unchanged—there is a conserved current whose form can be derived from the Lagrangian using Eq. (2.29).

There is a conserved *charge* associated with each conserved current that results from a symmetry of the Lagrangian. This is found by integrating the time component of J:

$$Q = \int d^3x \, J^0 \qquad (2.31)$$

We see that the translation symmetry in spacetime worked out in the previous section that led to the energy-momentum tensor is a special case of Noether's theorem, with the conserved "charges" being energy and momentum.

The Electromagnetic Field

The Maxwell or electromagnetic field tensor is given by

$$F^{\mu\nu} = \partial^\mu A^\nu - \partial^\nu A^\mu = \begin{pmatrix} 0 & -E_x & -E_y & -E_z \\ E_x & 0 & -B_z & B_y \\ E_y & B_z & 0 & -B_x \\ E_z & -B_y & B_x & 0 \end{pmatrix} \qquad (2.32)$$

The A^μ is the usual vector potential, but this is a 4-vector whose time component is the scalar potential and whose spatial component is the usual vector potential used to write down the magnetic field, that is, $A^\mu = (\psi, \vec{A})$. It can be shown that $F^{\mu\nu}$

satisfies or leads to Maxwell's equations. Note that $F^{\mu\nu}$ is antisymmetric; the sign flips when the indices are interchanged. That is,

$$F^{\mu\nu} = -F^{\nu\mu} \tag{2.33}$$

Without sources, the homogeneous Maxwell's equations can be written in terms of the electromagnetic field tensor as

$$\partial^\lambda F^{\mu\nu} + \partial^\nu F^{\lambda\mu} + \partial^\mu F^{\nu\lambda} \tag{2.34}$$

Meanwhile, the inhomogeneous Maxwell's equations can be written as

$$\partial_\mu F^{\mu\nu} - J^\nu = 0 \tag{2.35}$$

where J^ν are the current densities. It is an instructive exercise to derive Maxwell's equations using a variational procedure so that you can learn how to work with tensors of higher order, that is, vector fields. In the next example we derive Eq. (2.35) from a Lagrangian.

EXAMPLE 2.5
Show that the Lagrangian $\mathcal{L} = -\frac{1}{4} F_{\mu\nu} F^{\mu\nu} - J^\mu A_\mu$ leads to the inhomogeneous Maxwell's equations [Eq. (2.35)] if the potential A_μ is varied, leaving the current density constant.

SOLUTION
We begin as usual by writing the action as an integral of the Lagrangian density. The action in this case is

$$S = \int d^4x \left(-\frac{1}{4} F_{\mu\nu} F^{\mu\nu} - J^\mu A_\mu \right)$$

The variation we will compute is δA_μ. We have

$$\delta S = \int d^4x \left(-\frac{1}{4} (\delta F_{\mu\nu}) F^{\mu\nu} - \frac{1}{4} F_{\mu\nu} (\delta F^{\mu\nu}) - J^\mu \delta A_\mu \right)$$

Let's consider the first term. Using the definition of the electromagnetic field tensor Eq. (2.32), we can write $F_{\mu\nu} = \partial_\mu A_\nu - \partial_\nu A_\mu$ and so we have

$$-\frac{1}{4} (\delta F_{\mu\nu}) F^{\mu\nu} = -\frac{1}{4} (\partial_\mu \delta A_\nu - \partial_\nu \delta A_\mu) F^{\mu\nu}$$

Now we integrate by parts, transferring the derivatives from the δA_ν terms to $F^{\mu\nu}$. This allows us to write

$$-\frac{1}{4}(\delta F_{\mu\nu})F^{\mu\nu} = -\frac{1}{4}(\partial_\mu \delta A_\nu - \partial_\nu \delta A_\mu)F^{\mu\nu}$$

$$= \frac{1}{4}(\partial_\mu F^{\mu\nu}\delta A_\nu - \partial_\nu F^{\mu\nu}\delta A_\mu)$$

But, repeated indices are dummy indices. So let's swap μ and ν in the second term, and write this as

$$\frac{1}{4}(\partial_\mu F^{\mu\nu}\delta A_\nu - \partial_\mu F^{\nu\mu}\delta A_\nu)$$

Now we use the antisymmetry of the electromagnetic tensor under interchange of the indices in Eq. (2.23). This will get rid of the minus sign on the second term, giving

$$\frac{1}{4}(\partial_\mu F^{\mu\nu}\delta A_\nu - \partial_\mu F^{\nu\mu}\delta A_\nu) = \frac{1}{4}(\partial_\mu F^{\mu\nu}\delta A_\nu + \partial_\mu F^{\mu\nu}\delta A_\nu)$$

$$= \frac{1}{2}\partial_\mu F^{\mu\nu}\delta A_\nu$$

So we have the result

$$-\frac{1}{4}(\delta F_{\mu\nu})F^{\mu\nu} = \frac{1}{2}\partial_\mu F^{\mu\nu}\delta A_\nu \qquad (2.36)$$

Now let's tackle the next term $-\frac{1}{4}F_{\mu\nu}(\delta F^{\mu\nu})$. In this case we have

$$-\frac{1}{4}F_{\mu\nu}(\delta F^{\mu\nu}) = -\frac{1}{4}F_{\mu\nu}\delta(\partial^\mu A^\nu - \partial^\nu A^\mu) = -\frac{1}{4}F_{\mu\nu}(\partial^\mu \delta A^\nu - \partial^\nu \delta A^\mu)$$

We're going to have to lower and raise some indices using the metric (see Chap. 1) to get this in the form of Eq. (2.36). The first step is to raise the indices on the field tensor term

$$-\frac{1}{4}F_{\mu\nu}(\partial^\mu \delta A^\nu - \partial^\nu \delta A^\mu) = -\frac{1}{4}g_{\mu\rho}g_{\nu\sigma}F^{\rho\sigma}(\partial^\mu \delta A^\nu - \partial^\nu \delta A^\mu)$$

Now let's move $F^{\rho\sigma}$ inside the parentheses and integrate by parts to transfer the derivative onto it

$$-\frac{1}{4}g_{\mu\rho}g_{\nu\sigma}F^{\rho\sigma}(\partial^\mu \delta A^\nu - \partial^\nu \delta A^\mu) = \frac{1}{4}g_{\mu\rho}g_{\nu\sigma}[\partial^\mu (F^{\rho\sigma})\delta A^\nu - \partial^\nu (F^{\rho\sigma})\delta A^\mu]$$

$$= \frac{1}{4}g_{\mu\rho}g_{\nu\sigma}\partial^\mu (F^{\rho\sigma})\delta A^\nu - \frac{1}{4}g_{\mu\rho}g_{\nu\sigma}\partial^\nu (F^{\rho\sigma})\delta A^\mu$$

Now we lower the indices on the derivative operators to give

$$= \frac{1}{4}g_{\mu\rho}g_{\nu\sigma}\partial^\mu (F^{\rho\sigma})\delta A^\nu - \frac{1}{4}g_{\mu\rho}g_{\nu\sigma}\partial^\nu (F^{\rho\sigma})\delta A^\mu$$

$$= \frac{1}{4}g_{\nu\sigma}\partial_\rho (F^{\rho\sigma})\delta A^\nu - \frac{1}{4}g_{\mu\rho}\partial_\sigma (F^{\rho\sigma})\delta A^\mu$$

Next, do the same thing to the vector potential terms

$$\frac{1}{4}g_{\nu\sigma}\partial_\rho (F^{\rho\sigma})\delta A^\nu - \frac{1}{4}g_{\mu\rho}\partial_\sigma (F^{\rho\sigma})\delta A^\mu$$

$$= \frac{1}{4}\partial_\rho (F^{\rho\sigma})\delta A_\sigma - \frac{1}{4}\partial_\sigma (F^{\rho\sigma})\delta A_\rho$$

Once again, repeated indices are dummy indices so we can change the labels. Focusing on the second term, let's change $\rho \to \nu$, $\sigma \to \mu$. We obtain

$$\frac{1}{4}\partial_\rho (F^{\rho\sigma})\delta A_\sigma - \frac{1}{4}\partial_\sigma (F^{\rho\sigma})\delta A_\rho$$

$$= \frac{1}{4}\partial_\rho (F^{\rho\sigma})\delta A_\sigma - \frac{1}{4}\partial_\mu (F^{\nu\mu})\delta A_\nu$$

Now apply the antisymmetry of the electromagnetic tensor to the second term, to get rid of the minus sign as shown here.

$$\frac{1}{4}\partial_\rho (F^{\rho\sigma})\delta A_\sigma - \frac{1}{4}\partial_\mu (F^{\nu\mu})\delta A_\nu$$

$$= \frac{1}{4}\partial_\rho (F^{\rho\sigma})\delta A_\sigma + \frac{1}{4}\partial_\mu (F^{\nu\mu})\delta A_\nu$$

The relabeling procedure can also be applied to the first term. This time we let $\rho \rightarrow \mu$, $\sigma \rightarrow \nu$ and we get

$$\frac{1}{4}\partial_\rho(F^{\rho\sigma})\delta A_\sigma + \frac{1}{4}\partial_\mu(F^{\mu\nu})\delta A_\nu = \frac{1}{4}\partial_\mu(F^{\mu\nu})\delta A_\nu + \frac{1}{4}\partial_\mu(F^{\mu\nu})\delta A_\nu$$

$$= \frac{1}{2}\partial_\mu(F^{\mu\nu})\delta A_\nu$$

We can combine this result with Eq. (2.36), and we see that the variation in the action becomes

$$\delta S = \int d^4x \left(\frac{1}{2}\partial_\mu F^{\mu\nu}\delta A_\nu + \frac{1}{2}\partial_\mu F^{\mu\nu}\delta A_\nu - J^\mu \delta A_\mu \right)$$

$$= \int d^4x (\partial_\mu F^{\mu\nu} - J^\mu)\delta A_\nu$$

We require that the variation in the action vanish, that is, $\delta S = 0$. Since the variation is arbitrary, δA_ν cannot vanish. Once more we arrive at the conclusion that the integral will be 0 only if the integrand is 0 everywhere in the domain. This means the action will only vanish if Maxwell's equations are satisfied, that is,

$$\partial_\mu F^{\mu\nu} - J^\mu = 0$$

Gauge Transformations

In this section we will consider an extension of the idea of invariance, by introducing what is known as a *gauge transformation*. Here we're only going to provide a brief introduction to these ideas; they will be elaborated as we proceed through the book.

The idea of a gauge transformation follows from studies of electricity and magnetism where we can make changes to the scalar and vector potentials ψ and $\vec{A}$ without changing the field equations and hence the physical fields $\vec{E}$ and $\vec{B}$ themselves. For example, recall that the magnetic field $\vec{B}$ can be defined in terms of the vector potential $\vec{A}$ using the curl relation

$$\vec{B} = \vec{\nabla} \times \vec{A}$$

A rule from vector calculus tells us that $\nabla \cdot (\vec{\nabla} \times \vec{F}) = 0$ for any vector field $\vec{F}$. Hence the Maxwell's equation $\nabla \cdot \vec{B} = 0$ is still satisfied when we make the

definition $\vec{B} = \vec{\nabla} \times \vec{A}$. Now let f be some scalar function and define a new vector potential $\vec{A}'$ via

$$\vec{A}' = \vec{A} + \vec{\nabla}f$$

We also know from vector calculus that $\vec{\nabla} \times \vec{\nabla}f = 0$. Hence we can add a term of the form $\vec{\nabla}f$ to the vector potential with impunity if it's *mathematically convenient*. The magnetic field $\vec{B}$ is unchanged since

$$\vec{B} = \vec{\nabla} \times \vec{A}' = \vec{\nabla} \times (\vec{A} + \vec{\nabla}f) = \vec{\nabla} \times \vec{A} + \vec{\nabla} \times \vec{\nabla}f = \vec{\nabla} \times \vec{A}$$

Therefore the magnetic field, the physical quantity of interest, is unchanged by a transformation of the form $\vec{A}' = \vec{A} + \vec{\nabla}f$. We call this type of transformation in electrodynamics a *gauge transformation*. There are different choices that can be made when implementing a gauge transformation. For example, if we impose the requirement that $\nabla \cdot \vec{A} = 0$, we call this the *Coulomb gauge*. On the other hand, if $\nabla \cdot \vec{A} = -\mu_0 \varepsilon_0 \frac{\partial \psi}{\partial t}$, we have what is known as the *Lorentz gauge*.

In field theory, we arrive at a similar notion by considering transformations to the field that leave the Lagrangian invariant. To see how this works in field theory, let's consider a simple example, the Klein-Gordon Lagrangian with a complex field.

$$\mathcal{L} = \partial_\mu \varphi^\dagger \partial^\mu \varphi - m^2 \varphi^\dagger \varphi \tag{2.37}$$

Let U be a unitary transformation applied to the fields such that U does not, in any way, depend on spacetime. That is, we let

$$\varphi \to U\varphi \tag{2.38}$$

Then

$$\varphi^\dagger \to \varphi^\dagger U^\dagger \tag{2.39}$$

Since the transformation is unitary, we also know that $UU^\dagger = U^\dagger U = 1$. Let's see how this transformation affects the Lagrangian [Eq. (2.37)]. Looking at each term individually, we start with the first term where we have

$$\partial_\mu \varphi^\dagger \partial^\mu \varphi \to \partial_\mu (\varphi^\dagger U^\dagger) \partial^\mu (U\varphi)$$

But U does not depend on spacetime in any way, so the derivative operators do not affect it. Hence

$$\partial_\mu(\varphi^\dagger U^\dagger)\partial^\mu(U\varphi) = \partial_\mu(\varphi^\dagger)(U^\dagger U)\partial^\mu(\varphi) = \partial_\mu\varphi^\dagger\partial^\mu\varphi$$

Similarly for the second term in Eq. (2.37), we have

$$m^2\varphi^\dagger\varphi \to m^2(\varphi^\dagger U^\dagger)(U\varphi) = m^2\varphi^\dagger(U^\dagger U)\varphi = m^2\varphi^\dagger\varphi$$

Therefore we see that under the transformation [Eq. (2.38)], the Lagrangian [Eq. (2.37)] is invariant. Since U is a constant, we can write it in the form

$$U = e^{i\Lambda}$$

where Λ is a constant. However, in certain contexts, Λ can also be a matrix so long as it is Hermitian. Since it is a constant we say that the gauge transformation in this case is *global*, it does not depend on spacetime in any way.

LOCAL GAUGE TRANSFORMATIONS

The gauge transformations that are of interest are local transformations that do depend on spacetime. This type of transformation satisfies the requirements of special relativity—that no signal can travel faster than the speed of light.

Let's return to the transformation $\varphi \to U\varphi$. In the following, we still consider U to be a unitary transformation, however now we let it depend on spacetime so that $U = U(x)$. This means that terms like $\partial_\mu U$ will not vanish. Next consider how the Lagrangian is changed by the transformation $\varphi \to U\varphi$. We use the same Lagrangian we considered in the previous section, namely $\mathcal{L} = \partial_\mu\varphi^\dagger\partial^\mu\varphi - m^2\varphi^\dagger\varphi$. This time the second term in the Lagrangian remains invariant as shown here.

$$m^2\varphi^\dagger\varphi \to m^2\varphi^\dagger U^\dagger(x)U(x)\varphi = m^2\varphi^\dagger\varphi$$

The first term will change due to the spacetime dependence of $U = U(x)$, that is,

$$\partial_\mu\varphi^\dagger \to \partial_\mu(\varphi^\dagger U^\dagger) = (\partial_\mu\varphi^\dagger)U^\dagger + \varphi^\dagger\partial_\mu(U^\dagger)$$

Similarly we find that

$$\partial_\mu\varphi \to \partial_\mu(U\varphi) = (\partial_\mu U)\varphi + U\partial_\mu(\varphi)$$

We can write this in a more useful form by exploiting the fact that U is unitary, like

$$\partial_\mu \varphi \to \partial_\mu (U\varphi) = (\partial_\mu U)\varphi + U\partial_\mu(\varphi)$$
$$= UU^\dagger(\partial_\mu U)\varphi + U\partial_\mu(\varphi)$$
$$= U[\partial_\mu \varphi + (U^\dagger \partial_\mu U)\varphi]$$

To maintain invariance, we would like to cancel the extra term that has shown up here. In other words we want to get rid of

$$(U^\dagger \partial_\mu U)\varphi$$

We can do this by introducing a new object, a spacetime dependent field $A_\mu = A_\mu(x)$ called the *gauge potential*. It is given the label A_μ due to the analogy with electrodynamics—we are introducing a hidden field to keep the form of the Lagrangian invariant. As we will see later this has dramatic consequences, and this is one of the most important techniques in quantum field theory.

We introduce a *covariant derivative* that acts on the field as

$$D_\mu \varphi = \partial_\mu \varphi - iA_\mu \varphi \qquad (2.40)$$

(for readers of *Relativity Demystified*, notice the similarity to general relativity). Remember what the word *covariant* means—the form of the equations won't change. We introduce this derivative operator to keep the Lagrangian invariant under a local gauge transformation. Recalling the effect of a global gauge transformation $\varphi \to U\varphi$, we found that

$$\partial_\mu \varphi \to U\partial_\mu \varphi$$

The covariant derivative Eq. (2.40) will allow us to recover this result in the case of the local gauge transformation. That is, $\varphi \to U(x)\varphi$ will lead to $D_\mu \varphi \to U(x)D_\mu \varphi$. This can be accomplished if we define A_μ such that it obeys the similarity transformations.

$$A_\mu \to UA_\mu U^\dagger + iU\partial_\mu U^\dagger \qquad (2.41)$$

Note that some authors use the semicolon notation $D_\mu \varphi = \varphi_{;\mu}$ to represent the covariant derivative.

EXAMPLE 2.6
Consider a charged scalar particle of mass m with charge q and describe a suitable modification of the derivative operator $\partial_\mu \to \partial_\mu + qA_\mu$ that will yield a Lorentz invariant, real Lagrangian.

SOLUTION

The field equations for a complex field corresponding to a charged scalar particle are obtained using the covariant derivative in Eq. (2.40), in this case with the form

$$i\partial_\mu \to i\partial_\mu - qA_\mu$$

The Lagrangian is

$$\mathcal{L} = (i\partial_\mu + qA_\mu)\varphi^*(i\partial_\mu - qA_\mu)\varphi - m^2\varphi^*\varphi$$

Variation of this Lagrangian leads to the equation of motion as shown here.

$$(i\partial_\mu - qA_\mu)^2\varphi - m^2\varphi = 0$$

Summary

The Lagrangian is the difference of kinetic and potential energy given by $L = T - V$. It can be used to obtain the equations of motion for a system by applying variational calculus to the action S which is the integral of the Lagrangian. When extended to continuous systems, these techniques can be applied to fields to obtain the field equations. For problems with expected symmetries or conserved quantities, we require that the Lagrangian remains invariant under corresponding transformations. When a given transformation leaves the form of the Lagrangian invariant, we call that transformation a symmetry. Noether's theorem allows us to derive conservation laws, including conserved charges and currents from symmetries in the Lagrangian. Symmetries can be local, meaning that they are spacetime dependent, or they can be internal symmetries that are intrinsic to the system at hand. To maintain covariance of the equations, a covariant derivative must be introduced which necessitates the use of a gauge potential.

Quiz

1. Find the equation of motion for a forced harmonic oscillator with Lagrangian

$$L = \frac{1}{2}m\dot{x}^2 - \frac{1}{2}m\omega^2 x^2 + \alpha x$$

Here α is a constant.

2. Consider a Lagrangian given by

$$\mathcal{L} = -\frac{1}{2}\partial_\mu \varphi \partial^\mu \varphi - \frac{1}{2}m^2\varphi^2 - V(\varphi)$$

  (a) Write down the field equations for this system.

  (b) Find the canonical momentum density $\pi(x)$.

  (c) Write down the Hamiltonian.

3. Consider a free scalar field with Lagrangian $\mathcal{L} = \partial_\mu \varphi \partial^\mu \varphi$ and suppose that the field varies according to $\varphi \to \varphi + \alpha$, where α is a constant. Determine the conserved current.

4. Refer to the Lagrangian for a complex scalar field Eq. (2.37). Determine the equations of motion obeyed by the fields φ and $\varphi^\dagger$.

5. Refer to Eq. (2.37) and calculate the conserved charge.

6. Consider the action $S = \frac{1}{4}\int F_{\mu\nu}F^{\mu\nu}d^4x$. Vary the potential according to $A_\mu \to A_\mu + \partial_\mu \varphi$ where φ is a scalar field. Determine the variation in the action.

CHAPTER 3

An Introduction to Group Theory

An abstract branch of mathematics called *group theory* plays a fundamental role in modern particle physics. The reason it does so is because group theory is related to symmetry. For example, an important group in physics is the *rotation group,* which is related to the fact that the laws of physics don't change if you rotate your frame of reference. In general what we are after is a set of equations, or laws of physics, that keep the same mathematical form under various transformations. Group theory is related to this type of symmetry.

Definitions

We begin our discussion of group theory in a somewhat abstract manner, but later we will get to some more concrete material. Let's state what a group is and the four properties it must have.

A *group G* is a set of elements $\{a, b, c, \ldots\}$ which includes a "multiplication" or composition rule such that if $a \in G$ and $b \in G$, then the product is also a member of the group, that is,

$$ab \in G \tag{3.1}$$

We call this the *closure* property if $ab = ba,$ we say that the group is *abelian*. On the other hand if $ab \neq ba$, the group is *nonabelian*. The multiplication or composition rule is meant to convey a product in the abstract sense, the actual implementation of this rule can vary from group to group.

A group G must satisfy four axioms. These are

1. Associativity: The multiplication rule is associative, meaning $(ab)c = a(bc)$.

2. Identity element: The group has an identity element e that satisfies $ae = ea = a$. The identity element for the group is unique.

3. Inverse element: For every element $a \in G$, there exists an inverse which we denote by a^{-1} such that $aa^{-1} = a^{-1}a = e$.

4. Order: The order of the group is the number of elements that belong to G.

Representation of the Group

In particle physics we are often interested in what is called a *representation* of the group. Let's denote a representation by F. A representation is a mapping that takes group elements $g \in G$ into linear operators F that preserve the composition rule of the group in the sense that

- $F(a)F(b) = F(ab)$.

- The representation also preserves the identity, that is, $F(e) = I$.

Suppose that $a, b \in G$ and $f \in H$ where H is some other group. If the composition rule satisfies

$$f(a)f(b) = f(ab)$$

we say that G is *homomorphic* to H. This is a fancy way of saying that the two groups have a similar structure.

EXAMPLE 3.1A
Does the set of all integers form a group under addition?

SOLUTION

The set of all integers forms a group under addition. We can take the composition rule to be addition. Let $z_1 \in Z$ and $z_2 \in Z$. Clearly the sum

$$z_1 + z_2 \in Z$$

is another integer, so it belongs to the group. For the identity, we can take $e = 0$ since

$$z + 0 = 0 + z = z$$

for any $z \in Z$. Addition is commutative, that is,

$$z_1 + z_2 = z_2 + z_1$$

Therefore the group is abelian. The inverse of z is just $-z$, since

$$z + (-z) = e = 0$$

which satisfies $aa^{-1} = a^{-1}a = e$.

EXAMPLE 3.1

Does the set of all integers form a group under multiplication?

SOLUTION

The set of all integers does not form a group under multiplication. We can take the composition rule to be multiplication. Let $z_1 \in Z$ and $z_2 \in Z$. Clearly the product

$$z_1 z_2 \in Z$$

is another integer, so it belongs to the group. For the identity, we can take $e = 1$ since

$$z \times 1 = 1 \times z = z$$

for any $z \in Z$. Multiplication is commutative, that is,

$$z_1 \times z_2 = z_2 \times z_1$$

Therefore the group is abelian. The problem is the inverse: $1/z$ is not an integer.

$$z \times \frac{1}{z} = e = 1$$

So even though an inverse exists, the inverse is not in the group, and therefore the set of all integers does not form a group under multiplication.

Group Parameters

An ordinary function of position is specified by an input x, that is, we have $y = f(x)$. In an analogous way, a group can also be a function of one or more inputs that we call *parameters*.

Let a group G be such that individual elements $g \in G$ are specified by a finite set of parameters, say n of them. If we denote the set of parameters by

$$\{\theta_1, \theta_2, \ldots, \theta_n\}$$

The group element is then written as

$$g = G(\theta_1, \theta_2, \ldots, \theta_n)$$

The identity is the group element where the parameters are all set to 0.

$$e = G(0, 0, \ldots, 0)$$

Lie Groups

While there are discrete groups with a finite number of elements, most of the groups we will be concerned with have an infinite number of elements. However, they have a finite set of continuously varying parameters.

In the expression $g = G(\theta_1, \theta_2, \ldots, \theta_n)$ we have suggestively labeled the parameters as angles, since several important groups in physics are related to rotations. The angles vary continuously over a finite range $0 \ldots 2\pi$. In addition, the group is parameterized by a finite number of parameters, the angles of rotation.

So, if a group G

- Depends on a finite set of continuous parameters θ_i
- Derivatives of the group elements with respect to all the parameters exist

we call the group a *Lie group*. To simplify the discussion we begin with a group with a single parameter θ. We obtain the identity element by setting $\theta = 0$

$$g(\theta)\big|_{\theta=0} = e \tag{3.2}$$

By taking derivatives with respect to the parameters and evaluating the derivative at $\theta = 0$, we obtain the *generators* of the group. Let us denote an abstract generator by X. Then

$$X = \frac{\partial g}{\partial \theta}\Big|_{\theta=0} \tag{3.3}$$

More generally, if there are n parameters of the group, then there will be n generators such that each generator is given by

$$X_i = \frac{\partial g}{\partial \theta_i}\Big|_{\theta_i=0} \tag{3.4}$$

Rotations have a special property, in that they are length preserving (that is, rotate a vector and it maintains the same length). A rotation by $-\theta$ undoes a rotation by θ, hence rotations have an orthogonal or unitary representation. In the case of quantum theory, we seek a unitary representation of the group and choose the generators X_i to be Hermitian. In this case

$$X_i = -i\frac{\partial g}{\partial \theta_i}\Big|_{\theta_i=0} \tag{3.5}$$

For some finite θ, the generators allow us to define a representation of the group. Consider a small real number $\varepsilon > 0$ and use a Taylor expansion to form a representation of the group (which we denote by D)

$$D(\varepsilon\theta) \approx 1 + i\varepsilon\theta X$$

If $\theta = 0$, then clearly the representation gives the identity. You will also recall that the exponential function has a series expansion

$$e^x = 1 + x + \frac{1}{2!}x^2 + \cdots$$

So we can define the representation of the group in terms of the exponential using

$$D(\theta) = \lim_{n\to\infty}\left(1 + i\frac{\theta X}{n}\right)^n = e^{i\theta X} \tag{3.6}$$

Notice that if X is Hermitian, $X = X^\dagger$ and the representation of the group is unitary, since

$$D^\dagger(\theta) = (e^{i\theta X})^\dagger = e^{-i\theta X^\dagger} = e^{-i\theta X}$$
$$\Rightarrow D^\dagger(\theta)D(\theta) = (e^{-i\theta X})(e^{i\theta X}) = 1$$

One reason that the generators of a group are important is that they form a vector space. This means we can add two generators of the group together to obtain a third generator, and we can multiply generators by a scalar and still have a generator of the group. A complete vector space can be used as a basis for representing other vector spaces, hence the generators of a group can be used to represent other vector spaces. For example, the Pauli matrices from quantum mechanics can be used to describe any 2×2 matrix.

The character of the group is defined in terms of the generators in the following sense. The generators satisfy a commutation relation we write as

$$[X_i, X_j] = i f_{ijk} X_k \tag{3.7}$$

This is called the *Lie algebra* of the group. The quantities f_{ijk} are called the *structure constants* of the group. Looking at the commutation relation for the Lie algebra, Eq. (3.7), you should recognize the fact that you've already been working with group generators in your studies of non-relativistic quantum mechanics. You'll see this explicitly when we discuss the Pauli spin matrices later on.

The Rotation Group

The *rotation group* is the set of all rotations about the origin. A key feature is that rotations preserve the lengths of vectors. This mathematical property is expressed by saying the matrices are unitary. It is easy to see that the set of rotations forms a group. Let's check off each of the basic properties that a group must have.

The first is a group composition rule. Remember that if $a \in G$ and $b \in G$, if G is a group then $ab \in G$ as well. Now let R_1 and R_2 be two rotations. It is clear that the composition of these rotations, say by performing the rotation R_1 first followed by the rotation R_2, is itself just another rotation, as shown here.

$$R_3 = R_2 R_1$$

That is, performing the two rotations as described is the same as doing the single rotation R_3. Hence R_3 is a member of the rotation group. Next, we illustrate how two rotations in sequence are the same as a single rotation that is the sum of the two angles:

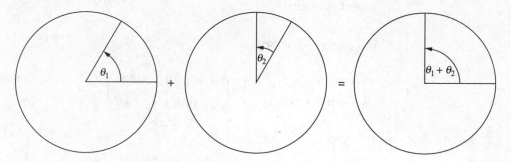

It is not the case that rotations commute. That is, in general

$$R_1 R_2 \neq R_2 R_1$$

To this, put a book on the table in front of you. Then rotate it about two different axes, and then try the experiment again doing the same rotations but in different order. You will see that the end results are not the same. Therefore, the rotation group is nonabelian. However, it's easy to see that rotations are associative, like

$$R_1 (R_2 R_3) = (R_1 R_2) R_3$$

The rotation group has an identity element—this is, simply doing no rotation at all. The inverse of a rotation is simply the rotation carried out in the opposite direction.

Representing Rotations

Let x_i be the coordinates of a two-dimensional vector and let x_i' be the coordinates of the vector rotated by an angle θ in the plane. The components of the two vectors are related by a transformation as

$$x_j' = R_{ij} x_i$$

where R_{ij} is a *rotation matrix*. This is a representation of the rotation group. Specifically, a rotation by an angle θ (in two dimensions) can be represented by the matrix

$$R(\theta) = \begin{pmatrix} \cos\theta & \sin\theta \\ -\sin\theta & \cos\theta \end{pmatrix}$$

So that

$$x_1' = \cos\theta\, x_1 + \sin\theta\, x_2$$

$$x_2' = -\sin\theta\, x_1 + \cos\theta\, x_2$$

$$\begin{pmatrix} x_1' \\ x_2' \end{pmatrix} = \begin{pmatrix} \cos\theta & \sin\theta \\ -\sin\theta & \cos\theta \end{pmatrix}\begin{pmatrix} x_1 \\ x_2 \end{pmatrix} = \begin{pmatrix} x_1\cos\theta + x_2\sin\theta \\ -x_1\sin\theta + x_2\cos\theta \end{pmatrix}$$

Let's write down the transpose of the rotation matrix as

$$R^T(\theta) = \begin{pmatrix} \cos\theta & \sin\theta \\ -\sin\theta & \cos\theta \end{pmatrix}$$

Notice that

$$R(\theta)R^T(\theta) = \begin{pmatrix} \cos\theta & \sin\theta \\ -\sin\theta & \cos\theta \end{pmatrix}\begin{pmatrix} \cos\theta & -\sin\theta \\ \sin\theta & \cos\theta \end{pmatrix}$$

$$RR^T = \begin{pmatrix} \cos\theta & \sin\theta \\ -\sin\theta & \cos\theta \end{pmatrix}\begin{pmatrix} \cos\theta & -\sin\theta \\ \sin\theta & \cos\theta \end{pmatrix}$$

$$= \begin{pmatrix} \cos^2\theta + \sin^2\theta & -\cos\theta\sin\theta + \cos\theta\sin\theta \\ -\sin\theta\cos\theta + \cos\theta\sin\theta & \cos^2\theta + \sin^2\theta \end{pmatrix}$$

$$= \begin{pmatrix} 1 & 0 \\ 0 & 1 \end{pmatrix}$$

This tells us that the transpose of the matrix is the inverse group element, since multiplying two matrices together gives the identity. You can see why this is true using basic trigonometry: $\sin(-\theta) = -\sin(\theta)$ and $\cos(-\theta) = \cos(\theta)$.

$$R^T(\theta) = R(-\theta) = \begin{pmatrix} \cos\theta & -\sin\theta \\ \sin\theta & \cos\theta \end{pmatrix}$$

So, the inverse of $R(\theta)$ is $R(-\theta)$: $\sin(\theta - \theta) = 0$ and $\cos(\theta - \theta) = 1$.

$$R(\theta)R^T(\theta) = R(\theta)R(-\theta) = R(\theta)R^{-1}(\theta) = \begin{pmatrix} 1 & 0 \\ 0 & 1 \end{pmatrix} = I$$

In group theory, various groups are classified according to the determinants of the matrices that represent the groups. Notice that in this case

$$\det R(\theta) = \det \begin{pmatrix} \cos\theta & \sin\theta \\ -\sin\theta & \cos\theta \end{pmatrix}$$
$$= \cos^2\theta + \sin^2\theta = 1$$

In general, the determinant will not be +1. However, when this condition is satisfied, the rotation matrix corresponds to a *proper rotation*.

A more readable notation is to make the angle implicit and to use a subscript. That is,

$$R(\theta_1) \rightarrow R_1$$

This is a better way to discuss problems to multiple angles.

Now, if $\det R_1 = 1$ and $\det R_2 = 1$, then the product is unity. We see this from the property of multiplication for determinants.

$$\det(R_1 R_2) = \det R_1 \det R_2 = (1)(1) = 1$$

The product of two rotations also has an inverse, since

$$R_1 R_2 (R_1 R_2)^T = R_1 R_2 R_2^T R_1^T = R_1 R_1^T = I$$

Another way to look at this is to use the closure property of this group. Call the successive rotations R_1 and R_2. Taken together they form the group element R_3:

$$R_1 R_2 = R_3$$

As before, the transpose is the inverse:

$$R_3^T = R_3^{-1}$$

In terms of the components, we have

$$R_3^T = (R_1 R_2)^T = R_2^T R_1^T \quad \text{and} \quad R_3^{-1} = (R_1 R_2)^{-1} = R_2^{-1} R_1^{-1}$$

We now see

$$R_3 R_3^T = I$$

as above.

So the matrix representation preserves the properties of the rotation group. Now that we've introduced the notion of a group, we will explore groups important for particle physics.

SO(N)

The group $SO(N)$ are special orthogonal $N \times N$ matrices. The term *special* is a reference to the fact that these matrices have determinant +1. A larger group, one that contains $SO(N)$ as a subgroup, is the group $O(N)$ which are orthogonal $N \times N$ matrices that can have arbitrary determinant. Generally speaking, rotations can be represented by orthogonal matrices, which themselves form a group. So the group $SO(3)$ is a representation of rotations in three dimensions, and the group consists of 3×3 orthogonal matrices with determinant +1.

A matrix O is called *orthogonal* if the transpose O^T is the inverse, that is,

$$OO^T = O^T O = I$$
$$\Rightarrow O \text{ is orthogonal} \tag{3.8}$$

As we stated above, a special orthogonal matrix is one with positive unit determinant

$$\det O = +1 \tag{3.9}$$

Now let's turn to a familiar case, $SO(3)$. This group has three parameters, the three angles defining rotations about the x, y, and z axes. Let these angles be denoted by ς, ϕ, and θ. Then

$$R_x(\varsigma) = \begin{pmatrix} 1 & 0 & 0 \\ 0 & \cos\varsigma & \sin\varsigma \\ 0 & -\sin\varsigma & \cos\varsigma \end{pmatrix} \tag{3.10}$$

$$R_y(\phi) = \begin{pmatrix} \cos\phi & 0 & \sin\phi \\ 0 & 1 & 0 \\ -\sin\phi & 0 & \cos\phi \end{pmatrix} \tag{3.11}$$

$$R_z(\theta) = \begin{pmatrix} \cos\theta & \sin\theta & 0 \\ -\sin\theta & \cos\theta & 0 \\ 0 & 0 & 1 \end{pmatrix} \tag{3.12}$$

These matrices are the representation of rotations in three dimensions. Rotations in three or more dimensions do not commute, and it is an easy although tedious exercise to show that the rotation matrices written down here do not commute either.

The task now is to find the generators for each group parameter. We do this using Eq. (3.5). Starting with $R_x(\varsigma)$, we have

$$\frac{dR_x}{d\varsigma} = \begin{pmatrix} 0 & 0 & 0 \\ 0 & -\sin\varsigma & \cos\varsigma \\ 0 & -\cos\varsigma & \sin\varsigma \end{pmatrix}$$

Now we let $\varsigma \to 0$ to obtain the generator

$$J_x = -i\frac{dR_x}{d\varsigma}\Big|_{\varsigma=0} = \begin{pmatrix} 0 & 0 & 0 \\ 0 & 0 & -i \\ 0 & i & 0 \end{pmatrix} \tag{3.13}$$

Next, we compute the generator for rotations about the y axis evaluated at $\phi = 0$.

$$J_y = -i\frac{dR_y}{d\phi} = \begin{pmatrix} 0 & 0 & -i \\ 0 & 0 & 0 \\ i & 0 & 0 \end{pmatrix} \tag{3.14}$$

And finally, for rotations about the z axis we find

$$J_z = -i\frac{dR_z}{d\theta}\Big|_{\theta=0} = \begin{pmatrix} 0 & -i & 0 \\ i & 0 & 0 \\ 0 & 0 & 0 \end{pmatrix}$$

These matrices are, of course, the familiar angular momentum matrices. So we've discovered the famous result that the angular momentum operators are the generators of rotations. We can use the generators to build infinitesimal rotations. For example, an infinitesimal rotation about the z axis by an angle $\varepsilon\theta$, where ε is a small positive parameter, is written as

$$R_z(\varepsilon\theta) = 1 + i J_z \varepsilon\theta$$

From quantum mechanics, you already know the algebra of the group. This is just the commutation relations satisfied by the angular momentum operators.

$$[J_i, J_j] = i\varepsilon_{ijk} J_k \tag{3.15}$$

The structure constants in this case are given by the Levi-Civita tensor, which takes on values of +1, −1, or 0 according to

$$\varepsilon_{123} = \varepsilon_{312} = \varepsilon_{231} = +1$$
$$\varepsilon_{132} = \varepsilon_{321} = \varepsilon_{213} = -1 \tag{3.16}$$

with all other combinations of the indices giving 0.

EXAMPLE 3.2
Show that the representation of the rotation group is of the form $e^{iJ_y \phi}$.

SOLUTION
What we need to show is that $R_y(\phi) = e^{iJ_y \phi}$. This is easy to do by just writing down the first few terms in the Taylor series expansion of the exponential. Remember Euler's formula which tells us that

$$e^{i\theta} = \cos\theta + i\sin\theta$$

From this we can extract the expansions for the sine and cosine.

$$\cos\theta = 1 - \frac{1}{2}\theta^2 + \frac{1}{4!}\theta^4 + \cdots$$

$$\sin\theta = \theta - \frac{1}{3!}\theta^3 + \frac{1}{5!}\theta^5 + \cdots$$

Now,

$$e^{iJ_y \phi} = 1 + iJ_y \phi - \frac{1}{2}J_y^2 \phi^2 - i\frac{1}{3!}J_y^3 \phi^3 + \cdots$$

What are the powers of the J_y matrix? A quick calculation shows that J_y^n takes one of two values depending upon whether n is odd or even.

$$J_y = J_y^3 = J_y^5 = \cdots = \begin{pmatrix} 0 & 0 & -i \\ 0 & 0 & 0 \\ i & 0 & 0 \end{pmatrix} \text{ and } J_y^2 = J_y^4 = J_y^6 = \cdots = \begin{pmatrix} 1 & 0 & 0 \\ 0 & 0 & 0 \\ 0 & 0 & 1 \end{pmatrix}$$

The expansion becomes

$$
e^{iJ_y\phi} =
\begin{pmatrix} 1 & 0 & 0 \\ 0 & 1 & 0 \\ 0 & 0 & 1 \end{pmatrix}
+ i\begin{pmatrix} 0 & 0 & i \\ 0 & 0 & 0 \\ -i & 0 & 0 \end{pmatrix}\phi
- \frac{1}{2}\begin{pmatrix} 0 & 0 & i \\ 0 & 0 & 0 \\ -i & 0 & 0 \end{pmatrix}^2\phi^2
- i\frac{1}{3!}\begin{pmatrix} 0 & 0 & i \\ 0 & 0 & 0 \\ -i & 0 & 0 \end{pmatrix}^3\phi^3 + \cdots
$$

$$
= \begin{pmatrix} 1 & 0 & 0 \\ 0 & 1 & 0 \\ 0 & 0 & 1 \end{pmatrix}
+ \begin{pmatrix} 0 & 0 & -\phi \\ 0 & 0 & 0 \\ \phi & 0 & 0 \end{pmatrix}
- \frac{1}{2}\begin{pmatrix} \phi^2 & 0 & 0 \\ 0 & 0 & 0 \\ 0 & 0 & \phi^2 \end{pmatrix}
- \frac{1}{3!}\begin{pmatrix} 0 & 0 & \phi^3 \\ 0 & 0 & 0 \\ \phi^3 & 0 & 0 \end{pmatrix} + \cdots
$$

$$
= \begin{pmatrix}
1-\frac{1}{2}\phi^2+\cdots & 0 & -\left(\phi-\frac{1}{3!}\phi^3+\cdots\right) \\
0 & 0 & 0 \\
\phi-\frac{1}{3!}\phi^3+\cdots & 0 & 1-\frac{1}{2}\phi^2+\cdots
\end{pmatrix}
$$

$$
= \begin{pmatrix} \cos\phi & 0 & -\sin\phi \\ 0 & 0 & 0 \\ \sin\phi & 0 & \cos\phi \end{pmatrix}
$$

So, rotations about the x, y, and z axes are represented by

$$ R_x(\varsigma)=e^{iJ_x\varsigma} \qquad R_y(\phi)=e^{iJ_y\phi} \qquad R_z(\theta)=e^{iJ_z\theta} \tag{3.17} $$

A rotation about an arbitrary axis defined by a unit vector $\vec{n}$ is given by

$$ R_n(\vec{\theta})=e^{i\vec{J}\cdot\vec{\theta}} \tag{3.18} $$

As stated earlier, orthogonal transformations (rotations) preserve the lengths of vectors. We say that the length of a vector is *invariant* under rotation. This means that given a vector $\vec{x}$ with length $\vec{x}^2=x^2+y^2+z^2$, when we transform it under a rotation $\vec{x}'=Rx$ we have

$$ \vec{x}'^2=\vec{x}^2 $$
$$ \Rightarrow x'^2+y'^2+z'^2=x^2+y^2+z^2 \tag{3.19} $$

The preservation of the lengths of vectors by orthogonal transformations will be important in establishing a correspondence between the unitary transformation $SU(2)$ (see in the next section, Unitary Groups) and rotations in three dimensions represented by $SU(3)$.

Unitary Groups

In particle physics unitary groups play a special role. This is due to the fact that unitary operators play an important role in quantum theory. Specifically, unitary operators preserve inner products—meaning that a unitary transformation leaves the probabilities for different transitions among the states unaffected. That is, quantum physics is invariant under a unitary transformation. As a result unitary groups play a special role in quantum field theory.

When the physical predictions of a theory are invariant under the action of some group, we can represent the group by a unitary operator U. Moreover, this unitary operator commutes with the Hamiltonian as shown here.

$$[U, H] = 0$$

The unitary group $U(N)$ consists of all $N \times N$ unitary matrices. Special unitary groups, denoted by $SU(N)$ are $N \times N$ unitary matrices with positive unit determinant. The dimension of $SU(N)$ and hence the number of generators, is given by $N^2 - 1$. Therefore,

- $SU(2)$ has $2^2 - 1 = 3$ generators.
- $SU(3)$ has $3^2 - 1 = 8$ generators.

The rank of $SU(N)$ is $N - 1$. So

- The rank of $SU(2)$ is $2 - 1 = 1$.
- The rank of $SU(3)$ is $3 - 1 = 2$.

The rank gives the number of operators in the algebra that can be simultaneously diagonalized.

The simplest unitary group is the group $U(1)$. A "1×1" matrix is just a complex number written in polar representation. In another way, we can say a $U(1)$ symmetry has a single parameter θ and is written as

$$U = e^{-i\theta}$$

where θ is a real parameter. It is completely trivial to see that $U(1)$ is abelian, since

$$U_1 U_2 = e^{-i\theta_1} e^{-i\theta_2} = e^{-i\theta_2} e^{-i\theta_1} = U_2 U_1$$

$$U_1 U_2 = e^{-i\theta_1} e^{-i\theta_2} = e^{-i(\theta_1 + \theta_2)} = e^{-i(\theta_2 + \theta_1)} = e^{-i\theta_2} e^{-i\theta_1} = U_2 U_1$$

We will see that many Lagrangians in field theory are invariant under a $U(1)$ transformation. When you look at the problem in the complex plane this invariance becomes obvious. Consider the arbitrary complex number $z = re^{i\alpha}$. When we multiply z by $e^{i\theta}$ we get

$$e^{i\theta}z = e^{i\theta}re^{i\alpha} = re^{i(\theta+\alpha)}$$

The new complex number has the same length, r, and the angle is increased by θ.

For example, the Lagrangian for a complex scalar field

$$L = \partial_\mu \varphi^* \partial^\mu \varphi - m^2 \varphi^* \varphi$$

is invariant under the transformation

$$\varphi \to e^{-i\theta}\varphi$$

As described in Chap. 2, when a Lagrangian is invariant under a transformation there is a symmetry, and in this case there is a $U(1)$ symmetry. Force-mediating particles, called *gauge bosons*, will be associated with unitary symmetries like this one. We will see later that when considering electrodynamics, the gauge boson associated with the $U(1)$ symmetry of quantum electrodynamics is the photon.

We will see in later chapters that a $U(1)$ symmetry also manifests itself in terms of the conservation of various quantum numbers. If there is a $U(1)$ symmetry associated with a quantum number a, then

$$U = e^{-ia\theta}$$

The importance of the $U(1)$ symmetry is that the Hamiltonian H is invariant under the transformation $e^{-ia\theta}He^{ia\theta}$, that is,

$$UHU^\dagger = H$$

Again we see that adjoint $U^\dagger$ (the Hermitian conjugate) is also the inverse.

$$U(\theta)U^\dagger(\theta) = U(\theta)U(-\theta) = 1$$

We will see that such symmetries are present in quantum field theory with conservation of lepton and baryon number, for example.

In summary, an element of the group $U(1)$ is a complex number of unit length written as

$$U = e^{-i\theta} \tag{3.20}$$

where θ is a number. This is the familiar unit circle.

Moving right along, the next non-trivial unitary group is $U(2)$, which is the set of all 2×2 unitary matrices. Being unitary these matrices satisfy

$$UU^\dagger = U^\dagger U = I \tag{3.21}$$

For physics, we are interested in a subgroup of $U(2)$, which is the set of all 2×2 unitary matrices with determinant +1. This group is called $SU(2)$. The generators of $SU(2)$ are the Pauli matrices, which we reproduce here for your convenience

$$\sigma_1 = \begin{pmatrix} 0 & 1 \\ 1 & 0 \end{pmatrix} \quad \sigma_2 = \begin{pmatrix} 0 & -i \\ i & 0 \end{pmatrix} \quad \sigma_3 = \begin{pmatrix} 1 & 0 \\ 0 & -1 \end{pmatrix} \tag{3.22}$$

Now we see how the rank of a unitary group comes into play. The rank of $SU(2)$ is 1, and there is one diagonalized operator σ_3 in the basis we have chosen. The generators of $SU(2)$ are actually taken to be $\frac{1}{2}\sigma_i$ and the Lie algebra is the familiar commutation relations that are satisfied by the Pauli matrices

$$\left[\frac{\sigma_i}{2}, \frac{\sigma_j}{2} \right] = i\varepsilon_{ijk} \frac{\sigma_k}{2} \tag{3.23}$$

The similar algebraic structure between $SU(2)$ as indicated in Eq. (3.23) and $SO(3)$ as indicated by Eq. (3.5) indicates that there will be a correspondence between these two groups. Since the Pauli matrices do not commute, $SU(2)$ is nonabelian, as seen in Eq. (3.23). Recalling that the rank of $SU(2)$ is 1, there is one diagonal generator which we have chosen to be σ_3.

An element of $SU(2)$ can be written as

$$U = e^{i\sigma_j \alpha_j/2} \tag{3.24}$$

where σ_i is one of the Pauli matrices and α_j is a number. To understand the presence of the factor ½ in Eqs. (3.23) and (3.24), let's explore the correspondence between $SU(2)$ and $SO(3)$. We saw in Eq. (3.19) that $SO(3)$ preserves the lengths of vectors. Since it is unitary, $SU(2)$ preserves the lengths of vectors as well. Let $\vec{r} = x\hat{x} + y\hat{y} + z\hat{z}$. We construct a matrix of the form $\vec{\sigma} \cdot \vec{r}$.

$$\vec{\sigma} \cdot \vec{r} = \sigma_x x + \sigma_y y + \sigma_z z$$

$$= \begin{pmatrix} 0 & x \\ x & 0 \end{pmatrix} + \begin{pmatrix} 0 & -iy \\ iy & 0 \end{pmatrix} + \begin{pmatrix} z & 0 \\ 0 & -z \end{pmatrix} \tag{3.25}$$

$$= \begin{pmatrix} z & x-iy \\ x+iy & -z \end{pmatrix}$$

If we take the determinant of this matrix, we obtain the length of the vector as shown here.

$$\det\begin{pmatrix} z & x-iy \\ x+iy & -z \end{pmatrix} = -x^2 - y^2 - z^2 = -\vec{x}^2$$

Also note that the matrix $\vec{\sigma}\cdot\vec{r}$ is Hermitian and has zero trace. Now consider a unitary transformation on this matrix. For example, we can take

$$U(\vec{\sigma}\cdot\vec{r})U^\dagger = \sigma_x(\vec{\sigma}\cdot\vec{r})\sigma_x$$

$$= \begin{pmatrix} 0 & 1 \\ 1 & 0 \end{pmatrix}\begin{pmatrix} z & x-iy \\ x+iy & -z \end{pmatrix}\begin{pmatrix} 0 & 1 \\ 1 & 0 \end{pmatrix}$$

$$= \begin{pmatrix} -z & x+iy \\ x-iy & z \end{pmatrix}$$

The transformed matrix still has zero trace, and is still Hermitian. Moreover, the determinant is preserved, and it again gives the length of the vector, that is,

$$\det\begin{pmatrix} -z & x+iy \\ x-iy & z \end{pmatrix} = -x^2 - y^2 - z^2 = -\vec{x}^2$$

The conclusion is that like $SO(3)$, $SU(2)$ preserves the lengths of vectors as shown here.

$$\vec{x}'^2 = \vec{x}^2$$
$$\Rightarrow x'^2 + y'^2 + z'^2 = x^2 + y^2 + z^2$$

The correspondence works by considering an $SU(2)$ transformation on a two component spinor, like

$$\psi = \begin{pmatrix} \alpha \\ \beta \end{pmatrix}$$

where

$$x = \frac{1}{2}(\beta^2 - \alpha^2) \qquad y = -\frac{i}{2}(\alpha^2 + \beta^2) \qquad z = \alpha\beta$$

Then an $SU(2)$ transformation on $\psi = \binom{\alpha}{\beta}$ is equivalent to an $SO(3)$ transformation on $\vec{x} = \binom{x}{y}_{z}$. As you can see from this equivalence, an $SU(2)$ transformation has three real parameters that correspond to the three angles of an $SO(3)$ transformation. Label the "angles" for the $SU(2)$ transformation by α, β, and γ. *Half* the rotation angle generated by $SU(2)$ corresponds to the rotation generated by $SO(3)$. For arbitrary angle α, a transformation generating a rotation about x in $SU(2)$ is given by

$$U = \begin{pmatrix} \cos\alpha/2 & i\sin\alpha/2 \\ i\sin\alpha/2 & \cos\alpha/2 \end{pmatrix}$$

(see quiz problem 1). This transformation corresponds to the rotation Eq. (3.10), a rotation about the x axis. Next, consider an $SU(2)$ transformation generating a rotation around the y axis. The unitary operator is

$$U = \begin{pmatrix} \cos\beta/2 & \sin\beta/2 \\ -\sin\beta/2 & \cos\beta/2 \end{pmatrix}$$

This corresponds to the $SO(3)$ transformation given in Eq. (3.11). Finally, for a rotation about the z axis we have the $SU(2)$ transformation

$$U = \begin{pmatrix} e^{i\gamma/2} & 0 \\ 0 & e^{-i\gamma/2} \end{pmatrix}$$

which corresponds to Eq. (3.12).

All 2×2 unitary matrices are specified by two parameters, complex numbers a and b where

$$U = \begin{pmatrix} a & b \\ -b^* & a^* \end{pmatrix}$$

For an element of $SU(2)$, $\det U = +1$ so we require that $|a|^2 + |b|^2 = 1$.

Later we will seek to define Lagrangians that are invariant under an $SU(2)$ symmetry. This symmetry will be of particular importance in the case of electroweak interactions. The gauge bosons corresponding to the $SU(2)$ symmetry will be the W and Z bosons that carry the weak interaction.

Next, we consider the unitary group $SU(3)$, which will be important in the study of quarks and the theory of quantum chromodynamics. Earlier we indicated that $SU(3)$ has eight generators. These are called the Gell-Mann matrices and are given by

$$\lambda_1 = \begin{pmatrix} 0 & 1 & 0 \\ 1 & 0 & 0 \\ 0 & 0 & 0 \end{pmatrix} \quad \lambda_2 = \begin{pmatrix} 0 & -i & 0 \\ i & 0 & 0 \\ 0 & 0 & 0 \end{pmatrix} \quad \lambda_3 = \begin{pmatrix} 1 & 0 & 0 \\ 0 & -1 & 0 \\ 0 & 0 & 0 \end{pmatrix}$$

$$\lambda_4 = \begin{pmatrix} 0 & 0 & 1 \\ 0 & 0 & 0 \\ 1 & 0 & 0 \end{pmatrix} \quad \lambda_5 = \begin{pmatrix} 0 & 0 & -i \\ 0 & 0 & 0 \\ i & 0 & 0 \end{pmatrix} \quad \lambda_6 = \begin{pmatrix} 0 & 0 & 0 \\ 0 & 0 & 1 \\ 0 & 1 & 0 \end{pmatrix}$$

$$\lambda_7 = \begin{pmatrix} 0 & 0 & 0 \\ 0 & 0 & -i \\ 0 & i & 0 \end{pmatrix} \quad \lambda_8 = \frac{1}{\sqrt{3}}\begin{pmatrix} 1 & 0 & 0 \\ 0 & 1 & 0 \\ 0 & 0 & -2 \end{pmatrix}$$

Notice two of the matrices are diagonal, λ_3 and λ_8, as we would expect from the rank of the group. The Gell-Mann matrices are traceless and they satisfy the commutation relations

$$\left[\lambda_i, \lambda_j\right] = 2i\sum_{k=1}^{8} f_{ijk}\lambda_k \tag{3.26}$$

This defines the algebraic structure of $SU(3)$. The nonzero structure constants are

$$f_{123} = 1 \qquad f_{147} = f_{165} = f_{246} = f_{257} = f_{345} = f_{376} = \frac{1}{2}$$

$$f_{458} = f_{678} = \frac{\sqrt{3}}{2} \tag{3.27}$$

We will see more of $SU(3)$ when we examine the standard model.

Casimir Operators

A *casimir operator* is a nonlinear function of the generators of a group that commutes with all of the generators. The number of casimir operators for a group is given by the rank of the group.

Considering $SU(3)$ as an example, the generators are the angular momentum operators. A casimir operator in this case is

$$\vec{J}^2 = J_x^2 + J_y^2 + J_z^2$$

A casimir operator is an invariant. In this case, the invariance suggests J^2 is a multiple of the group identity element.

Summary

Group theory plays an important role in physics because groups are used to describe symmetries. The structure of a group is defined by the algebra among its generators. If two groups have the same algebra, they are related. Unitary transformations preserve the probabilities of state transitions in quantum theory. As a result, the most important groups in quantum field theory are the unitary groups, specifically $U(1)$, $SU(2)$, and $SU(3)$.

Quiz

1. Consider an element of $SU(2)$ given by $U = e^{i\sigma_x \alpha/2}$. By writing down the power series expansion, write U in terms of trigonometry functions.

2. Consider $SU(3)$ and calculate $tr(\lambda_i \lambda_j)$.

3. How many casimir operators are there for $SU(2)$?

4. Write down the casimir operators for $SU(2)$.

   A Lorentz transformation can be described by boost matrices with rapidity defined by $\tanh\phi = v/c$. A boost in the x direction is represented by the matrix

$$\begin{pmatrix} \cosh\phi & \sinh\phi & 0 & 0 \\ \sinh\phi & \cosh\phi & 0 & 0 \\ 0 & 0 & 1 & 0 \\ 0 & 0 & 0 & 1 \end{pmatrix}$$

5. Find the generator K_x.

6. Knowing that $[K_x, K_y] = -iJ_z$, where J_z is the angular momentum operator written in four dimensions as

$$J_z = -i \begin{pmatrix} 0 & 0 & 0 & 0 \\ 0 & 0 & 1 & 0 \\ 0 & -1 & 0 & 0 \\ 0 & 0 & 0 & 0 \end{pmatrix}$$

find K_y.

7. Do pure Lorentz boosts constitute a group?

CHAPTER 4

Discrete Symmetries and Quantum Numbers

In Chap. 3 we examined *continuous symmetries,* that is, symmetries with continuously varying parameters such as rotations. Now we consider a different kind of symmetry, a *discrete symmetry.* There are three important discrete symmetries in particle physics: *parity, charge conjugation,* and *time reversal.*

Additive and Multiplicative Quantum Numbers

A *quantum number* is some quantity (a quantized property of a particle) that is conserved in a particle reaction (decay, collision, etc.). An *additive quantum number*

n is one such that if $n_1, n_2, \ldots, n_i, \ldots$ are the quantum numbers before the reaction, and $m_1, m_2, \ldots, m_i, \ldots$ are the quantum numbers after the reaction, then the sum is preserved.

$$\sum_i n_i = \sum_i m_i \tag{4.1}$$

Or, if we have a composite system with quantum numbers $n_1, n_2, \ldots, n_i, \ldots$, and if the quantum number is additive, then the quantum number of the composite system is

$$\sum_i n_i$$

A *multiplicative quantum number* is one such that if $n_1, n_2, \ldots, n_i, \ldots$ are the quantum numbers before the reaction, and $m_1, m_2, \ldots, m_i, \ldots$ are the quantum numbers after the reaction, then the product is preserved.

$$\prod_i n_i = \prod_i m_i \tag{4.2}$$

Or, if we have a composite system, and if a quantum number is multiplicative, then

$$\prod_i n_i$$

is the quantum number for the composite system. If a quantum number is conserved, then it represents a symmetry of the system.

Parity

We begin our discussion of parity by examining nonrelativistic quantum mechanics. Consider a potential V that is symmetric about the origin and therefore $V(-x) = V(x)$. This implies that if $\psi(x)$ is a solution of the Schrödinger equation, then so is $\psi(-x)$, and it solves the equation with the *same eigenvalue*. This is because

$$-\frac{\hbar^2}{2m}\frac{d^2\psi(-x)}{dx^2} + V(-x)\psi(-x) = E\psi(-x)$$

$$\Rightarrow -\frac{\hbar^2}{2m}\frac{d^2\psi(-x)}{dx^2} + V(x)\psi(-x) = E\psi(-x)$$

when $V(-x) = V(x)$. If $\psi(x)$ and $\psi(-x)$ both solve the Schrödinger equation with the same eigenvalue E, then they must be related to each other as

$$\psi(x) = \alpha\psi(-x) \tag{4.3}$$

If we let $x \to -x$, then we obtain

$$\psi(-x) = \alpha\psi(x)$$

Inserting this into Eq. (4.3) gives

$$\psi(x) = \alpha\psi(-x) = \alpha\left[\alpha\psi(x)\right] = \alpha^2\psi(x)$$

$$\Rightarrow 1 = \alpha^2$$

This tells us that $\alpha = \pm 1$. Then either

$$\psi(-x) = \psi(x)$$

in which case we say that the wave function has *even parity* or

$$\psi(-x) = -\psi(x)$$

in which case we say the wave function has *odd parity*. This leads us to the concept of the *parity operator P*. The parity operator causes a change in sign when $x \to -x$ in the wave function.

$$P\psi(x) = \psi(-x) \tag{4.4}$$

As seen here in the illustrations, even powers lead to functions with even parity $\left[\psi(-x) = \psi(x)\right]$ while odd powers lead to functions with odd parity $\left[\psi(-x) = -\psi(x)\right]$.

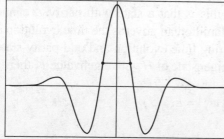

An example of an even function where $\psi(-x) = \psi(x)$. Monomials with even powers $(x^2, x^4, x^6, \ldots)$ are even functions.

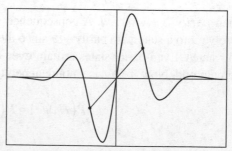

An example of an odd function where $\psi(-x) = -\psi(x)$. Monomials with odd powers $(x^1, x^3, x^5, \ldots)$ are odd functions.

Obviously, applying the parity operator twice in succession gives the original wave function back as shown here.

$$P^2 \psi(x) = P\psi(-x) = \psi(x)$$

Think of the parity operator as a reflection through the y axis. If ψ is even, we see the same function values; if ψ is odd, we see the negative of the function values. In either case, another reflection through the y axis brings us back to our initial state. The reflection of the reflection is the image.

It follows that

$$P^2 = I \tag{4.5}$$

The eigenstates of parity are ± 1:

$$P|\psi\rangle = \pm|\psi\rangle \tag{4.6}$$

As we have seen, true reflections preserve length.

If $|\psi\rangle$ is an angular momentum state with angular momentum L, that is, $|\psi\rangle = |L, m_z\rangle$, then the parity operator acts as

$$P|L, m_z\rangle = (-1)^L |L, m_z\rangle \tag{4.7}$$

We showed above that $\alpha = \pm 1$ when $\psi(x)$ and $\psi(-x)$ both solve the Schrödinger equation with the same eigenvalue E. We can generalize this by saying that if the Parity operator and Hamiltonian commute

$$[P, H] = 0 = PH - HP$$

then *parity is conserved*. A consequence of this is that a state with parity α cannot evolve into a state with parity $-\alpha$ since the Hamiltonian governs the time evolution of the states. Even parity states remain even during time evolution and odd parity states remain odd. Now if $|\psi\rangle$ is a nondegenerate eigenstate of H with eigenvalue E, then

$$P(H|\psi\rangle) = P(E|\psi\rangle) = EP|\psi\rangle$$

But if $[P, H] = 0 = PH - HP$, then

$$P(H|\psi\rangle) = H(P|\psi\rangle) = E(P|\psi\rangle)$$

So the eigenstates of H are also eigenstates of the parity operator. Also, it follows that the eigenvalues of P are $\alpha = \pm 1$. Notice that this precludes states of mixed, or indefinite parity. This is a powerful constraint mathematically and physically it means that particles are either fermions or bosons.

In quantum field theory, the eigenvalue of parity α is a property of particles called the *intrinsic parity* of the particle. Following our discussion of parity and wave functions at the beginning of this section, if $\alpha = +1$ for a given particle, we say that the particle has even parity. If $\alpha = -1$, the particle has odd parity.

Parity for fermions is assigned as follows:

- Particles with spin-1/2 have positive parity. Hence an electron and a quark both have $\alpha = +1$.

- Antiparticles with spin-1/2 have negative parity. Therefore a positron has $\alpha = -1$.

Bosons have the same intrinsic parity for both particles and antiparticles.

Parity is a multiplicative quantum number. Let $|\psi\rangle = |a\rangle|b\rangle$ define a composite system. If the parities of $|a\rangle$ and $|b\rangle$ are P_a and P_b respectively, then the parity of the composite system is the product of the individual parities, that is,

$$P_\psi = P_a P_b$$

We can construct new parity operators by combining P with one of the conserved charges of the standard model. These are

- The electric charge operator Q
- Lepton number L
- Baryon number B

Earlier we said something about the conservation of parity. Parity *is not always conserved*, and there are specific cases, like

- Parity *is conserved* in the electromagnetic and strong interactions.

- Parity *is not conserved* in the weak interaction.

Particles are often labeled as follows

$$J^P \equiv \text{spin}^{\text{parity}} \tag{4.8}$$

A spin-0 particle with negative parity is called a *pseudoscalar*. Examples of pseudoscalar particles include the π and K mesons. Using the notation in Eq. (4.8) we write 0^- to indicate a pseudoscalar particle.

A spin-0 particle with positive parity is called a *scalar*. We denote a scalar by 0^+. An example of a scalar particle is the *Higgs boson*, the particle corresponding to the field believed to be responsible for mass generation. The elusive Higgs hasn't been detected at the time of writing, but may be found soon when the Large Hadron Collider (LHC) begins operation.

A *vector boson* has spin-1 and negative parity (1^-). The most famous vector boson is the photon. A pseudovector has unit spin and positive parity 1^+.

Parity is conserved in the electromagnetic and strong interactions, so the total parity of a system before an electromagnetic or strong interaction is the same as the parity after the interaction. In the 1950s, two physicists named Lee and Yang proposed that parity conservation is violated in the weak interaction.

This is called *parity violation*. This was demonstrated experimentally by observing weak decays of Cobalt-60, leading to the Nobel prize for Lee and Yang. Parity violation also became apparent in the weak decays of two particles called the θ and τ mesons. They decay as

$$\theta^+ \to \pi^+ \pi^0$$
$$\tau^+ \to \pi^+ \pi^- \pi^+$$

The final states of these two decays have opposite parity and therefore physicists believed the θ^+ and the τ^+ to be different particles. However, successively refined measurements of the θ^+ and τ^+ mass and lifetimes suggested they were actually the same particle. The discovery of parity violation in the weak interaction resolved this dilemma and today we call this particle the K^+ meson.

Charge Conjugation

We now consider *charge conjugation C*, an operator which converts particles into antiparticles. Let $|\psi\rangle$ represent a particle state and $|\bar{\psi}\rangle$ represent the antiparticle state. Then the charge conjugation operator acts as

$$C|\psi\rangle = |\bar{\psi}\rangle \tag{4.9}$$

Charge conjugation also acts on antiparticle states, turning them into particle states.

$$C|\bar{\psi}\rangle = |\psi\rangle \tag{4.10}$$

It follows that

$$C^2|\psi\rangle = CC|\psi\rangle = C|\bar{\psi}\rangle = |\psi\rangle$$

We can use this relation to determine the eigenvalues of charge conjugation. It is apparent that like parity, they must be $C = \pm 1$. Charge conjugation is also like parity in that it is a multiplicative quantum number. Since charge conjugation converts particles into antiparticles and vice versa, it reverses the sign of all quantum numbers (and also changes the sign of magnetic moment). Consider a proton $|p\rangle$. It has positive charge q

$$Q|p\rangle = q|p\rangle$$

and Baryon number $B = +1$. If we operate on the proton with the charge conjugation operator, then $C|p\rangle = |\bar{p}\rangle$ and since

$$Q|\bar{p}\rangle = -q|\bar{p}\rangle$$

the charge has been reversed. The baryon number has also been changed to $B = -1$. Notice that the proton state cannot be an eigenstate of charge conjugation, since the result of $C|p\rangle = |\bar{p}\rangle$ is a state with different quantum numbers—a different quantum state. The eigenstates of the charge conjugation operator are eigenstates with 0 charge, that is, neutral particles. More generally, the eigenstates of C must have all additive quantum numbers equal to 0. An example is a neutral pion π^0. In this case, π^0 is its own antiparticle and so

$$C|\pi^0\rangle = \alpha|\pi^0\rangle$$

for some α. Applying charge conjugation twice

$$C^2|\pi^0\rangle = \alpha C|\pi^0\rangle = \alpha^2|\pi^0\rangle$$

$$\Rightarrow \alpha = \pm 1$$

We can find the charge conjugation properties of the photon in the following way, and hence determine the eigenvalue α for the π^0. First charge conjugation will reverse the sign of the charge density J as shown here.

$$CJC^{-1} = -J$$

Now, the interaction part of the electromagnetic Lagrangian can be used to determine the charge conjugation properties of the photon. We need to find the action of C on $J_\mu A^\mu$. We have

$$CJ_\mu A^\mu C^{-1} = CJ_\mu C^{-1} CA^\mu C^{-1}$$

$$= -J_\mu CA^\mu C^{-1}$$

This can only be invariant if

$$CA^\mu C^{-1} = -A^\mu$$

$$\Rightarrow CJ_\mu A^\mu C^{-1} = J_\mu A^\mu$$

Since A^μ is the electromagnetic vector potential, this tells us that the eigenvalue of charge conjugation for the photon is $\alpha = -1$. Therefore if there are n photons, the charge conjugation is $(-1)^n$. The π^0 decays into two photons as

$$\pi^0 \rightarrow \gamma + \gamma$$

The two photon state has $\alpha = (-1)^2 = +1$, therefore we conclude that

$$C|\pi^0\rangle = (+1)|\pi^0\rangle$$

$$\Rightarrow \alpha = +1 \text{ for the } \pi^0$$

Charge conjugation proceeds in the same way as parity, that is,

- Charge conjugation C is conserved in the strong and electromagnetic interactions.
- Charge conjugation is not conserved in the weak interaction.

CP Violation

The fact that charge conjugation and parity are each individually violated in weak interactions led to the hope that the combination of charge conjugation and parity would be conserved. It almost is, but there is a slight violation that can be seen from the decay of neutral K mesons.

The neutral $|K^0\rangle$ meson is an interesting particle which is observed in a linear combination of states with its antiparticle. This is because the $|K^0\rangle$ spontaneously transitions into its antiparticle and vice versa as shown here.

$$|K^0\rangle \leftrightarrow |\bar{K}^0\rangle$$

The $|K^0\rangle$ and its antiparticles are pseudoscalars 0^- and so have negative parity, that is

$$P|K^0\rangle = -|K^0\rangle$$

$$P|\bar{K}^0\rangle = -|\bar{K}^0\rangle$$

Charge conjugation, is of course the operation that transforms the $\left|K^0\right\rangle$ into its antiparticle, that is,

$$C\left|K^0\right\rangle = \left|\bar{K}^0\right\rangle$$

$$C\left|\bar{K}^0\right\rangle = \left|K^0\right\rangle$$

Taken together, CP acts on the states as

$$CP\left|K^0\right\rangle = -C\left|K^0\right\rangle = -\left|\bar{K}^0\right\rangle$$

$$CP\left|\bar{K}^0\right\rangle = -C\left|\bar{K}^0\right\rangle = -\left|K^0\right\rangle$$

We see from this relation that $\left|K^0\right\rangle$ and $\left|\bar{K}^0\right\rangle$ *are not* eigenstates of CP. To see if CP is violated, we need to construct states that are eigenstates of CP out of $\left|K^0\right\rangle$ and $\left|\bar{K}^0\right\rangle$. The states that do this are

$$\left|K_1\right\rangle = \frac{\left|K^0\right\rangle - \left|\bar{K}^0\right\rangle}{\sqrt{2}} \qquad \left|K_2\right\rangle = \frac{\left|K^0\right\rangle + \left|\bar{K}^0\right\rangle}{\sqrt{2}}$$

It is helpful to think of particle states in terms of vector spaces. In terms of a rotation by $\pi/4$ we see that

$$\begin{pmatrix} \left|K_1\right\rangle \\ \left|K_2\right\rangle \end{pmatrix} = R\left(-\frac{\pi}{4}\right)\begin{pmatrix} \left|K_0\right\rangle \\ \left|\bar{K}_0\right\rangle \end{pmatrix}$$

which demonstrates an advantage of the vector space view of particle.

So we have

$$CP\left|K_1\right\rangle = +\left|K_1\right\rangle$$

$$CP\left|K_2\right\rangle = -\left|K_2\right\rangle$$

Fortunately, these states can be created in the laboratory. It turns out that they both decay into π mesons. Now, if CP is conserved, then each of these states will decay into a state with the same value of CP. That is,

$$\left|K_1\right\rangle \rightarrow \text{ decays into a state with } CP = +1$$

$$\left|K_2\right\rangle \rightarrow \text{ decays into a state with } CP = -1$$

A neutral K meson $\left| K^0 \right\rangle$ can decay into a state with two α mesons or into a state with three π mesons. The charge conjugation and parity eigenvalues of these states are

$$2\pi \text{ mesons} : C = +1, P = +1,$$

$$\Rightarrow CP = +1$$

$$3\pi \text{ mesons} : C = +1, P = -1,$$

$$\Rightarrow CP = -1$$

If CP is conserved then, we would have

$$\left| K_1 \right\rangle \rightarrow \text{only decays into } 2\pi \text{ mesons}$$

$$\left| K_1 \right\rangle \rightarrow \text{only decays into } 3\pi \text{ mesons}$$

This is *not* what is observed experimentally. It is found that a small fraction of the time

$$\left| K_2 \right\rangle \rightarrow \text{decays into } 2\pi \text{ mesons}$$

Hence we have a transition from $CP = -1 \rightarrow CP = +1$. It turns out that the long lived K meson state is

$$\left| K_L \right\rangle = \frac{\left| K_2 \right\rangle + \varepsilon \left| K_1 \right\rangle}{\sqrt{1 + \left| \varepsilon \right|^2}}$$

The parameter ε is a measure of the amount of violation of CP conservation. Experimental evidence indicates that

$$\varepsilon = 2.3 \times 10^{-3}$$

A small number indeed, but *not zero*. *CP* violation happens because a small fraction of the time, the long lived neutral K meson state, is found in $\left| K_1 \right\rangle$, giving the unexpected decays.

The CPT Theorem

To restore invariance, we have to bring in one more symmetry, *time reversal*. This is another discrete transformation on states, turning a state $\left| \psi \right\rangle$ into a state $\left| \psi' \right\rangle$ that evolves with time flowing in the negative direction. Momentums change sign:

linear momentum $p \rightarrow -p$ and angular momentum $L \rightarrow -L$, but all other quantities maintain the same sign. The time-reversal operator acts to transform the states as

$$T|\psi\rangle = |\psi'\rangle$$

The time-reversal operator is *antiunitary* and *antilinear*. To say that it is antilinear, we mean that

$$T(\alpha|\psi\rangle + \beta|\phi\rangle) = \alpha^*|\psi'\rangle + \beta^*|\phi'\rangle$$

While a unitary operator U preserves inner products,

$$\langle U\phi|U\psi\rangle = \langle\phi|\psi\rangle$$

an antiunitary operator does not, but instead gives the complex conjugate as shown here.

$$\langle A\phi|A\psi\rangle = \langle\phi|\psi\rangle^*$$

The time-reversal operator is antiunitary and can be written as a product of an operator K that converts states into their complex conjugates (note that K does not refer to the K meson of the previous section, but in this context is an operator)

$$K\psi = \psi^*$$

and a unitary operator U

$$T = UK$$

If the time-reversal operator commutes with the Hamiltonian $[T,H] = 0$, and if $|\psi\rangle$ satisfies the Schrödinger equation, $T|\psi\rangle$ will also satisfy the Schrödinger equation when $t \rightarrow -t$. Hence the name time-reversal operator. If the laws of physics are unchanged under time reversal, then they are a symmetry of the system.

The *CPT theorem* considers the three symmetries C, P, and T taken together. According to this theorem, if charge conjugation, parity reversal, and time reversal are taken together we have an exact symmetry and so the laws of physics are invariant. More colloquially, the theorem means that if matter is replaced by antimatter (charge conjugation), momentum is reversed with spatial inversion (parity conjugation), and time is reversed, the result would be a universe indistinguishable from the one we live in. For the *CPT* theorem to be valid, all three symmetries must be valid or if one or more symmetries are violated, another symmetry must be violated to negate the first violation. However we have seen that in the weak interaction, there is *CP* violation. To compensate for the problem of *CP* violation, there must be *T* violation as well.

Now let's look at this (loosely) in the context of quantum field theory. To satisfy special relativity, we need Lorentz invariance. This means that we implement a Lorentz transformation with

$$\Lambda^{\mu}{}_{\nu} = \begin{pmatrix} \cosh\phi & -\sinh\phi & 0 & 0 \\ -\sinh\phi & \cosh\phi & 0 & 0 \\ 0 & 0 & 1 & 0 \\ 0 & 0 & 0 & 1 \end{pmatrix}$$

and the theory is the same. Quantum theory allows for ϕ to be *complex*. If we take $\phi = i\pi$ and

$$\Lambda^{\mu}{}_{\nu} = \begin{pmatrix} \cosh\phi & \sinh\phi & 0 & 0 \\ \sinh\phi & \cosh\phi & 0 & 0 \\ 0 & 0 & 1 & 0 \\ 0 & 0 & 0 & 1 \end{pmatrix}$$

then by setting $\phi = i\pi$ we get

$$\Lambda^{\mu}{}_{\nu} = \begin{pmatrix} -1 & 0 & 0 & 0 \\ 0 & -1 & 0 & 0 \\ 0 & 0 & 1 & 0 \\ 0 & 0 & 0 & 1 \end{pmatrix}$$

giving time reversal $t \to -t$ and space inversion $x \to -x$. This is a *PT* invariant theory. If the particles are charged then you recover the entire *CPT* invariance.

Summary

In this chapter we have examined some discrete symmetries, including parity, charge conjugation, and time reversal. Interesting physics arises because these discrete symmetries are not conserved in weak interactions. The symmetry that is always conserved in all interactions is *CPT*, which is stated in the *CPT* theorem.

Quiz

1. Angular momentum states transform under the parity operator as

   (a) $P|L,m_z\rangle = -|L,m_z\rangle$

   (b) $P|L,m_z\rangle = L|L,m_z\rangle$

   (c) $P|L,m_z\rangle = (-1)^L|L,m_z\rangle$

   (d) $P|L,m_z\rangle = |L,m_z\rangle$

2. The interaction Lagrangian of electromagnetism is invariant under charge conjugation if

   (a) $CA^\mu C^{-1} = -A^\mu$

   (b) It is not invariant under charge conjugation

   (c) $CJ^\mu C^{-1} = J^\mu$

   (d) $CA^\mu C^{-1} = A^\mu$

3. Parity is

   (a) Conserved in weak and electromagnetic interactions, but is violated in the strong interaction

   (b) Conserved in strong interactions, but is violated in weak and electromagnetic interactions

   (c) Not conserved

   (d) Conserved in the strong and electromagnetic interactions, but is violated in the weak interaction

4. The eigenvalues of charge conjugation are

   (a) $c = \pm 1$

   (b) $c = 0, \pm 1$

   (c) $c = \pm q$

   (d) $c = 0, \pm q$

5. An operator is antiunitary if

   (a) $\langle A\phi | A\psi \rangle = -\langle \phi | \psi \rangle$

   (b) $\langle A\phi | A\psi \rangle = \langle \phi | \psi \rangle$

   (c) $\langle A\phi | A\psi \rangle = \langle \phi | \psi \rangle^*$

   (d) $\langle A\phi | A\psi \rangle = -\langle \phi | \psi \rangle^*$

CHAPTER 5

The Dirac Equation

In the next chapter we will see that scientists began with an attempt to arrive at a relativistic wave equation that put time and space on an equal footing by "promoting" derivatives with respect to time to second order. This was done because the Schrödinger equation has second-order spatial derivatives. The equation that results is called the *Klein-Gordon equation*. Unfortunately, this leads to several problems as far as quantum theory is concerned. In particular, it gives us a probability density that can be negative and it leads to negative energy states.

While these problems can be resolved by reinterpreting the resulting Klein-Gordon equation, a fruitful line of inquiry results if we try to tackle the first problem head on by taking a different approach. This is exactly what Dirac did in deriving his now famous equation. The Dirac equation applies to spin-1/2 fields, and puts time and space on equal ground in the equation by considering first-order spatial derivatives, rather than increasing the order of the time derivatives.

The Classical Dirac Field

We begin by looking at the Dirac equation in terms of classical field theory. Again, we approach the problem with the goal of satisfying the tenets of special relativity, hence we want time and space to appear in the equation in a similar fashion. As discussed in Chap. 4, derivatives with respect to time in the Schrödinger equation are first order while derivatives with respect to spatial coordinates are second order. The Klein-Gordon equation attempts to deal with this discrepancy by using second-order derivatives with respect to time. With the Dirac equation, we are going to take the opposite approach and consider using first-order derivatives for the spatial coordinates, while simultaneously keeping the derivatives with respect to time first order as well. The reason for doing this is to avoid the negative probability distributions that we saw arise from the Klein-Gordon equation. In Chap. 4 we saw that this is *due* to the fact that the equation contained second-order derivatives with respect to time. So we will attempt to avoid that problem by keeping time derivatives first order. The result is an equation that beautifully describes spin-1/2 particles.

First let's remind ourselves about the Schrödinger equation.

$$i\hbar \frac{\partial \psi}{\partial t} = -\frac{\hbar^2}{2m}\nabla^2 \psi + V\psi \tag{5.1}$$

Let's write it in a more suggestive form using the operator definition $\hat{H} = -\frac{\hbar^2}{2m}\nabla^2 + V$. Then,

$$i\hbar \frac{\partial \psi}{\partial t} = \hat{H}\psi \tag{5.2}$$

Why is this suggestive? Well, the Dirac equation can be thought of as a type of Schrödinger equation if we just change what the Hamiltonian operator $\hat{H}$ is, that is, applied to the wave function. The form of the Hamiltonian is chosen so that the requirements of special relativity can be satisfied. Assuming that the particle in equation has rest mass m, the form of the Hamiltonian operator used in the Dirac equation is

$$\hat{H} = c\vec{\alpha}\cdot(-i\hbar\nabla) + \beta mc^2 \tag{5.3}$$

We will explain in a bit what $\vec{\alpha}$ and β are. For now, using Eq. (5.3) in Eq. (5.2) gives us the Dirac equation.

$$i\hbar \frac{\partial \psi}{\partial t} = \left[c\vec{\alpha}\cdot(-i\hbar\nabla) + \beta mc^2 \right]\psi \tag{5.4}$$

This is a relativistically covariant equation. Time and space have been put on the same footing since they both appear in the equation in terms of first-order derivatives. The new terms in the equation, $\vec{\alpha}$ and β, are actually 4×4 *matrices*. Before writing them down, we are going to rewrite Eq. (5.4) using what physicists call the *Dirac matrices,* or you can call them the gamma matrices if you like. First let's hold on a moment and recall the gradient operator, which is a vector operator

$$\vec{\nabla} = \frac{\partial}{\partial x}\hat{x} + \frac{\partial}{\partial y}\hat{y} + \frac{\partial}{\partial z}\hat{z}$$

or, using the notation of Chap. 1

$$\vec{\nabla} = \frac{\partial}{\partial x^1}\hat{e}_1 + \frac{\partial}{\partial x^2}\hat{e}_2 + \frac{\partial}{\partial x^3}\hat{e}_3$$

Writing this out we see the first-order spatial derivatives that appear in the Dirac equation [Eq. (5.4)] explicitly. We are taking $\vec{\alpha}$ to be a vector as well, whose components are matrices

$$\vec{\alpha} = \alpha_1\hat{e}_1 + \alpha_2\hat{e}_2 + \alpha_3\hat{e}_3$$

Now we define the gamma matrices (or Dirac matrices) in terms of $\vec{\alpha}$ and β in the following way.

$$\gamma^0 = \beta \tag{5.5}$$

$$\gamma^i = \beta\alpha_i \tag{5.6}$$

Adding Quantum Theory

At this point the first hint of quantum theory comes into play. Well of course since they are matrices, we don't necessarily expect the Dirac matrices to commute. In other words it's not necessarily true that $\gamma^1\gamma^2 = \gamma^2\gamma^1$, say. In fact, the Dirac matrices obey an important *anticommutation* rule. The anticommutator of two matrices A and B is

$$\{A,B\} = AB + BA \tag{5.7}$$

Note that in many texts the anticommutator is denoted by $\left[A,B\right]_+$, but we will stick to the notation used in Eq. (5.7). The relationship for the Dirac matrices actually connects them to the spacetime metric (perhaps a connection to quantum gravity here?). It is given by

$$\{\gamma^{\mu},\gamma^{\nu}\} = \gamma^{\mu}\gamma^{\nu} + \gamma^{\nu}\gamma^{\mu} = 2g^{\mu\nu} \tag{5.8}$$

Using the Dirac matrices, we can write down the Dirac equation in a fancy relativistic notation sure to impress all of your friends. It is

$$i\hbar\gamma^\mu \frac{\partial\psi}{\partial x^\mu} - mc\psi = 0 \tag{5.9}$$

Better yet, we work in units where we set $\hbar = c = 1$ and use $\frac{\partial}{\partial x^\mu} = \partial_\mu$ to write the Dirac equation in the compact form

$$i\gamma^\mu \partial_\mu \psi - m\psi = 0 \tag{5.10}$$

The correct way to interpret this equation, when the quantum theory is brought in, is that it applies to the *Dirac field* whose quanta are spin-1/2 particles—electrons.

EXAMPLE 5.1
Show that the Dirac field ψ also satisfies the Klein-Gordon equation.

SOLUTION
That it should ought not to surprise you, the Klein-Gordon equation is nothing other than a restatement of Einstein's relation between energy, mass, and momentum in special relativity derived using the quantum substitutions $E \to i\hbar \frac{\partial}{\partial t}$ and $\vec{p} \to -i\hbar \frac{\partial}{\partial x}$. Since $E^2 = p^2 + m^2$ is an absolutely fundamental relation that applies to everything, all particles and fields, including the Dirac field, must satisfy the Klein-Gordon equation. It turns out that in a sense the Dirac equation is the "square root" of the Klein-Gordon equation. Let's see how to derive the Klein-Gordon equation directly from the Dirac equation. We start with the Dirac equation

$$i\gamma^\mu \partial_\mu \psi - m\psi = 0$$

Now multiply from the left by $i\gamma_\nu \partial^\nu$. This gives

$$-\gamma_\nu \gamma^\mu \partial^\nu \partial_\mu \psi - im\gamma_\nu \partial^\nu \psi = 0$$

Take a look at the second term. From the Dirac equation itself, we know that

$$i\gamma^\mu \partial_\mu \psi = m\psi$$

So we can write

$$im\gamma_\nu \partial^\nu \psi = m^2 \psi$$

Using this and moving everything to the other side to get rid of the minus sign

$$-\gamma_\nu \gamma^\mu \partial^\nu \partial_\mu \psi - im\gamma_\nu \partial^\nu \psi = 0$$

becomes

$$\gamma_\nu \gamma^\mu \partial^\nu \partial_\mu \psi + m^2 \psi = 0$$

Now we apply the anticommutation relation obeyed by the Dirac matrices, Eq. (5.8). Restating it here so you don't have to flip through the pages of the book it is

$$\{\gamma^\mu, \gamma^\nu\} = \gamma^\mu \gamma^\nu + \gamma^\nu \gamma^\mu = 2g^{\mu\nu}$$

This can be applied to write the first term in $\gamma_\nu \gamma^\mu \partial^\nu \partial_\mu \psi + m^2 \psi = 0$ in a symmetric form. This is done as follows:

$$\gamma_\nu \gamma^\mu = g_{\nu\sigma} \gamma^\sigma \gamma^\mu = g_{\nu\sigma} \frac{1}{2}(\gamma^\mu \gamma^\sigma + \gamma^\sigma \gamma^\mu)$$

But we have

$$g_{\nu\sigma} \frac{1}{2}\left(\gamma^\mu \gamma^\sigma + \gamma^\sigma \gamma^\mu\right) = g_{\nu\sigma} \frac{1}{2}\left(2g^{\mu\sigma}\right)$$
$$= g_{\nu\sigma} g^{\mu\sigma}$$
$$= \delta_\nu^\mu$$

Therefore,

$$0 = \gamma_\nu \gamma^\mu \partial^\nu \partial_\mu \psi + m^2 \psi$$
$$= \delta_\nu^\mu \partial^\nu \partial_\mu \psi + m^2 \psi$$
$$= \partial^\mu \partial_\mu \psi + m^2 \psi$$

This is nothing other than our old friend, the Klein-Gordon equation! This ought to be a really satisfying result—we have *derived* the Klein-Gordon equation directly from the Dirac equation. This shows that Dirac fields (particles . . .) also satisfy the Klein-Gordon equation and hence automatically satisfy the relativistic relation between energy, mass, and momentum.

The Form of the Dirac Matrices

It will be hard to get anywhere if we don't know how to explicitly write down the Dirac matrices. There are actually a couple of different ways to do it. One way, which we introduce first, is the Dirac-Pauli representation, and it's pretty straightforward. Keep in mind that these matrices are 4×4 matrices. The first Dirac matrix is an extension of the unassuming identity matrix, which has 1s all along the diagonal. In the 2×2 case,

$$I = \begin{pmatrix} 1 & 0 \\ 0 & 1 \end{pmatrix}$$

In the Dirac-Pauli representation, we can write the first gamma matrix as

$$\gamma_0 = \begin{pmatrix} I & 0 \\ 0 & -I \end{pmatrix} \tag{5.11}$$

where I is the 2×2 identity matrix shown above. The 0s are 2×2 blocks of 0s. So this matrix is actually

$$\gamma_0 = \begin{pmatrix} 1 & 0 & 0 & 0 \\ 0 & 1 & 0 & 0 \\ 0 & 0 & -1 & 0 \\ 0 & 0 & 0 & -1 \end{pmatrix}$$

Now, an aside, note that since the Dirac matrices are 4×4 matrices (think operators), the Dirac field ψ must be a four component vector that they can act on, since they appear in the Dirac equation. We call this vector a *spinor* and write

$$\psi(x) = \begin{pmatrix} \psi^1(x) \\ \psi^2(x) \\ \psi^3(x) \\ \psi^4(x) \end{pmatrix} \tag{5.12}$$

Now that the aside is over, let's turn to the other Dirac matrices. They are written in terms of the Pauli matrices that you should be familiar with ordinary nonrelativistic quantum mechanics. To review, they are

$$\sigma_1 = \begin{pmatrix} 0 & 1 \\ 1 & 0 \end{pmatrix} \qquad \sigma_2 = \begin{pmatrix} 0 & -i \\ i & 0 \end{pmatrix} \qquad \sigma_3 = \begin{pmatrix} 1 & 0 \\ 0 & -1 \end{pmatrix} \tag{5.13}$$

In the Dirac-Pauli representation, then

$$\vec{\gamma} = \begin{pmatrix} 0 & \vec{\sigma} \\ -\vec{\sigma} & 0 \end{pmatrix} \tag{5.14}$$

so that

$$\gamma_1 = \begin{pmatrix} 0 & \sigma_1 \\ -\sigma_1 & 0 \end{pmatrix} \qquad \gamma_2 = \begin{pmatrix} 0 & \sigma_2 \\ -\sigma_2 & 0 \end{pmatrix} \qquad \gamma_3 = \begin{pmatrix} 0 & \sigma_3 \\ -\sigma_3 & 0 \end{pmatrix}$$

Let's write down one example explicitly as shown here.

$$\gamma_1 = \begin{pmatrix} 0 & \sigma_1 \\ -\sigma_1 & 0 \end{pmatrix} = \begin{pmatrix} 0 & 0 & 0 & 1 \\ 0 & 0 & 1 & 0 \\ 0 & -1 & 0 & 0 \\ -1 & 0 & 0 & 0 \end{pmatrix}$$

The Dirac matrices can also be written down using what is known as the *chiral representation*. In this case,

$$\gamma^0 = \begin{pmatrix} 0 & I \\ I & 0 \end{pmatrix} \qquad \gamma^i = \begin{pmatrix} 0 & \sigma_i \\ -\sigma_i & 0 \end{pmatrix} \tag{5.15}$$

Some Tedious Properties of the Dirac Matrices

Regardless of representation used, the Dirac matrices satisfy several tedious but important relations that are useful when doing calculations. First we define yet another matrix, the mysterious gamma five matrix.

$$\gamma^5 = i\gamma^0\gamma^1\gamma^2\gamma^3 \tag{5.16}$$

It is Hermitian

$$\left(\gamma^5\right)^\dagger = \gamma^5 \tag{5.17}$$

and it squares to the identity

$$\left(\gamma^5\right)^2 = I \tag{5.18}$$

In the chiral representation we write it as

$$\gamma^5 = \begin{pmatrix} -I & 0 \\ 0 & I \end{pmatrix} = \begin{pmatrix} -1 & 0 & 0 & 0 \\ 0 & -1 & 0 & 0 \\ 0 & 0 & 1 & 0 \\ 0 & 0 & 0 & 1 \end{pmatrix} \tag{5.19}$$

but in the Dirac representation we write it as

$$\gamma^5 = \begin{pmatrix} 0 & I \\ I & 0 \end{pmatrix} = \begin{pmatrix} 0 & 0 & 1 & 0 \\ 0 & 0 & 0 & 1 \\ 1 & 0 & 0 & 0 \\ 0 & 1 & 0 & 0 \end{pmatrix} \tag{5.20}$$

(notice that these matrices are traceless). Now we also have $(\gamma^\mu)^2 = I$. Hence,

$$\gamma^\mu \gamma_\mu = 4I \tag{5.21}$$

We can show this explicitly:

$$\gamma^\mu \gamma_\mu = \gamma^1 \gamma_1 + \gamma^2 \gamma_2 + \gamma^3 \gamma_3 + \gamma^4 \gamma_4 = 4 \begin{pmatrix} 1 & 0 & 0 & 0 \\ 0 & 1 & 0 & 0 \\ 0 & 0 & 1 & 0 \\ 0 & 0 & 0 & 1 \end{pmatrix}$$

$$\gamma^\mu \gamma^\nu \gamma_\mu = -2\gamma^\nu \tag{5.22}$$

Again, we see this by writing out the implied summation:

$$\gamma^\mu \gamma^\nu \gamma_\mu = \gamma^1 \gamma^\nu \gamma_1 + \gamma^2 \gamma^\nu \gamma_2 + \gamma^3 \gamma^\nu \gamma_3 + \gamma^4 \gamma^\nu \gamma_4 = -2\gamma^\nu$$

You should repeat these computations yourself to ensure you understand how to work with these matrices.

EXAMPLE 5.2
Find the anticommutator $\{\gamma^5, \gamma^0\}$.

SOLUTION
We are helped by the fact that $\{\gamma^i, \gamma^0\} = 0$. This is because we apply Eq. (5.8) and use the fact that $g^{i0} = 0$. For instance,

$$\{\gamma^3, \gamma^0\} = \gamma^3 \gamma^0 + \gamma^0 \gamma^3 = 2g^{03} = 0$$

Using this fact, that is $\{\gamma^i, \gamma^0\} = 0$, we can write

$$\gamma^i \gamma^0 = -\gamma^0 \gamma^i$$

Therefore,

$$\{\gamma^5, \gamma^0\} = \gamma^5\gamma^0 + \gamma^0\gamma^5$$
$$= (i\gamma^0\gamma^1\gamma^2\gamma^3)\gamma^0 + \gamma^0(i\gamma^0\gamma^1\gamma^2\gamma^3)$$
$$= -(i\gamma^0\gamma^1\gamma^2\gamma^0\gamma^3) + \gamma^0(i\gamma^0\gamma^1\gamma^2\gamma^3)$$
$$= -(i\gamma^0\gamma^1\gamma^2\gamma^0\gamma^3) + (\gamma^0)^2(i\gamma^1\gamma^2\gamma^3)$$
$$= -(i\gamma^0\gamma^1\gamma^2\gamma^0\gamma^3) + (i\gamma^1\gamma^2\gamma^3)$$
$$= (i\gamma^0\gamma^1\gamma^0\gamma^2\gamma^3) + (i\gamma^1\gamma^2\gamma^3)$$
$$= -(\gamma^0)^2(i\gamma^1\gamma^2\gamma^3) + (i\gamma^1\gamma^2\gamma^3)$$
$$= -(i\gamma^1\gamma^2\gamma^3) + (i\gamma^1\gamma^2\gamma^3) = 0$$

We conclude that $\{\gamma^5, \gamma^0\} = 0$.

EXAMPLE 5.3
Find $tr(\gamma^5)$.

SOLUTION
We use $(\gamma^\mu)^2 = I$ together with the result derived in Example 5.2, which means that $\gamma^5\gamma^0 = -\gamma^0\gamma^5$. We have

$$tr(\gamma^5) = tr(I\gamma^5)$$
$$= tr(\gamma^0\gamma^0\gamma^5)$$
$$= tr(-\gamma^0\gamma^5\gamma^0)$$

Now let's recall a basic property of the trace. Remember that the trace operation is cyclic, meaning that

$$tr(ABC) = tr(CBA) = tr(BCA) = tr(BAC)$$

We can also pull scalars (numbers) right outside the trace, that is, $tr(\alpha A) = \alpha tr(A)$ Hence,

$$tr(\gamma^5) = tr(-\gamma^0\gamma^5\gamma^0)$$
$$= -tr(\gamma^0\gamma^5\gamma^0)$$
$$= -tr(\gamma^0\gamma^0\gamma^5)$$
$$= -tr(\gamma^5)$$

We have found that

$$tr(\gamma^5) = -tr(\gamma^5)$$

This can only be true if

$$tr(\gamma^5) = 0$$

which is easily verified by looking at the explicit representation of this matrix in Eq. (5.19).

Adjoint Spinors and Transformation Properties

The adjoint spinor of ψ is not simply $\psi^\dagger$, but turns out to be

$$\bar{\psi} = \psi^\dagger \gamma^0 \tag{5.23}$$

We can form composite objects among the fields $\psi, \bar{\psi},$ and the Dirac matrices. Each of these objects transforms in a different way, so we can construct vectors, tensors, and pseudovectors, for example. We can form the following Lorentz scalar

$$\bar{\psi}\psi \tag{5.24}$$

Using the gamma matrices, we can construct a pseudoscalar which means it is a quantity that changes sign under either parity or space inversion. This pseudoscalar is

$$\bar{\psi}\gamma_5\psi \tag{5.25}$$

Taking an arbitrary gamma matrix, we get an object that transforms as a four vector.

$$\bar{\psi}\gamma^\mu\psi \tag{5.26}$$

Another scalar that can be constructed is

$$\bar{\psi}\gamma^\mu\partial_\mu\psi \tag{5.27}$$

Since $\bar{\psi}\psi$ is a scalar and the mass m is a scalar, we can use this to write down the Lagrangian that can be used to derive the Dirac equation using the usual methods. Remember that the Lagrangian must transform as a scalar. The Lagrangian that works is

$$\mathcal{L} = i\bar{\psi}\gamma^\mu\partial_\mu\psi - m\bar{\psi}\psi \tag{5.28}$$

Notice that both terms in this Lagrangian are scalars. By varying $\bar{\psi}$ we obtain the Dirac equation as the equation of motion for this to Lagrangian.

Slash Notation

In quantum field theory texts and papers you will often see a shorthand notation developed by Feynman called *slash notation*. Slash notation is used to indicate a contraction between a 4-vector and a gamma matrix. Let a_μ be some 4-vector. Then,

$$\not{a} = \gamma^\mu a_\mu = \gamma^0 a_0 + \gamma^1 a_1 + \gamma^2 a_2 + \gamma^3 a_3 \tag{5.29}$$

So for momentum, we have

$$\not{p} = \gamma^\mu p_\mu \tag{5.30}$$

In fact, this can be written as the 4×4 matrix.

$$\not{p} = \begin{pmatrix} E & -\vec{\sigma} \cdot \vec{p} \\ \vec{\sigma} \cdot \vec{p} & -E \end{pmatrix} \tag{5.31}$$

Solutions of the Dirac Equation

Let's introduce two new components of the Dirac field so that we can write down a spinor as a two-component object. We call these components u and v where

$$\psi(x) = \begin{pmatrix} \psi^1(x) \\ \psi^2(x) \\ \psi^3(x) \\ \psi^4(x) \end{pmatrix} = \begin{pmatrix} u^1(x) \\ u^2(x) \\ v^1(x) \\ v^2(x) \end{pmatrix} = \begin{pmatrix} u \\ v \end{pmatrix} \tag{5.32}$$

It will be easy to find solutions of the Dirac equation if we consider momentum space, since the single spatial derivative will be converted into momentum and we will arrive at an algebraic relationship. So consider the Fourier expansion of the Dirac field as

$$\psi(x) = \int \frac{d^4 k}{(2\pi)^4} \, \psi(k) e^{-ik_\mu x^\mu} \tag{5.33}$$

Now we return to the Dirac equation, which for your convenience we reproduce here as

$$i\gamma^\mu \partial_\mu \psi - m\psi = 0$$

Using the Fourier expansion of the field Eq. (5.33), we have

$$i\gamma^\nu \partial_\nu \psi = i\gamma^\mu \partial_\mu \int \frac{d^4 k}{(2\pi)^4} \psi(k) e^{-ik_\mu x^\mu}$$

$$= \int \frac{d^4 k}{(2\pi)^4} i\gamma^\nu \psi(k) \partial_\nu e^{-ik_\mu x^\mu}$$

$$= \int \frac{d^4 k}{(2\pi)^4} i\gamma^\nu (-ik_\nu) \psi(k) e^{-ik_\mu x^\mu}$$

$$= \int \frac{d^4 k}{(2\pi)^4} [\gamma^\nu k_\nu \psi(k)] e^{-ik_\mu x^\mu}$$

The other piece of the equation is

$$m\psi(x) = m \int \frac{d^4 k}{(2\pi)^4} \psi(k) e^{-ik_\mu x^\mu}$$

$$= \int \frac{d^4 k}{(2\pi)^4} m \psi(k) e^{-ik_\mu x^\mu}$$

Putting these terms together gives

$$0 = \int \frac{d^4 k}{(2\pi)^4} [\gamma^\nu k_\nu \psi(k)] e^{-ik_\mu x^\mu} - \int \frac{d^4 k}{(2\pi)^4} m \psi(k) e^{-ik_\mu x^\mu}$$

$$= \int \frac{d^4 k}{(2\pi)^4} [\gamma^\nu k_\nu \psi(k) - m \psi(k)] e^{-ik_\mu x^\mu}$$

Once more, the only way for this integral to be 0 is for the integrand to be 0. So it must be true that

$$\gamma^\nu k_\nu \psi(k) - m \psi(k) = 0$$

Well of course you could have arrived at this, which is just the momentum space equivalent of the Dirac equation, using the usual representation of momentum as a derivative in position space. To see how to proceed toward a solution, let's write down the Dirac representation of the gamma matrices again. We have

$$\gamma_0 = \begin{pmatrix} I & 0 \\ 0 & -I \end{pmatrix}$$

together with

$$\gamma_1 = \begin{pmatrix} 0 & \sigma_1 \\ -\sigma_1 & 0 \end{pmatrix} \qquad \gamma_2 = \begin{pmatrix} 0 & \sigma_2 \\ -\sigma_2 & 0 \end{pmatrix} \qquad \gamma_3 = \begin{pmatrix} 0 & \sigma_3 \\ -\sigma_3 & 0 \end{pmatrix}$$

Let's take a look at the individual terms in the equation $\gamma^\nu k_\nu \psi(k) - m\psi(k) = 0$. First we have

$$\gamma^0 k_0 \psi = k_0 \begin{pmatrix} I & 0 \\ 0 & -I \end{pmatrix} \begin{pmatrix} u \\ v \end{pmatrix} = \begin{pmatrix} k_0 u \\ -k_0 v \end{pmatrix}$$

and

$$m\psi = m \begin{pmatrix} u \\ v \end{pmatrix} = \begin{pmatrix} mu \\ mv \end{pmatrix}$$

Next we have

$$\gamma^1 k_1 \psi = k_1 \begin{pmatrix} 0 & \sigma_1 \\ -\sigma_1 & 0 \end{pmatrix} \begin{pmatrix} u \\ v \end{pmatrix} = k_1 \begin{pmatrix} \sigma_1 v \\ -\sigma_1 u \end{pmatrix}$$

Similarly,

$$\gamma^2 k_2 \psi = k_2 \begin{pmatrix} 0 & \sigma_2 \\ -\sigma_2 & 0 \end{pmatrix} \begin{pmatrix} u \\ v \end{pmatrix} = k_2 \begin{pmatrix} \sigma_2 v \\ -\sigma_2 u \end{pmatrix}$$

and

$$\gamma^3 k_3 \psi = k_3 \begin{pmatrix} 0 & \sigma_3 \\ -\sigma_3 & 0 \end{pmatrix} \begin{pmatrix} u \\ v \end{pmatrix} = k_3 \begin{pmatrix} \sigma_3 v \\ -\sigma_3 u \end{pmatrix}$$

Putting everything together, these relations and the Dirac equation in momentum space can be written in matrix form as

$$\begin{pmatrix} k_0 - m & -\vec{k} \cdot \vec{\sigma} \\ \vec{k} \cdot \vec{\sigma} & -(k_0 + m) \end{pmatrix} \begin{pmatrix} u \\ v \end{pmatrix} = 0 \tag{5.34}$$

So there are two coupled equations

$$(k_0 - m)u - (\vec{k} \cdot \vec{\sigma})v = 0$$
$$(k_0 + m)v - (\vec{k} \cdot \vec{\sigma})u = 0 \tag{5.35}$$

For the sake of simplicity, let's identify the matrix in Eq. (5.34) as

$$K = \begin{pmatrix} k_0 - m & -\vec{k} \cdot \vec{\sigma} \\ \vec{k} \cdot \vec{\sigma} & -(k_0 + m) \end{pmatrix}$$

For a solution to this system to exist, it must be true that K has vanishing determinant. That is,

$$\det \begin{pmatrix} k_0 - m & -\vec{k} \cdot \vec{\sigma} \\ \vec{k} \cdot \vec{\sigma} & -(k_0 + m) \end{pmatrix} = 0$$

The determinant works out to be

$$\det K = \det \begin{pmatrix} k_0 - m & -\vec{k} \cdot \vec{\sigma} \\ \vec{k} \cdot \vec{\sigma} & -(k_0 + m) \end{pmatrix}$$
$$= -(k_0 - m)(k_0 + m) + (\vec{k} \cdot \vec{\sigma})^2$$

To work out this result, we'll need to calculate $(\vec{k} \cdot \vec{\sigma})^2$. From ordinary quantum mechanics recall that the Pauli matrices satisfy $\sigma_j^2 = I$. In addition, they satisfy the anticommutation relation.

$$\{\sigma_i, \sigma_j\} = 2\delta_{ij} \qquad (5.36)$$

This greatly simplifies the calculation of $(\vec{k} \cdot \vec{\sigma})^2$. Writing out the terms, it's going to be

$$(\vec{k} \cdot \vec{\sigma})^2 = (k_1\sigma_1 + k_2\sigma_2 + k_3\sigma_3)(k_1\sigma_1 + k_2\sigma_2 + k_3\sigma_3)$$

Since the Pauli matrices satisfy Eq. (5.36), mixed terms in this expression will vanish. As a specific example consider

$$k_1 k_2 \sigma_1 \sigma_2 + k_2 k_1 \sigma_2 \sigma_1 = k_1 k_2 (\sigma_1 \sigma_2 + \sigma_2 \sigma_1)$$
$$= k_1 k_2 2\delta_{12}$$
$$= 0$$

So we are left with

$$(\vec{k} \cdot \vec{\sigma})^2 = (k_1\sigma_1 + k_2\sigma_2 + k_3\sigma_3)(k_1\sigma_1 + k_2\sigma_2 + k_3\sigma_3)$$
$$= k_1^2\sigma_1^2 + k_2^2\sigma_2^2 + k_3^2\sigma_3^2$$
$$= k_1^2 + k_2^2 + k_3^2$$
$$= \vec{k}^2$$

Hence,

$$\det K = -(k_0 - m)(k_0 + m) + (\vec{k} \cdot \vec{\sigma})^2$$
$$= -k_0^2 + m^2 + \vec{k}^2$$

This is just another way to write down our old friend, the relativistic relation between energy, mass, and momentum. Since we are working in units where $\hbar = 1$, $\vec{k} = \vec{p}$, and $k_0 = p_0 = E$, we can also write the energy in terms of a frequency since as you recall $E = \hbar\omega$.

Remember for a solution to exist, this quantity must vanish $(\det K = 0)$. That is,

$$k_0^2 = \vec{k}^2 + m^2$$

Taking the square root, we see that the possible energies are

$$\omega_k = k_0 = \pm\sqrt{\vec{k}^2 + m^2}$$

This means that the Dirac equation is still plagued by negative energies. If we take the positive solution, so that $E = \omega_k > 0$, we call the solution a *positive frequency* solution. The solution with $E = \omega_k < 0$ is the *negative energy* solution. We get out of this problem the same way we did with the Klein-Gordon equation—we interpret the positive energy solutions as corresponding to particles with positive energy, and the negative energy solutions correspond to antiparticles with positive energy. Dirac predicted their existence with the notion of a "sea" of negative energy states. However his logic used to predict the existence of antiparticles is wrong, so we will not discuss it. There is no Dirac sea. You can read about this in most quantum field theory texts.

Free Space Solutions

If a particle is at rest (or let's rephrase that and say you are observing it in its rest frame), it has no motion and hence no momentum. This means we can disregard the spatial derivatives in the Dirac equation. Let's use this case to develop the free space solution for a particle at rest. Once you have that, you can always do a Lorentz boost to find the solution for a particle with arbitrary momentum.

In the case where the particle is at rest, the Dirac equation reduces to

$$i\gamma^0 \frac{\partial \psi}{\partial t} - m\psi = 0 \qquad (5.37)$$

Once again we take $\psi = \binom{u}{v}$ where u and v each have two components and we work with the Dirac-Pauli representation and take

$$\gamma^0 = \begin{pmatrix} I & 0 \\ 0 & -I \end{pmatrix}$$

The first term works out to be

$$i\gamma^0 \frac{\partial \psi}{\partial t} = i \begin{pmatrix} I & 0 \\ 0 & -I \end{pmatrix} \frac{\partial}{\partial t} \binom{u}{v}$$

$$= i \begin{pmatrix} I & 0 \\ 0 & -I \end{pmatrix} \binom{\dot{u}}{\dot{v}}$$

$$= i \binom{\dot{u}}{-\dot{v}}$$

where $\dot{u} = \frac{\partial u}{\partial t}$ and similarly for v. The minus sign that appears here will once again lead to negative energies. The other term in the Dirac equation is

$$m\psi = m\binom{u}{v}$$

So we have the system

$$0 = i\gamma^0 \frac{\partial \psi}{\partial t} - m\psi$$

$$= i\binom{\dot{u}}{-\dot{v}} - m\binom{u}{v}$$

This leads to the two elementary differential equations

$$i\frac{\partial u}{\partial t} = mu$$

$$i\frac{\partial v}{\partial t} = -mv$$

with the solutions

$$u(t) = u(0)e^{-imt}$$
$$v(t) = v(0)e^{imt}$$

Thinking back to nonrelativistic quantum mechanics, a free space solution has a time dependence of the form e^{-iEt}. So we make a comparison in each case to determine the energy, which in this case is the rest energy. For u we have the correspondence

$$e^{-imt} \sim e^{-iEt}$$

So for u we have the pleasing relationship that $E = m$. Since we're working in units where $c = 1$, this is just the statement that the rest-mass energy is $E = mc^2$. Making the same comparison for v, we see that

$$e^{imt} \sim e^{-iEt}$$

This time we have $E = -m$. Yet again negative-energy states have reared their ugly head. Therefore we conclude that u is a two-component spinor representing a *particle*, while v is a two-component spinor representing an *antiparticle*. These are two-component objects because this equation describes spin-1/2 particles. Recall that the column vector representation of a spin-1/2 particle in quantum mechanics is

$$\phi = \begin{pmatrix} \alpha \\ \beta \end{pmatrix}$$

where α and β are the probability amplitudes to find the particle spin-up or spin-down, respectively.

Summarizing, we've learned that using the two component spinors u and v, the Dirac spinor can be thought of as

$$\psi = \begin{pmatrix} u \\ v \end{pmatrix} = \begin{pmatrix} \text{particle} \\ \text{antiparticle} \end{pmatrix} \tag{5.38}$$

Now let's consider a particle moving with arbitrary momentum p. We will use the scalar product

$$p \cdot x = Et - \vec{p} \cdot \vec{x} \tag{5.39}$$

Free space solutions are going to be plane waves, and we can immediately use the rest frame solutions to guess at a solution for a particle with momentum p. We use Eq. (5.39) and take

$$u \propto e^{-ip \cdot x}$$

$$v \propto e^{ip \cdot x}$$

Notice that if $u \propto e^{-ip \cdot x}$, then

$$i\gamma^\mu \partial_\mu u = i\gamma^\mu \partial_\mu e^{-ip \cdot x}$$

$$= \gamma^\mu p_\mu e^{-ip \cdot x}$$

$$= p\!\!\!/ \psi$$

For $v \propto e^{ip \cdot x}$ we find

$$i\gamma^\mu \partial_\mu v = i\gamma^\mu \partial_\mu e^{ip \cdot x}$$

$$= -\gamma^\mu p_\mu e^{ip \cdot x}$$

$$= -p\!\!\!/ \psi$$

The Dirac equation, as you should have ingrained in your mind by now, is $i\gamma^\mu \partial_\mu \psi - m\psi = 0$. Using the above results we have two algebraic relations for u and v.

$$(p\!\!\!/ - m)u = 0 \tag{5.40}$$

$$(p\!\!\!/ + m)v = 0 \tag{5.41}$$

Notice that

$$(p\!\!\!/ - m)(p\!\!\!/ + m) = (\gamma^\mu p_\mu - m)(\gamma^\nu p_\nu + m)$$

$$= \gamma^\mu \gamma^\nu p_\mu p_\nu + \gamma^\nu \gamma^\mu p_\mu p_\nu + m\gamma^\mu p_\mu - m\gamma^\nu p_\nu - m^2$$

When an index in an expression is repeated, it is a dummy index so we can relabel it. This allows us to get rid of two of the terms in this expression, because we have

$$m\gamma^\mu p_\mu - m\gamma^\nu p_\nu = m\gamma^\mu p_\mu - m\gamma^\mu p_\mu = 0$$

This leaves

$$(p\!\!\!/ - m)(p\!\!\!/ + m) = \gamma^\mu \gamma^\nu p_\mu p_\nu + \gamma^\nu \gamma^\mu p_\mu p_\nu - m^2$$

$$= (\gamma^\mu \gamma^\nu + \gamma^\nu \gamma^\mu) p_\mu p_\nu - m^2$$

Using a now familiar friend, the anticommutation relation for the gamma matrices Eq. (5.8), this simplifies to

$$(p - m)(p + m) = (\gamma^\mu \gamma^\nu + \gamma^\nu \gamma^\mu) p_\mu p_\nu - m^2$$
$$= p_\mu p^\mu - m^2$$
$$= p^2 - m^2$$
$$= E^2 - \vec{p}^2 - m^2 = m^2 - m^2 = 0$$

We take the solutions to be of the form

$$u(p) = (\not{p} + m)u(0) \tag{5.42}$$

$$v(p) = (\not{p} - m)v(0) \tag{5.43}$$

Notice that

$$(p - m)u(p) = (p - m)(p + m)u(0)$$
$$= (p^2 - m^2)u(0)$$
$$= 0$$

The Dirac equation, written as Eq. (5.40), is satisfied. The initial states are given by ordinary spin-up and spin-down states, that we are free to choose. For example, we can take

$$u(0) = \begin{pmatrix} 1 \\ 0 \end{pmatrix} \qquad v(0) = \begin{pmatrix} 0 \\ 1 \end{pmatrix}$$

Boosts, Rotations, and Helicity

We have already written down the Dirac matrices and their defining anticommutation relation as

$$\{\gamma^\mu, \gamma^\nu\} = 2g^{\mu\nu}$$

Now we need to figure out how to use them to generate boosts and rotations for the spin-1/2 case. The so-called Lorentz algebra requires that we seek operators $J^{\mu\nu}$ that satisfy

$$[J^{\mu\nu}, J^{\alpha\beta}] = i(g^{\nu\alpha}J^{\mu\beta} - g^{\mu\alpha}J^{\nu\beta} - g^{\nu\beta}J^{\mu\alpha} + g^{\mu\beta}J^{\nu\alpha})$$

This will work if we define the tensor

$$S^{\mu\nu} = \frac{i}{4}[\gamma^\mu, \gamma^\nu] \tag{5.44}$$

For a spin-1/2 particle, the generator of a Lorentz boost in the jth direction is

$$S^{0j} = \frac{i}{4}[\gamma^0, \gamma^j] \tag{5.45}$$

The generator of a rotation for a spin-1/2 particle is

$$S^{ij} = \frac{i}{4}[\gamma^i, \gamma^j] = \frac{1}{2}\varepsilon^{ijk}\begin{pmatrix} \sigma_k & 0 \\ 0 & \sigma_k \end{pmatrix} \tag{5.46}$$

Now let

$$\vec{\Sigma} = \begin{pmatrix} \vec{\sigma} & 0 \\ 0 & \vec{\sigma} \end{pmatrix} \tag{5.47}$$

so that, for example

$$\Sigma_1 = \begin{pmatrix} \sigma_1 & 0 \\ 0 & \sigma_1 \end{pmatrix}$$

This can be used to define the *helicity operator*. Helicity tells us how closely aligned the spin of a particle is with its direction of motion. The direction of a particle's motion is given by its spatial momentum vector $\vec{p}$, so we can write the helicity operator as

$$h = \frac{\vec{\Sigma} \cdot \vec{p}}{|\vec{p}|} \tag{5.48}$$

Looking at Eq. (5.47) you can see that a simple way to write the helicity operator is $\vec{\sigma} \cdot \vec{p}$.

Weyl Spinors

Let's return to the chiral representation, which is useful when considering a special type of spinor known as a *Weyl spinor*. In the chiral representation, the Dirac matrices are represented by

$$\gamma^0 = \begin{pmatrix} 0 & I \\ I & 0 \end{pmatrix} \qquad \gamma^i = \begin{pmatrix} 0 & \sigma_i \\ -\sigma_i & 0 \end{pmatrix}$$

$$\gamma_5 = \begin{pmatrix} -I & 0 \\ 0 & I \end{pmatrix}$$

In the chiral representation, we have

$$\not{p} = \gamma^\mu p_\mu = \gamma^0 p_0 - \gamma^1 p_1 - \gamma^2 p_2 - \gamma^3 p_3$$

$$= \begin{pmatrix} 0 & I \\ I & 0 \end{pmatrix} p_0 - \begin{pmatrix} 0 & \sigma_1 \\ -\sigma_1 & 0 \end{pmatrix} p_1 - \begin{pmatrix} 0 & \sigma_2 \\ -\sigma_2 & 0 \end{pmatrix} p_2 - \begin{pmatrix} 0 & \sigma_3 \\ -\sigma_3 & 0 \end{pmatrix} p_3$$

$$= \begin{pmatrix} 0 & E - \vec{p}\cdot\vec{\sigma} \\ E + \vec{p}\cdot\vec{\sigma} & 0 \end{pmatrix}$$

where $p_0 = E$. Considering the simple massless case, we write the Dirac spinor as

$$\psi = \begin{pmatrix} \psi_L \\ \psi_R \end{pmatrix} \tag{5.49}$$

We call Eq. (5.49) a Weyl spinor. The components ψ_L and ψ_R will be left-handed and right-handed spinors. Like u and v, these are two component spinors. We also use the term chiral representation because these components are eigenstates of the helicity operator.

When $m = 0$, the Dirac equation reduces to the pleasingly simple form like

$$\not{p}\psi = 0$$

or in matrix form

$$\begin{pmatrix} 0 & E - \vec{p}\cdot\vec{\sigma} \\ E + \vec{p}\cdot\vec{\sigma} & 0 \end{pmatrix}\begin{pmatrix} \psi_L \\ \psi_R \end{pmatrix} = 0$$

This produces two equations.

$$\left(E + \vec{p}\cdot\vec{\sigma}\right)\psi_L = 0 \tag{5.50}$$

$$\left(E - \vec{p}\cdot\vec{\sigma}\right)\psi_R = 0 \tag{5.51}$$

Let's take a look at the first equation. Using $(\vec{p}\cdot\vec{\sigma})^2 = |\vec{p}|^2$, we have

$$0 = (E - \vec{p}\cdot\vec{\sigma})(E + \vec{p}\cdot\vec{\sigma})\psi_L$$

$$= (E^2 - |\vec{p}|^2)\psi_L$$

$$\Rightarrow E^2 - |\vec{p}|^2 = 0,\ E = |\vec{p}|$$

A similar relation with a sign change can be found for the right-handed spinor. Writing the helicity operator as $\vec{\sigma} \cdot \vec{p}$, the left- and right-handed spinors satisfy the eigenvalue equations.

$$(\vec{\sigma} \cdot \vec{p})\psi_L = -E\psi_L = -|\vec{p}|\psi_L \tag{5.52}$$

$$(\vec{\sigma} \cdot \vec{p})\psi_R = E\psi_R = |\vec{p}|\psi_R \tag{5.53}$$

The Weyl spinors are also eigenstates of γ_5, as we show in Example 5.4.

EXAMPLE 5.4

Consider massless Weyl spinors and show that they are eigenstates of the γ_5 matrix if we take $\psi_L = \frac{1}{2}(I - \gamma_5)\psi$, $\psi_R = \frac{1}{2}(I + \gamma_5)\psi$.

SOLUTION

In the chiral representation we have

$$\gamma_5 = \begin{pmatrix} -I & 0 \\ 0 & I \end{pmatrix}$$

Applying this to the Weyl spinor, we get

$$\gamma_5\psi = \begin{pmatrix} -I & 0 \\ 0 & I \end{pmatrix}\begin{pmatrix} \psi_L \\ \psi_R \end{pmatrix} = \begin{pmatrix} -\psi_L \\ \psi_R \end{pmatrix}$$

Now of course

$$I\psi = \begin{pmatrix} I & 0 \\ 0 & I \end{pmatrix}\begin{pmatrix} \psi_L \\ \psi_R \end{pmatrix} = \begin{pmatrix} \psi_L \\ \psi_R \end{pmatrix}$$

Therefore,

$$(I - \gamma_5)\psi = \begin{pmatrix} \psi_L \\ \psi_R \end{pmatrix} - \begin{pmatrix} -\psi_L \\ \psi_R \end{pmatrix} = 2\begin{pmatrix} \psi_L \\ 0 \end{pmatrix}$$

This leads to the relationship

$$\psi_L = \frac{1}{2}(I - \gamma_5)\psi$$

Similarly, we have

$$\psi_R = \frac{1}{2}(I + \gamma_5)\psi$$

Notice that

$$\gamma_5^2 = \begin{pmatrix} -I & 0 \\ 0 & I \end{pmatrix}\begin{pmatrix} -I & 0 \\ 0 & I \end{pmatrix} = \begin{pmatrix} I & 0 \\ 0 & I \end{pmatrix} = I$$

So we have

$$\gamma_5 \psi_L = \frac{1}{2}(\gamma_5 I - \gamma_5^2)\psi$$
$$= \frac{1}{2}(\gamma_5 - I)\psi$$
$$= -\frac{1}{2}(I - \gamma_5)\psi$$
$$= -\psi_L$$

This shows that the left-handed Weyl spinor is an eigenstate of γ_5 with eigenvalue -1. Similarly,

$$\gamma_5 \psi_R = \frac{1}{2}(\gamma_5 I + \gamma_5^2)\psi$$
$$= \frac{1}{2}(\gamma_5 + I)\psi$$
$$= \frac{1}{2}(I + \gamma_5)\psi$$
$$= \psi_R$$

That is, ψ_R is an eigenstate of γ_5 with eigenvalue $+1$.

Summary

In this chapter we have introduced the Dirac equation. This equation was derived by Dirac in an attempt to get a relativistic equation with time and space on the same footing, while avoiding the negative probability densities associated with the Klein-Gordon equation. The equation still has solutions with negative energy, which are the result of the fact that it describes antiparticles, as well as particles.

Quiz

1. Given the Lagrangian

$$\mathcal{L} = \bar{\psi}(x)[i\gamma^{\mu}\partial_{\mu} - m]\psi$$

   vary $\psi(x)$ to find the equation of motion obeyed by $\bar{\psi}(x)$.

2. Calculate $\{\gamma_5, \gamma^{\mu}\}$.

3. Consider the solution of the Dirac equation with $E = \omega_k > 0$. Find a relationship between the u and v components of the Dirac field.

4. Find the normalization of the free space solutions of the Dirac equation using the density $\bar{\psi}\gamma^0\psi$.

5. Find S^{01}, the generator of a boost in the x direction.

6. We can introduce an electromagnetic field with a vector potential A_{μ}. Let the source charge be q. Using the substitution $p_{\mu} \rightarrow p_{\mu} - qA_{\mu}$, determine the form of the Dirac equation in the presence of an electromagnetic field.

CHAPTER 6

Scalar Fields

The first attempts to merge the theory of relativity with quantum mechanics involved what you might think of as relativistic generalizations of the Schrödinger equation that were imagined to apply to a single particle. In fact, Schrödinger himself derived a relativistic equation—one that we first learned in Chap. 2—the Klein-Gordon equation, before coming up with his famous nonrelativistic wave equation. He ended up discarding the Klein-Gordon equation as the correct one for quantum mechanics for three main reasons:

- It appeared to have solutions with negative energy.
- It appeared to lead to a negative probability distribution.
- It gave an incorrect spectrum for the hydrogen atom.

Looking at these factors he ended up discarding what we now know as the Klein-Gordon equation in favor of what is now known as the Schrödinger equation. But as we'll see later, the main problem with the Klein-Gordon equation is a problem in interpretation.

We can begin our path to a relativistic wave equation by thinking about what we learned in Chap. 1. Relativity treats time and space in a similar fashion. In a wave equation, this implies that the derivatives applied to the time and spatial coordinates must be of the same order. In the nonrelativistic Schrödinger equation, there is a first-order derivative with respect to time but derivatives with respect to the spatial coordinates are of second order. Let's write down the Schrödinger equation in the case of one spatial dimension to remind ourselves of this explicitly.

$$i\hbar \frac{\partial \psi}{\partial t} = -\frac{\hbar^2}{2m}\frac{\partial^2 \psi}{\partial x^2} + V\psi \tag{6.1}$$

This equation cannot be relativistic since we have a first-order derivative with respect to time $\frac{\partial \psi}{\partial t}$ on the left hand side, while we have a second-order derivative with respect to the spatial coordinate $\frac{\partial^2 \psi}{\partial x^2}$ on the right-hand side. To incorporate special relativity into quantum theory, we would expect symmetry. This situation is rectified in the Klein-Gordon equation where second-order derivatives are applied to both the time and spatial coordinates. In contrast, Dirac, when deriving his famous equation that applies to spin-1/2 particles, insisted that first-order derivatives be applied to both the spatial and time coordinates. We will see later why Dirac decided to "demote" the spatial derivative to first order to get the symmetry in time and space we are looking for when we see how a second-order time derivative causes problems in the Klein-Gordon equation. In this chapter we will discuss the Klein-Gordon equation, which applies to scalar fields.

Arriving at the Klein-Gordon Equation

We begin our examination of relativistic wave equations by returning to an equation we learned briefly in the second chapter, the *Klein-Gordon equation*. In Chap. 2 we saw how it could be derived from a Lagrangian which was given to us, but the ultimate origin of this equation may have seemed mysterious. What we are going to see in a moment is that the Klein-Gordon equation follows from the application of two fundamental principles—one taken from special relativity and the other taken from quantum mechanics. These are

- The relativistic relation between energy, mass, and momentum derived by Einstein
- The promotion of measurable quantities ("observables") to mathematical operators in quantum mechanics

Now let's go forward and see how Schrödinger, Klein, and Gordon (my apologies to any other dead people who derived this equation I am leaving out) derived the equation. The Klein-Gordon equation is very easy to derive in two steps. We start by writing down the fundamental relation between energy, momentum, and mass used in special relativity.

$$E^2 = p^2 c^2 + m^2 c^4 \qquad (6.2)$$

Now we turn immediately to quantum mechanics. In quantum theory, observables turn into mathematical operators using a specific prescription you are no doubt very familiar with. We can see how this is done looking at the nonrelativistic Schrödinger equation [Eq. (6.1)]. You remember that the time-independent version of this equation is given by

$$E\psi = -\frac{\hbar^2}{2m}\frac{\partial^2 \psi}{\partial x^2} + V\psi \qquad (6.3)$$

So it might occur to you that the Schrödinger equation can be thought of as a statement of the nonrelativistic definition of energy. Hence we make the following substitution for energy, promoting it to an operator that takes the derivative with respect to time.

$$E \rightarrow i\hbar \frac{\partial}{\partial t} \qquad (6.4)$$

We also recall that in ordinary quantum mechanics, momentum p is given by a spatial derivative, that is,

$$p \rightarrow -i\hbar \frac{\partial}{\partial x} \qquad (6.5)$$

Generalizing to three dimensions, the relation is

$$\vec{p} \rightarrow -i\hbar \nabla \qquad (6.6)$$

To derive the Klein-Gordon equation, all we do is put the substitutions [Eqs. (6.4) and (6.6)] into the Einstein relation for energy, momentum, and mass Eq. (6.2) and apply it to a wave function φ. Using Eq. (6.4) we see that

$$E^2 \rightarrow -\hbar^2 \frac{\partial^2}{\partial t^2}$$

Now, using Eq. (6.6) we have

$$p^2 = -\hbar^2 \nabla^2$$

Therefore, in terms of operators, the Einstein relation between energy, momentum, and mass Eq. (6.2) can be written as

$$-\hbar^2 \frac{\partial^2}{\partial t^2} = -\hbar^2 c^2 \nabla^2 + m^2 c^4$$

This isn't going to be of much use or make any sense unless we *do something* with it. So we'll apply this operator to a function of space and time $\varphi = \varphi(\vec{x}, t)$. Doing this and rearranging terms a little gives us the Klein-Gordon equation.

$$\hbar^2 \frac{\partial^2 \varphi}{\partial t^2} - \hbar^2 c^2 \nabla^2 \varphi + m^2 c^4 \varphi = 0 \qquad (6.7)$$

As discussed in Chap. 1, in particle physics we typically work in units where $\hbar = c = 1$ (*natural units*) so this becomes

$$\frac{\partial^2 \varphi}{\partial t^2} - \nabla^2 \varphi + m^2 \varphi = 0 \qquad (6.8)$$

We can simplify the appearance of the equation a little further by using different notation. In fact we'll write it in two different ways. The first is to recall the D'Alembertian operator in Minkowski space as

$$\Box = \frac{\partial^2}{\partial t^2} - \nabla^2$$

This allows us to write Eq. (6.8) in the following simplified way:

$$(\Box + m^2)\varphi = 0$$

This is a nice way to write the equation for the following reason. We know that $\Box$ is a relativistic invariant, that is, it is the same in all inertial reference frames because it transforms as a scalar. The mass m is of course a scalar so the operator given by

$$\Box + m^2$$

is also a scalar. What this tells us is that the Klein-Gordon equation will be covariant provided that the function φ—which we will interpret later as a field—also transforms as a scalar. In Chap. 1 we learned that the coordinates x^μ transform as

$$x'^\mu = \Lambda^\mu{}_\nu x^\nu \tag{6.9}$$

under a Lorentz transformation. If a field $\varphi(x)$ is a *scalar* field, then it transforms as

$$\varphi'(x) = \varphi(\Lambda^{-1}x) \tag{6.10}$$

We are led to the first important characteristics of the Klein-Gordon equation.

- It applies to scalar particles (actually scalar fields).
- These particles are spin-0 particles.

We can also write Eq. (6.7) in a nice, compact style using the notation developed in Chap. 1. Using $\partial_\mu\partial^\mu = \frac{\partial^2}{\partial t^2} - \nabla^2$ it becomes

$$(\partial_\mu\partial^\mu + m^2)\varphi = 0 \tag{6.11}$$

As it is written, Eq. (6.11) describes a free particle. The free particle solution is given by

$$\varphi(\vec{x},t) = e^{-ip\cdot x}$$

Remember, we are applying special relativity here so p and x are 4-vectors given by $p = (E,\vec{p})$ and $x = (t,\vec{x})$, respectively. The scalar product in the exponent is

$$p\cdot x = p_\mu x^\mu = Et - \vec{p}\cdot\vec{x} \tag{6.12}$$

The free particle solution implies the relativistic relation between energy, mass, and momentum. This is very easy to show, so let's do it. For simplicity, we consider one spatial dimension only. Since

$$\frac{\partial\varphi}{\partial t} = \frac{\partial}{\partial t}e^{-i(Et-px)} = -iEe^{-i(Et-px)} = -iE\varphi$$

and

$$\frac{\partial\varphi}{\partial x} = \frac{\partial}{\partial x}e^{-i(Et-px)} = ipe^{-i(Et-px)} = ip\varphi$$

Therefore, we have

$$\frac{\partial^2 \varphi}{\partial t^2} - \frac{\partial^2 \varphi}{\partial x^2} = -E^2 \varphi + p^2 \varphi$$

Hence, applying the full Klein-Gordon equation [Eq. (6.8)] we have

$$(E^2 - p^2)\varphi = m^2 \varphi$$

Canceling the wave function and rearranging terms gives $E^2 = p^2 + m^2$, the desired result. Solving for the energy, we take the square root, being careful to include both positive and negative square roots.

$$E = \pm\sqrt{p^2 + m^2} \tag{6.13}$$

This is a dramatic result which is one reason Schrödinger discarded the Klein-Gordon equation. The solution for the energy of the particle tells us that it is possible to have both positive and negative energy states—a nonphysical result. How do we get around this?

Finding negative energy states was the first indication that the *interpretation* of the Klein-Gordon equation as a single particle wave equation is incorrect. It turns out we deal with the negative energy states in the following way. Solutions for particles with negative energy are actually solutions describing *antiparticles,* that is, particles with the same mass but opposite charge, with positive energy.

However as we mentioned in the introduction, there are more problems with the Klein-Gordon equation. The second problem we will see is the fact that the time derivatives are second order leads to the problem of *negative probability densities*, which is nonsense. At least if you are constrained by the ideas of nonrelativistic quantum mechanics. The way around this is that we will decide the equation is not describing a single particle wave function the way the nonrelativistic Schrödinger equation does. Let's see how a negative probability arises from the free particle solution in Example 6.1.

EXAMPLE 6.1
Show that the Klein-Gordon equation leads to a negative probability density in the free particle case. For simplicity, consider one spatial dimension.

SOLUTION
We begin by assuming that the *probability current* assumes the same form as it does in ordinary quantum mechanics. Keeping $\hbar = 1$ we define the probability current as

$$J = -i\varphi^* \frac{\partial \varphi}{\partial x} + i\varphi \frac{\partial \varphi^*}{\partial x} \tag{6.14}$$

Now,

$$\frac{\partial J}{\partial x} = -i\frac{\partial \varphi^*}{\partial x}\frac{\partial \varphi}{\partial x} - i\varphi^*\frac{\partial^2 \varphi}{\partial x^2} + i\frac{\partial \varphi}{\partial x}\frac{\partial \varphi^*}{\partial x} + i\varphi\frac{\partial^2 \varphi^*}{\partial x^2}$$

$$= -i\varphi^*\frac{\partial^2 \varphi}{\partial x^2} + i\varphi\frac{\partial^2 \varphi^*}{\partial x^2}$$

We can transform the derivatives with respect to the spatial coordinate into derivatives with respect to the time coordinate using the Klein-Gordon equation [Eq. (6.8)]. Sticking to one spatial dimension, we rearrange terms a bit and find that

$$\frac{\partial^2 \varphi}{\partial x^2} = \frac{\partial^2 \varphi}{\partial t^2} + m^2 \varphi \tag{6.15}$$

So we see that

$$\frac{\partial J}{\partial x} = -i\varphi^*\frac{\partial^2 \varphi}{\partial x^2} + i\varphi\frac{\partial^2 \varphi^*}{\partial x^2}$$

$$= -i\varphi^*\left(\frac{\partial^2 \varphi}{\partial t^2} + m^2 \varphi\right) + i\varphi\left(\frac{\partial^2 \varphi^*}{\partial t^2} + m^2 \varphi^*\right)$$

$$\Rightarrow \frac{\partial J}{\partial x} = -i\left(\varphi^*\frac{\partial^2 \varphi}{\partial t^2} - \varphi\frac{\partial^2 \varphi^*}{\partial t^2}\right)$$

Now we recall another fundamental result from ordinary quantum mechanics. The probability current and probability density ρ satisfy a conservation equation, called the *conservation of probability*, which becomes, when working in one spatial dimension

$$\frac{\partial \rho}{\partial t} + \frac{\partial J}{\partial x} = 0 \tag{6.16}$$

Hence we find that

$$\frac{\partial \rho}{\partial t} = i\left(\varphi^*\frac{\partial^2 \varphi}{\partial t^2} - \varphi\frac{\partial^2 \varphi^*}{\partial t^2}\right)$$

This equation will be satisfied if

$$\rho = i\left(\varphi^*\frac{\partial \varphi}{\partial t} - \varphi\frac{\partial \varphi^*}{\partial t}\right) \tag{6.17}$$

Let's summarize what we have found. The fact that the Klein-Gordon equation includes second-order derivatives with respect to time leads to a probability density Eq. (6.17) that has a very different form than it takes in ordinary quantum mechanics. In fact you probably remember the probability density is defined in terms of the wave function ψ as

$$\rho = |\psi|^2 = \psi^* \psi$$

This is a direct result of the expression we found for the probability current, namely

$$\frac{\partial J}{\partial x} = -i \left(\varphi^* \frac{\partial^2 \varphi}{\partial t^2} - \varphi \frac{\partial^2 \varphi^*}{\partial t^2} \right) \tag{6.18}$$

together with the Klein-Gordon equation. Now let's see what happens when we consider the free particle solution. The presence of the first-order time derivatives in the probability density Eq. (6.17) together with the solutions for the energy Eq. (6.13) leads to problems. Remembering the free particle solution

$$\varphi(\vec{x}, t) = e^{-ip \cdot x} = e^{-i(Et - px)}$$

the time derivatives are

$$\frac{\partial \varphi}{\partial t} = -iE e^{-i(Et - px)} \qquad \frac{\partial \varphi^*}{\partial t} = iE e^{i(Et - px)}$$

We have

$$\varphi^* \frac{\partial \varphi}{\partial t} = e^{i(Et - px)} [-iE e^{-i(Et - px)}] = -iE$$

$$\varphi \frac{\partial \varphi^*}{\partial t} = e^{-i(Et - px)} [iE e^{i(Et - px)}] = iE$$

So the probability density Eq. (6.17) is

$$\rho = i \left(\varphi^* \frac{\partial \varphi}{\partial t} - \varphi \frac{\partial \varphi^*}{\partial t} \right) = i(-iE - iE) = 2E$$

Looks good so far—except for those pesky negative energy solutions. Remember that

$$E = \pm \sqrt{p^2 + m^2}$$

In the case of the negative energy solution

$$\rho = 2E = -2\sqrt{p^2 + m^2} < 0$$

which is a negative probability density, something which simply does not make sense. Why doesn't it make sense? A probability of 1 means the particle is in hand. Probability of ½ means you might find it. A probability of 0 means you can't find the particle. A negative probability, say −1, has no consistent interpretation.

Reinterpreting the Field

The solution to the problem of negative probability density with the Klein-Gordon equation involves reinterpreting what the equation represents. Instead of imagining that the equation governs the wave function of a scalar particle, we instead imagine that φ is a field. We promote φ to an operator $\varphi \to \hat{\varphi}$ that includes creation and annihilation operators that create and destroy quanta of the field (the particles), and we will force φ to obey the usual/respected/canonical commutation relations.

Field Quantization of Scalar Fields

We now turn to the task of quantizing a given field $\varphi(x)$. The process of quantization which basically means creating a quantum theory from a classical one is based on imposing commutation relations. *Canonical quantization* refers to the process of imposing the fundamental commutation relation on the position and momentum operators.

$$[\hat{x}, \hat{p}] = i \tag{6.19}$$

In total, the quantization procedure is to

- Promote position and momentum functions to operators
- Impose the commutation relation Eq. (6.19)

We will follow a similar procedure for quantizing a classical field theory. In this case, the procedure is called *second quantization*.

118

Quantum Field Theory Demystified

SECOND QUANTIZATION

In quantum field theory, we quantize the fields themselves rather than quantizing dynamical variables like position. Once again we are faced with the problem of having to put space and time on an equal footing. In nonrelativistic quantum mechanics, position and momentum are operators. The position operator acts on a wave function according to

$$\hat{X}\psi(x) = x\psi(x)$$

The momentum operator acts as

$$\hat{p}\,\psi(x) = -i\hbar\frac{\partial\psi}{\partial x}$$

On the other hand, time t is nothing but a parameter in nonrelativistic quantum mechanics. Clearly it is treated differently than position, as there is no operator that acts as

$$\hat{T}\psi(x,t) = t\psi(x,t)$$

Maybe you could try to construct a theory based on promoting time to such an operator, but that is not what is done in quantum field theory. What happens in quantum field theory is that we actually take the opposite approach, and demote position and momentum from their lofty status as operators. In quantum field theory, time t and position x are just parameters that label a position in spacetime for a field as shown here.

$$\varphi(x,t)$$

To quantize the theory, we are going to take a different approach and treat the fields themselves as operators. The procedure of second quantization is therefore to

- Promote the fields to operators, and
- Impose *equal time* commutation relations on the fields and their conjugate momenta

Since we are quantizing the fields rather than the position and momentum, we call this procedure second quantization—the type of quantization used in ordinary quantum mechanics is first quantization.

This is important so let's summarize. In quantum field theory,

- Position x and momentum p are not operators—they are just numbers like in classical physics.
- The fields $\varphi(x,t)$ and their conjugate momentum fields $\pi(x,t)$ are operators.
- Canonical commutation relations are imposed on the fields.

The fields are operators in the following sense. We have quantum states as we do in quantum mechanics, but these are states of the field. The field operators act on these states to destroy or create particles. This is important because in special relativity,

- Particle number is not fixed. Particles can be created and destroyed.
- To create a particle, we need at least twice the rest-mass energy $E = mc^2$.

The mathematics that describe a quantum theory with changing particle number has its roots in the simple harmonic oscillator, one of the few exactly solvable models. We will briefly review this now.

The Simple Harmonic Oscillator

The Hamiltonian for a simple harmonic oscillator in nonrelativistic quantum mechanics is

$$\hat{H} = \frac{\hat{p}^2}{2m} + \frac{m\omega^2}{2}\hat{x}^2 \tag{6.20}$$

We define two non-Hermitian operators, which are known as the *annihilation* and *creation* operators, respectively.

$$\hat{a} = \sqrt{\frac{m\omega}{2}}\left(\hat{x} + \frac{i}{m\omega}\hat{p}\right) \tag{6.21}$$

$$\hat{a}^\dagger = \sqrt{\frac{m\omega}{2}}\left(\hat{x} - \frac{i}{m\omega}\hat{p}\right) \tag{6.22}$$

It is straightforward to show, using $[\hat{x},\hat{p}] = i$, that

$$[\hat{a},\hat{a}^\dagger] = 1 \tag{6.23}$$

The Hamiltonian can be written in terms of these operators. It is given by

$$\hat{H} = \omega\left(\hat{a}^\dagger\hat{a} + \frac{1}{2}\right) \tag{6.24}$$

If we define the number operator as

$$\hat{N} = \hat{a}^\dagger\hat{a} \tag{6.25}$$

then the Hamiltonian takes on the wonderfully simple form

$$\hat{H} = \omega\left(\hat{N} + \frac{1}{2}\right) \tag{6.26}$$

The eigenstates of the Hamiltonian satisfy

$$\hat{H}|n\rangle = \omega\left(n + \frac{1}{2}\right)|n\rangle \tag{6.27}$$

This tells us that the energy of the state $|n\rangle$ is

$$E_n = \omega\left(n + \frac{1}{2}\right) \tag{6.28}$$

We call the states $|n\rangle$ the *number states*. They are eigenstates of the number operator

$$\hat{N}|n\rangle = n|n\rangle \tag{6.29}$$

The number n is an integer. The number operator obeys the following commutation relations with the annihilation and creation operators.

$$[\hat{N},\hat{a}] = -\hat{a} \tag{6.30}$$

$$[\hat{N},\hat{a}^\dagger] = \hat{a}^\dagger \tag{6.31}$$

The annihilation operator drops n by one unit.

$$\hat{a}|n\rangle = \sqrt{n}|n-1\rangle \tag{6.32}$$

The creation operator increases n by one unit.

$$\hat{a}^{\dagger}\left|n\right\rangle = \sqrt{n+1}\left|n+1\right\rangle \qquad (6.33)$$

There is a lowest lying state, otherwise the system would be able to degenerate into negative energy states. We call the lowest energy state the *ground state* and denote it by $\left|0\right\rangle$. In quantum field theory, we will often refer to this state as the *vacuum state*. The vacuum state is annihilated by the annihilation operator, rather it's destroyed by it.

$$\hat{a}\left|0\right\rangle = 0 \qquad (6.34)$$

Meanwhile, $\hat{a}^{\dagger}$ raises the energy of the system so that $\left|n\right\rangle \rightarrow \left|n+1\right\rangle$ without limit. The state $\left|n\right\rangle$ is obtained from the ground state through repeated applications of $\hat{a}^{\dagger}$.

$$\left|n\right\rangle = \frac{(\hat{a}^{\dagger})^{n}}{\sqrt{n!}}\left|0\right\rangle \qquad (6.35)$$

These ideas carry over to quantum field theory, but with a different interpretation. In quantum mechanics we are talking about a single particle with state $\left|n\right\rangle$ and energy levels $E_n = \omega(n + \frac{1}{2})$. The creation and annihilation operators move the state of the particle up and down in energy from the ground.

In quantum field theory, we take the notion of "number operator" literally. The state $\left|n\right\rangle$ is not a state of a single particle, rather it is a state of the field with n particles present. The ground state which is also the lowest energy state is a state of the field with 0 particles (but the field is still there). The creation operator $\hat{a}^{\dagger}$ adds a single quantum (a particle) to the field, while the annihilation operator $\hat{a}$ destroys a single quantum (removes a single particle) from the field. As we will see, in general there will be creation operators and annihilation operators for particles as well as for antiparticles.

These operators will be functions of momentum. The fields will become operators which will be written as sums over annihilation and creation operators.

SCALAR FIELD QUANTIZATION

The best way to learn about quantizing fields is to consider the simplest case first, the real scalar field that satisfies the Klein-Gordon equation as shown here.

$$\partial_{\mu}\partial^{\mu}\varphi + m^{2}\varphi = \frac{\partial^{2}\varphi}{\partial t^{2}} - \nabla^{2}\varphi + m^{2}\varphi = 0$$

We saw that the free field solution of the Klein-Gordon equation is of the form

$$\varphi(x,t) \sim e^{-i(Et - \vec{p} \cdot \vec{x})}$$

Let's write this using the wave number k and let $E \to k_0 = \omega_k, \vec{p} \to \vec{k}$

$$\varphi(x) \sim e^{-i(\omega_k x^0 - \vec{k} \cdot \vec{x})}$$

where we have also adopted the relativistic notation for position. The reason we are doing this is so that we can write down the general solution of the Klein-Gordon equation in terms of a Fourier expansion.

$$\varphi(x) = \int \frac{d^3 k}{(2\pi)^{3/2} \sqrt{2\omega_k}} \left[\varphi(\vec{k}) e^{-i(\omega_k x^0 - \vec{k} \cdot \vec{x})} + \varphi^*(\vec{k}) e^{i(\omega_k x^0 - \vec{k} \cdot \vec{x})} \right] \qquad (6.36)$$

Now we apply step one of the quantization process—we promote the field $\varphi(x)$ to an operator. This is done by replacing the Fourier transforms of the field $\varphi(\vec{k})$ and $\varphi^*(\vec{k})$ by annihilation and creation operators, associated with each mode. That is,

$$\varphi(\vec{k}) \to \hat{a}(\vec{k})$$

$$\varphi^*(\vec{k}) \to \hat{a}^\dagger(\vec{k})$$

Now that the field is an operator, we will add a caret to it and write it as $\hat{\varphi}(x)$ to remind us of that fact. In terms of the creation and annihilation operators, the field is written as

$$\hat{\varphi}(x) = \int \frac{d^3 k}{(2\pi)^{3/2} \sqrt{2\omega_k}} \left[\hat{a}(\vec{k}) e^{-i(\omega_k x^0 - \vec{k} \cdot \vec{x})} + \hat{a}^\dagger(\vec{k}) e^{i(\omega_k x^0 - \vec{k} \cdot \vec{x})} \right] \qquad (6.37)$$

To have a quantum theory, we need to have a conjugate momentum to the field so that we can impose commutation relations. Reminding ourselves of what that is, we repeat the definition starting with the Lagrangian

$$\mathcal{L} = \frac{1}{2} \partial_\mu \partial^\mu \varphi - \frac{1}{2} m^2 \varphi^2$$

We showed that the conjugate momentum to the field is

$$\pi(x) = \frac{\partial \mathcal{L}}{\partial(\partial_0 \varphi)} = \partial_0 \varphi$$

Now

$$\partial_0 \hat{\varphi}(x) = \partial_0 \int \frac{d^3k}{(2\pi)^{3/2}\sqrt{2\omega_k}} \left[\hat{a}(\vec{k}) e^{-i(\omega_k x^0 - \vec{k}\cdot\vec{x})} + \hat{a}^\dagger(\vec{k}) e^{i(\omega_k x^0 - \vec{k}\cdot\vec{x})} \right]$$

$$= \int \frac{d^3k}{(2\pi)^{3/2}\sqrt{2\omega_k}} \left[\hat{a}(\vec{k}) \partial_0 (e^{-i(\omega_k x^0 - \vec{k}\cdot\vec{x})}) + \hat{a}^\dagger(\vec{k}) \partial_0 (e^{i(\omega_k x^0 - \vec{k}\cdot\vec{x})}) \right]$$

$$= \int \frac{d^3k}{(2\pi)^{3/2}\sqrt{2\omega_k}} \left[\hat{a}(\vec{k})(-i\omega_k) e^{-i(\omega_k x^0 - \vec{k}\cdot\vec{x})} + \hat{a}^\dagger(\vec{k})(+i\omega_k) e^{i(\omega_k x^0 - \vec{k}\cdot\vec{x})} \right]$$

$$= -i \int \frac{d^3k}{(2\pi)^{3/2}} \sqrt{\frac{\omega_k}{2}} \left[\hat{a}(\vec{k}) e^{-i(\omega_k x^0 - \vec{k}\cdot\vec{x})} - \hat{a}^\dagger(\vec{k}) e^{i(\omega_k x^0 - \vec{k}\cdot\vec{x})} \right]$$

Therefore, the conjugate momentum to the field in Eq. (6.37) is

$$\hat{\pi}(x) = -i \int \frac{d^3k}{(2\pi)^{3/2}} \sqrt{\frac{\omega_k}{2}} \left[\hat{a}(\vec{k}) e^{-i(\omega_k x^0 - \vec{k}\cdot\vec{x})} - \hat{a}^\dagger(\vec{k}) e^{i(\omega_k x^0 - \vec{k}\cdot\vec{x})} \right] \qquad (6.38)$$

The commutation relations we impose follow from the canonical commutation relations used in ordinary quantum mechanics. For Cartesian coordinates x_i we have

$$[x_i, p_j] = i\delta_{ij}$$

$$[x_i, x_j] = [p_i, p_j] = 0$$

where δ_{ij} is the Kronecker delta function. Going to the continuum, with spatial locations $\vec{x}$ and $\vec{y}$ we let

$$\delta_{ij} \to \delta(\vec{x} - \vec{y})$$

Now we consider commutators between the fields evaluated at *the same time*. We say that the fields obey *equal time commutation relations*. The fields are evaluated at different spatial locations $\vec{x}$ and $\vec{y}$, but $x^0 = y^0$. Then we have

$$[\hat{\varphi}(x), \hat{\pi}(y)] = i\delta(\vec{x} - \vec{y}) \qquad (6.39)$$

$$[\hat{\varphi}(x), \hat{\varphi}(y)] = 0 \qquad (6.40)$$

$$[\hat{\pi}(x), \hat{\pi}(y)] = 0 \qquad (6.41)$$

EXAMPLE 6.2

Suppose that a real scalar field is given by

$$\varphi(x) = \int \frac{d^3p}{\sqrt{(2\pi)^3 2p^0}} \left[a(\vec{p})e^{ipx} + a^\dagger(\vec{p})e^{-ipx} \right]$$

Compute the equal time commutator

$$[\varphi(x), \pi(y)]$$

where $x^0 = y^0$.

SOLUTION

The momentum is

$$\pi(x) = \frac{\partial \varphi}{\partial x^0}$$

$$= \frac{\partial}{\partial x^0} \int \frac{d^3p}{\sqrt{(2\pi)^3 2p^0}} \left[a(\vec{p})e^{ipx} + a^\dagger(\vec{p})e^{-ipx} \right]$$

$$= i \int \frac{d^3p}{\sqrt{(2\pi)^3}} \sqrt{\frac{p^0}{2}} \left[a(\vec{p})e^{ipx} - a^\dagger(\vec{p})e^{-ipx} \right]$$

Now the commutator is

$$[\varphi(x), \pi(y)] = \varphi(x)\pi(y) - \pi(y)\varphi(x)$$

where $x^0 = y^0$ (equal time commutation relation). Looking at the first term, we have

$$\varphi(x)\pi(y) = \int \frac{d^3p}{\sqrt{(2\pi)^3 2p^0}} \left[a(\vec{p})e^{ipx} + a^\dagger(\vec{p})e^{-ipx} \right] i \int \frac{d^3p'}{\sqrt{(2\pi)^3}} \sqrt{\frac{p'^0}{2}} \left[a(\vec{p}')e^{ip'y} - a^\dagger(\vec{p}')e^{-ip'y} \right]$$

$$= i \int \underbrace{\frac{d^3p}{\sqrt{(2\pi)^3}} \frac{d^3p'}{\sqrt{(2\pi)^3}} \frac{1}{2} \sqrt{\frac{p'^0}{p^0}}}_{\text{Phase space factors}} \underbrace{\left[a(\vec{p})e^{ipx} + a^\dagger(\vec{p})e^{-ipx} \right]}_{\text{x terms}} \underbrace{\left[a(\vec{p}')e^{ip'y} - a^\dagger(\vec{p}')e^{-ip'y} \right]}_{\text{y terms}}$$

Unfortunately in this case, there is only one way to proceed in order to complete the calculation—and that is to use brute force. Multiplying out term by term we get

$$\varphi(x)\pi(y) = i \int \frac{d^3 p}{\sqrt{(2\pi)^3 2 p^0}} \frac{d^3 p'}{\sqrt{(2\pi)^3}} \frac{1}{2} \sqrt{\frac{p'^0}{p^0}} \{ a(p)a(p') e^{ipx} e^{ip'y}$$

$$- a(p)a^\dagger(p') e^{ipx} e^{-ip'y} + a^\dagger(p)a(p') e^{-ipx} e^{ip'y}$$

$$\times a^\dagger(p)a^\dagger(p') e^{-ipx} e^{-ip'y} \}$$

Now we compute the other term in the commutator, which turns out to be

$$\pi(y)\varphi(x) = i \int \frac{d^3 p'}{\sqrt{(2\pi)^3}} \frac{d^3 p}{\sqrt{(2\pi)^3}} \frac{1}{2} \sqrt{\frac{p'^0}{p^0}} \{ a(p')a(p) e^{ipx} e^{ip'y}$$

$$- a^\dagger(p')a(p) e^{ipx} e^{-ip'y} + a(p')a^\dagger(p) e^{-ipx} e^{ip'y}$$

$$\times a^\dagger(p')a^\dagger(p) e^{-ipx} e^{-ip'y} \}$$

The next step is to take the difference and to collect terms using the creation and annihilation operators. Not surprisingly, they obey similar commutation relations to the creation and annihilation operators used with the simple harmonic oscillator—we simply generalize to the continuous case. The relevant relations are

$$\left[a(\vec{p}), a^\dagger(\vec{p}') \right] = \delta(\vec{p} - \vec{p}') \tag{6.42}$$

$$\left[a(\vec{p}), a(\vec{p}') \right] = 0 \tag{6.43}$$

$$\left[a^\dagger(\vec{p}), a^\dagger(\vec{p}') \right] = 0 \tag{6.44}$$

Taking the difference of the first term we computed for $\varphi(x)\pi(y)$ with the first term we computed for $\pi(y)\varphi(x)$, we get

$$a(p)a(p') e^{ipx} e^{ip'y} - a(p')a(p) e^{ipx} e^{ip'y} = \left[a(p), a(p') \right] e^{ipx} e^{ip'y}$$

$$= 0$$

using Eq. (6.43).

In a similar fashion, if we take the difference of the last terms in each expression, we also get 0 since $\left[a^\dagger(\vec{p}), a^\dagger(\vec{p}') \right] = 0$. Now let's look at the second term in each expression. Taking the difference of the second term in $\varphi(x)\pi(y)$ and the second term in $\pi(y)\varphi(x)$, we get

$$
\begin{aligned}
-a(p)a^\dagger(p')e^{ipx}e^{-ip'y} + a^\dagger(p')a(p)e^{ipx}e^{-ip'y} &= -\left(a(p)a^\dagger(p') - a^\dagger(p')a(p)\right)e^{ipx}e^{-ip'y} \\
&= -\left[a(p), a^\dagger(p')\right]e^{ipx}e^{-ip'y} \\
&= -\delta(\vec{p} - \vec{p}')e^{ipx}e^{-ip'y} \\
&= -\delta(\vec{p} - \vec{p}')e^{i\vec{p}(\vec{x}-\vec{y})}
\end{aligned}
$$

To get to the last step, we used the fact that

$$
\delta(p - p')f(p) = \delta(p - p')f(p')
$$

together with the fact that $x^0 = y^0$ to get rid of the time component. Now we apply the same procedure to the difference of the third terms in each expression. The result is

$$
-\delta(\vec{p} - \vec{p}')e^{i\vec{p}(\vec{x}-\vec{y})}
$$

Putting everything together, we obtain

$$
\begin{aligned}
\left[\varphi(x), \pi(y)\right] &= \varphi(x)\pi(y) - \pi(y)\varphi(x) \\
&= i\int \frac{d^3p}{\sqrt{(2\pi)^3}} \frac{d^3p'}{\sqrt{(2\pi)^3}} \sqrt{\frac{p'^0}{p^0}} \frac{1}{2}\left[-\delta(\vec{p} - \vec{p}')e^{i\vec{p}(\vec{x}-\vec{y})} - \delta(\vec{p} - \vec{p}')e^{-i\vec{p}(\vec{x}-\vec{y})}\right] \\
&= -i\int \frac{d^3p}{\sqrt{(2\pi)^3}} \frac{1}{2}\left[e^{i\vec{p}(\vec{x}-\vec{y})} + e^{-i\vec{p}(\vec{x}-\vec{y})}\right]
\end{aligned}
$$

But one definition of the Dirac delta function is

$$
\delta(\vec{x} - \vec{y}) = \int \frac{d^3p}{(2\pi)^3} e^{i(\vec{x}-\vec{y})p} \tag{6.45}
$$

and from the symmetry of the delta functions we know $\delta(\vec{x} - \vec{y}) = \delta(\vec{y} - \vec{x})$, so

$$
\begin{aligned}
\left[\varphi(x), \pi(y)\right] = \varphi(x)\pi(y) - \pi(y)\varphi(x) &= -i\int \frac{d^3p}{\sqrt{(2\pi)^3}} \frac{1}{2}\left[e^{i\vec{p}(\vec{x}-\vec{y})} + e^{-i\vec{p}(\vec{x}-\vec{y})}\right] \\
&= -i\frac{1}{2}\left[\delta(\vec{x} - \vec{y}) + \delta(\vec{y} - \vec{x})\right] \\
&= -i\delta(\vec{x} - \vec{y})
\end{aligned}
$$

States in Quantum Field Theory

Now that we know how to write down the scalar field in terms of creation and annihilation operators, we are ready to see how the operators act on the states of the field. We already have some idea of how they act by the analogy with the simple harmonic oscillator. As always let's start with the simplest case, the state of lowest energy or the ground state, which is commonly referred to as the *vacuum (or vacuum state)* in quantum field theory. The vacuum, represented by $|0\rangle$, is destroyed by the annihilation operator.

$$\hat{a}(\vec{k})|0\rangle = 0 \tag{6.46}$$

Now notice that the creation and annihilation operators entered the field via the Fourier expansion. Therefore, we have been labeling them by the momentum $\vec{p}$ or wave number $\vec{k}$. States can be denoted by momentum, so we can step up from the vacuum to a state $|\vec{k}\rangle$ with an application of the creation operator.

$$|\vec{k}\rangle = \hat{a}^{\dagger}(\vec{k})|0\rangle \tag{6.47}$$

This describes a one-particle state. We can apply multiple creation operators of different modes $\vec{k}_1, \vec{k}_2, \ldots$ and so on. For example, the two-particle state $|\vec{k}_1, \vec{k}_2\rangle$ is created by

$$|\vec{k}_1, \vec{k}_2\rangle = \hat{a}^{\dagger}(\vec{k}_1)\hat{a}^{\dagger}(\vec{k}_2)|0\rangle$$

By extension, we can create an *n*-particle state using

$$|\vec{k}_1, \vec{k}_2, \ldots, \vec{k}_n\rangle = \hat{a}^{\dagger}(\vec{k}_1)\hat{a}^{\dagger}(\vec{k}_2) \ldots \hat{a}^{\dagger}(\vec{k}_n)|0\rangle \tag{6.48}$$

Each creation operator $\hat{a}^{\dagger}(\vec{k}_i)$ creates a single particle with momentum $\hbar\vec{k}_i$ and energy $\hbar\omega_{k_i}$ (we are restoring the $\hbar$'s for the moment, for clarity) where

$$\omega_{k_i} = \sqrt{\vec{k}_i^{\,2} + m^2}$$

An annihilation operator $\hat{a}(\vec{k}_i)$ destroys a particle with the said momentum and energy.

Positive and Negative Frequency Decomposition

We can decompose the field into two parts, a positive frequency part and a negative frequency part. The positive frequency part consists of annihilation operators and is written as

$$\hat{\varphi}^+(x) = \int \frac{d^3k}{(2\pi)^{3/2}\sqrt{2\omega_k}} \hat{a}(\vec{k}) e^{-i(\omega_k x^0 - \vec{k}\cdot\vec{x})} \tag{6.49}$$

The negative frequency part of the field is composed of creation operators.

$$\hat{\varphi}^-(x) = \int \frac{d^3k}{(2\pi)^{3/2}\sqrt{2\omega_k}} \hat{a}^\dagger(\vec{k}) e^{i(\omega_k x^0 - \vec{k}\cdot\vec{x})} \tag{6.50}$$

Therefore, since $\hat{a}(\vec{k})|0\rangle = 0$, the positive frequency part of the field annihilates the vacuum.

$$\hat{\varphi}^+(x)|0\rangle = 0 \tag{6.51}$$

And the negative frequency part creates particles.

$$\hat{\varphi}^-(x)|0\rangle = \int \frac{d^3k}{(2\pi)^{3/2}\sqrt{2\omega_k}} e^{i(\omega_k x^0 - \vec{k}\cdot\vec{x})} \hat{a}^\dagger(\vec{k})|0\rangle$$

$$= \int \frac{d^3k}{(2\pi)^{3/2}\sqrt{2\omega_k}} e^{i(\omega_k x^0 - \vec{k}\cdot\vec{x})} |\vec{k}\rangle \tag{6.52}$$

Number Operators

We can construct a number operator from the creation and annihilation operators.

$$\hat{N}(\vec{k}) = \hat{a}^\dagger(\vec{k})\hat{a}(\vec{k}) \tag{6.53}$$

The eigenvalues of the number operator are called *occupation numbers*. These are integers

$$n(\vec{k}) = 0, 1, 2, \ldots \qquad (6.54)$$

which tell us how many particles there are of momentum $\vec{k}$ for a given state. The state

$$\left| \vec{k}_1, \vec{k}_2, \ldots, \vec{k}_n \right\rangle = \hat{a}^\dagger(\vec{k}_1) \hat{a}^\dagger(\vec{k}_2) \ldots \hat{a}^\dagger(\vec{k}_n) | 0 \rangle$$

consists of n particles, with a single particle with momentum $\vec{k}_1$, a single particle with momentum $\vec{k}_2$, a single particle with momentum $\vec{k}_3$, and so on. However, we can have states where there are multiple particles with the same momentum. Suppose that we have two particles with momentum $\vec{k}_1$ and a single particle with momentum $\vec{k}_2$. We can write the state as

$$\left| \vec{k}_1, \vec{k}_1, \vec{k}_2 \right\rangle = \frac{\hat{a}^\dagger(\vec{k}_1) \hat{a}^\dagger(\vec{k}_1)}{\sqrt{2}} \hat{a}^\dagger(\vec{k}_2) | 0 \rangle$$

We can also write this state as

$$\left| \vec{k}_1, \vec{k}_1, \vec{k}_2 \right\rangle = \left| n(\vec{k}_1) n(\vec{k}_2) \right\rangle$$

where $n(\vec{k}_1) = 2$, $n(\vec{k}_2) = 1$. From the vacuum state, we have

$$\left| n(\vec{k}_1) n(\vec{k}_2) \right\rangle = \frac{\hat{a}^\dagger(\vec{k}_1)^{n(\vec{k}_1)}}{\sqrt{n(k_1)!}} \frac{\hat{a}^\dagger(\vec{k}_2)^{n(\vec{k}_2)}}{\sqrt{n(k_2)!}} | 0 \rangle$$

In general,

$$\left| n(\vec{k}_1) n(\vec{k}_2) \ldots n(\vec{k}_m) \right\rangle = \prod_j \frac{\hat{a}^\dagger(\vec{k}_j)^{n(\vec{k}_j)}}{\sqrt{n(k_j)!}} | 0 \rangle$$

As it is written, the number operator Eq. (6.53) is actually a density. It tells us the number density of particles in a given state, so to get the total number of particles we have to integrate over all of the states in momentum space. Doing so one obtains

$$\hat{N} = \int d^3k \, \hat{a}^\dagger(\vec{k}) \hat{a}(\vec{k})$$

EXAMPLE 6.3
Find $\hat{N}|\vec{k}'\rangle$.

SOLUTION
Since

$$|\vec{k}'\rangle = \hat{a}^\dagger(\vec{k}')|0\rangle$$

and

$$\left[\hat{a}(\vec{k}),\hat{a}^\dagger(\vec{k}')\right] = \hat{a}(\vec{k})\hat{a}^\dagger(\vec{k}') - \hat{a}^\dagger(\vec{k}')\hat{a}(\vec{k}) = \delta(\vec{k} - \vec{k}')$$

we have

$$\begin{aligned}
\hat{a}^\dagger(\vec{k})\hat{a}(\vec{k})|\vec{k}'\rangle &= \hat{a}^\dagger(\vec{k})\hat{a}(\vec{k})\hat{a}^\dagger(\vec{k}')|0\rangle \\
&= \hat{a}^\dagger(\vec{k})\left[\hat{a}^\dagger(\vec{k}')\hat{a}(\vec{k}) + \delta(\vec{k} - \vec{k}')\right]|0\rangle \\
&= \hat{a}^\dagger(\vec{k})\delta(\vec{k} - \vec{k}')|0\rangle \\
&= \delta(\vec{k} - \vec{k}')\hat{a}^\dagger(\vec{k}')|0\rangle = \delta(\vec{k} - \vec{k}')|\vec{k}'\rangle
\end{aligned}$$

To get from the second to the third line, remember that $\hat{a}|0\rangle = 0$. So we find that

$$\begin{aligned}
\hat{N}|\vec{k}'\rangle &= \int d^3k\,\hat{a}^\dagger(\vec{k})\hat{a}(\vec{k})|\vec{k}'\rangle \\
&= \left\{\int d^3k\,\delta(\vec{k} - \vec{k}')\right\}|\vec{k}'\rangle \\
&= |\vec{k}'\rangle
\end{aligned}$$

Hence we've found that the single particle state $|\vec{k}'\rangle$ has $n(\vec{k}') = 1$.

Normalization of the States

An important question that always comes up in quantum theory is the normalization of a given state. How do we tackle it here? First we start with the premise that the vacuum is normalized to unity.

$$\langle 0|0\rangle = 1 \tag{6.55}$$

Then, to compute the normalization of an arbitrary state $\left| \vec{k} \right\rangle$, we proceed by using the commutation relation Eq. (6.42). This is shown in the Example 6.4.

EXAMPLE 6.4
Compute the normalization of the state $\left| \vec{k} \right\rangle$ by considering the inner product $\left\langle \vec{k} \middle| \vec{k}' \right\rangle$.

SOLUTION
We proceed using the fact that $\hat{a}^{\dagger}(\vec{k})|0\rangle = \left| \vec{k} \right\rangle$ and that the adjoint of this expression is $\langle 0| = \left\langle \vec{k} \middle| \hat{a}(\vec{k}) \right.$. Then,

$$\left\langle \vec{k} \middle| \vec{k}' \right\rangle = \langle 0| \hat{a}(\vec{k})\hat{a}^{\dagger}(\vec{k}')|0\rangle$$

$$= \langle 0| \hat{a}^{\dagger}(\vec{k}')\hat{a}(\vec{k}) + \delta(\vec{k} - \vec{k}')|0\rangle$$

$$= \langle 0| \hat{a}^{\dagger}(\vec{k}')\hat{a}(\vec{k})|0\rangle + \langle 0| \delta(\vec{k} - \vec{k}')|0\rangle$$

$$= \delta(\vec{k} - \vec{k}')\langle 0|0\rangle$$

$$= \delta(\vec{k} - \vec{k}')$$

$$\Rightarrow \left\langle \vec{k} \middle| \vec{k}' \right\rangle = \delta(\vec{k} - \vec{k}')$$

Bose-Einstein Statistics

The theory being developed in this chapter applies to *bosons*, which are indistinguishable particles of integral spin (or spin-0 in this case). To see this, we note that we can interchange the order of creation operators as applied to a state. So

$$\left| \vec{k}_1, \vec{k}_2 \right\rangle = \hat{a}^{\dagger}(\vec{k}_1)\hat{a}^{\dagger}(\vec{k}_2)|0\rangle$$

but

$$\left| \vec{k}_1, \vec{k}_2 \right\rangle = \hat{a}^{\dagger}(\vec{k}_1)\hat{a}^{\dagger}(\vec{k}_2)|0\rangle$$

$$= \hat{a}^{\dagger}(\vec{k}_2)\hat{a}^{\dagger}(\vec{k}_1)|0\rangle$$

$$= \left| \vec{k}_2, \vec{k}_1 \right\rangle$$

$$\Rightarrow \left| \vec{k}_1, \vec{k}_2 \right\rangle = \left| \vec{k}_2, \vec{k}_1 \right\rangle$$

This tells us we are dealing with a theory that describes bosons. If we had fermions, there would have been a sign change in this calculation.

ENERGY AND MOMENTUM

Next we turn to the question of computing the energy and momentum of the field. Starting with the operator expansion of the field

$$\hat{\varphi}(x) = \int \frac{d^3k}{(2\pi)^{3/2}\sqrt{2\omega_k}} \left[\hat{a}(\vec{k})e^{-i(\omega_k x^0 - \vec{k}\cdot\vec{x})} + \hat{a}^\dagger(\vec{k})e^{i(\omega_k x^0 - \vec{k}\cdot\vec{x})} \right]$$

and using the number operator $\hat{N} = \hat{a}^\dagger(\vec{k})\hat{a}(\vec{k})$, it can be shown that the Hamiltonian operator is

$$\hat{H} = \int d^3k\, \omega_k \left[\hat{N}(\vec{k}) + \frac{1}{2} \right] \tag{6.56}$$

The momentum in the field is

$$\hat{P} = \int d^3k\, \vec{k} \left[\hat{N}(\vec{k}) + \frac{1}{2} \right] \tag{6.57}$$

EXAMPLE 6.5
For the real scalar field, find the energy of the vacuum.

SOLUTION
The solution to this example is the famous infinite energy of the vacuum, which may or may not be a problem depending on your point of view. To find the energy of the vacuum, we need to compute

$$\langle 0 | \hat{H} | 0 \rangle \tag{6.58}$$

We have

$$\langle 0 | \hat{H} | 0 \rangle = \langle 0 | \int d^3k\, \omega_k \left(\hat{N}(\vec{k}) + \frac{1}{2} \right) | 0 \rangle$$

$$= \langle 0 | \int d^3k\, \omega_k \left(\hat{a}^\dagger(\vec{k})\hat{a}(\vec{k}) + \frac{1}{2} \right) | 0 \rangle$$

$$= \langle 0| \int d^3k \, \omega_k \left(\hat{a}^\dagger(\vec{k}) \hat{a}(\vec{k}) \right) |0\rangle + \langle 0| \int d^3k \, \omega_k \left(\frac{1}{2} \right) |0\rangle$$

$$= \frac{\omega_k}{2} \int d^3k \, \langle 0|0\rangle$$

$$= \frac{\omega_k}{2} \int d^3k$$

This solution is reminiscent of the energy of the harmonic oscillator in ordinary quantum mechanics. In that case the energy of the ground state is $\frac{1}{2}\hbar\omega$. We have found a similar term here, but the integral blows up, since we're integrating over all momentum space as shown here.

$$\int d^3k \to \infty$$

This result can be ignored, or swept under the rug depending on your point of view. The usual explanation is that we only measure energy *differences*, so energy is measured relative to the ground state and this term falls out. The end result is we just throw it in the trash and say that the energy is 0. We simply subtract the infinity and say we are "renormalizing" the theory. This trick works but you have to think about the fact that we have to resort to a mathematical sleight of hand to make the theory work—perhaps it's an indicator that things are not quite right.

The *renormalized Hamiltonian* is constructed by subtracting off the term that gives rise to the infinite energy. Thus,

$$\hat{H}_R = \hat{H} - \int d^3k$$

$$= \int d^3k \, \omega_k \hat{N}(\vec{k}) = \int d^3k \, \omega_k \hat{a}^\dagger(\vec{k}) \hat{a}(\vec{k}) \tag{6.59}$$

EXAMPLE 6.6
Find the energy of the state $|\vec{k}\rangle$ using the renormalized Hamiltonian.

SOLUTION
We have

$$\langle \vec{k}| \hat{H}_R |\vec{k}\rangle = \langle \vec{k}| \int d^3k' \omega_{k'} \, \hat{a}^\dagger(\vec{k}') \hat{a}(\vec{k}') |\vec{k}\rangle$$

$$= \langle \vec{k}| \int d^3k' \omega_{k'} \, \delta(\vec{k} - \vec{k}') |\vec{k}'\rangle$$

$$= \langle \vec{k}| \omega_k |\vec{k}\rangle$$

$$= \omega_k$$

Normal and Time-Ordered Products

In quantum field theory we often find it desirable to write expressions in a way such that all creation operators are to *the left* of all annihilation operators. When an expression is written in this way, we say that we are using *normal ordering*. When normal ordering is applied to an expression, we denote this by enclosing it in two colons, so the normal ordering of ψ is denoted by writing : ψ :. Since normal ordering means move all creation operators to the left of all annihilation operators, then

$$: \hat{a}(\vec{k})\hat{a}^\dagger(\vec{k}): = \hat{a}^\dagger(\vec{k})\hat{a}(\vec{k}) \tag{6.60}$$

The normal ordering of a scalar field can be written down using the positive and negative frequency parts. Recall that

$$\hat{\varphi}^+(x) = \int \frac{d^3k}{(2\pi)^{3/2}\sqrt{2\omega_k}} \hat{a}(\vec{k})e^{-i(\omega_k x^0 - \vec{k}\cdot\vec{x})}$$

while

$$\hat{\varphi}^-(x) = \int \frac{d^3k}{(2\pi)^{3/2}\sqrt{2\omega_k}} \hat{a}^\dagger(\vec{k})e^{i(\omega_k x^0 - \vec{k}\cdot\vec{x})}$$

Normal ordering puts creation operators to the left, so we expect the normal ordered field to have negative frequency parts to the left of positive frequency components. Explicitly

$$: \varphi(x)\varphi(y): = \varphi^+(x)\varphi^+(y) + \varphi^-(x)\varphi^+(y) + \varphi^-(y)\varphi^+(x) + \varphi^-(x)\varphi^-(y)$$

A *time-ordered product* is a mathematical representation of the physical fact that a particle has to be created before it gets destroyed. Time ordering is accomplished using the time-ordering operator which acts on the product $\varphi(t_1)\psi(t_2)$ as

$$T[\varphi(t_1)\psi(t_2)] = \begin{cases} \varphi(t_1)\psi(t_2) \text{ if } t_1 > t_2 \\ \psi(t_2)\varphi(t_1) \text{ if } t_2 > t_1 \end{cases} \tag{6.61}$$

Remember that the fields are operators. Operators act in a right to left order. So a product of operators $\hat{A}\hat{B}$ acts on a state $|\psi\rangle$ in such a way that $\hat{B}$ acts on the state first, and then $\hat{A}$ acts on the result. Therefore if $t_1 > t_2$ which means that t_1 is later in time, $\psi(t_2)$ acts on the state first, followed by the action of $\varphi(t_1)$. The order is reversed if $t_2 > t_1$.

The Complex Scalar Field

Now let's quantize the complex scalar field. This is a good step forward because the complex scalar field represents particles with charge q and antiparticles with charge $-q$, so we will be able to tackle a relatively simple case and see how antiparticles can be represented in quantum field theory.

When we are dealing with antiparticles, the field is expanded in terms of positive frequency modes (annihilation operators) for particles and negative frequency modes (creation operators) for antiparticles. It is common to use the creation and annihilation operators $\hat{a}^\dagger$, $\hat{a}$ for particles

$$\hat{a}^\dagger(\vec{k}) \qquad \hat{a}(\vec{k}) \quad \text{(particles)}$$

We use $\hat{b}^\dagger, \hat{b}$ to represent the creation and annihilation operators for antiparticles.

$$\hat{b}^\dagger(\vec{k}) \qquad \hat{b}(\vec{k}) \quad \text{(antiparticles)}$$

Hence $\hat{a}^\dagger(\vec{k})$ creates a *particle* of momentum $\hbar k$ and energy $\hbar \omega_k$, while $\hat{b}^\dagger(\vec{k})$ creates an *antiparticle* of momentum $\hbar k$ and energy $\hbar \omega_k$. To write the field operator, we sum up positive frequency parts for particles together with negative frequency parts for antiparticles to get

$$\hat{\varphi}(x) = \int \frac{d^3 k}{(2\pi)^{3/2} \sqrt{2\omega_k}} \hat{a}(\vec{k}) e^{-i(\omega_k x^0 - \vec{k}\cdot\vec{x})} + \hat{b}^\dagger(\vec{k}) e^{i(\omega_k x^0 - \vec{k}\cdot\vec{x})} \qquad (6.62)$$

There is an adjoint field (not surprising since it's a *complex* field) given by

$$\hat{\varphi}^\dagger(x) = \int \frac{d^3 k}{(2\pi)^{3/2} \sqrt{2\omega_k}} \hat{a}^\dagger(\vec{k}) e^{i(\omega_k x^0 - \vec{k}\cdot\vec{x})} + \hat{b}(\vec{k}) e^{-i(\omega_k x^0 - \vec{k}\cdot\vec{x})} \qquad (6.63)$$

We still require that $\left[\hat{a}(\vec{k}), \hat{a}^\dagger(\vec{k}')\right] = \delta(\vec{k} - \vec{k}')$, and similarly for the creation and annihilation operators for antiparticles

$$\left[\hat{b}(\vec{k}), \hat{b}^\dagger(\vec{k}')\right] = \delta(\vec{k} - \vec{k}') \qquad (6.64)$$

There are two conjugate momenta corresponding to the field and its adjoint. For example,

$$\hat{\pi}(x) = \partial_0 \hat{\varphi}(x)$$

$$= \int \frac{d^3k}{(2\pi)^{3/2}\sqrt{2\omega_k}} (-i\omega_k)\hat{a}(\vec{k})e^{-i(\omega_k x^0 - \vec{k}\cdot\vec{x})} + (i\omega_k)\hat{b}^\dagger(\vec{k})e^{i(\omega_k x^0 - \vec{k}\cdot\vec{x})}$$

$$= -i\int \frac{d^3k}{(2\pi)^{3/2}} \sqrt{\frac{\omega_k}{2}} \left[\hat{a}(\vec{k})e^{-i(\omega_k x^0 - \vec{k}\cdot\vec{x})} + \hat{b}^\dagger(\vec{k})e^{i(\omega_k x^0 - \vec{k}\cdot\vec{x})} \right] \qquad (6.65)$$

In the case of the charged complex field, we have two number operators. The first is the familiar number operator that corresponds to the number of particles.

$$\hat{N}_{\hat{a}} = \int d^3k\, \hat{a}^\dagger(\vec{k})\hat{a}(\vec{k}) \qquad (6.66)$$

The second is the number operator that represents the number of antiparticles.

$$\hat{N}_{\hat{b}} = \int d^3k\, \hat{b}^\dagger(\vec{k})\hat{b}(\vec{k}) \qquad (6.67)$$

The total energy in the field is expressed as the energy of the particles added to the energy of the antiparticles.

$$\hat{H} = \int d^3k\, \omega_k \left[\hat{a}^\dagger(\vec{k})\hat{a}(\vec{k}) + \hat{b}^\dagger(\vec{k})\hat{b}(\vec{k}) \right] \qquad (6.68)$$

Notice that the energy density is the number density of particles added to the number density of antiparticles multiplied by the energy ω_k. Then, to get the total energy we integrate over all modes of the field. Next, the total momentum is the momentum due to particles added to the momentum due to antiparticles.

$$\hat{P} = \int d^3k\, \vec{k} \left[\hat{a}^\dagger(\vec{k})\hat{a}(\vec{k}) + \hat{b}^\dagger(\vec{k})\hat{b}(\vec{k}) \right] \qquad (6.69)$$

A complex field corresponds to a charged field. Particles and antiparticles have opposite charge. The total charge is found by subtracting the charge due to antiparticles from the charge due to particles. The charge operator is

$$\hat{Q} = \int d^3k \left[\hat{a}^\dagger(\vec{k})\hat{a}(\vec{k}) + \hat{b}^\dagger(\vec{k})\hat{b}(\vec{k}) \right]$$

$$= \hat{N}_{\hat{a}} - \hat{N}_{\hat{b}} \qquad (6.70)$$

Finally, the fields and the conjugate momenta satisfy a series of commutation relations. Once again we consider equal time commutation relations such that $x^0 = y^0$. Then,

$$\left[\hat{\varphi}(x),\hat{\pi}(y)\right]=\left[\hat{\varphi}^{\dagger}(x),\hat{\pi}^{\dagger}(y)\right]=i\delta(\vec{x}-\vec{y}) \tag{6.71}$$

All the equal time commutators vanish, leaving the following, which is *not* an equal time commutator.

$$\left[\hat{\varphi}(x),\hat{\varphi}^{\dagger}(y)\right]=i\Delta(x-y) \tag{6.72}$$

At equal times,

$$\left[\hat{\varphi}(x),\hat{\varphi}^{\dagger}(y)\right]=\left[\hat{\varphi}(x),\hat{\varphi}(y)\right]=\left[\hat{\varphi}^{\dagger}(x),\hat{\varphi}^{\dagger}(y)\right]=0$$

The commutator $\left[\hat{\varphi}(x),\hat{\varphi}^{\dagger}(y)\right]=i\Delta(x-y)\partial$ represents a new function called the *propagator,* which we will explore in detail in the next chapter.

Summary

The Klein-Gordon equation results from a straightforward substitution of the quantum mechanical operators for energy and momentum into the Einstein relation for energy, momentum, and mass from special relativity. This leads to inconsistencies such as negative probabilities and negative energy states. We can get around the inconsistencies by reinterpreting the equation. Rather than viewing it as a single particle wave equation, we instead apply it to a field that includes creation and annihilation operators similar to the harmonic oscillator of quantum mechanics. There is one difference, however, in that the creation and annihilation operators now create and destroy particles, rather than changing the energy level of an individual particle.

Quiz

1. Compute $\left[\hat{N}(\vec{k}),\hat{N}^{\dagger}(\vec{k}')\right]$ for the real scalar field.
2. Find $\hat{N}(\vec{k})\hat{a}^{\dagger}(\vec{k})\left|n(\vec{k})\right\rangle$.

3. Find $\hat{N}|0\rangle$.

4. Consider the complex scalar field. Determine if charge is conserved by examining the Heisenberg equation of motion for the charge operator $\hat{Q}$.

$$\dot{Q} = [H, Q]$$

Do this computation by writing out the operators using Eqs. (6.68) and (6.70) and using the commutation relations Eq. (6.64).

CHAPTER 7

The Feynman Rules

The tricky mathematics of quantum field theory have been distilled into a series of operations called the Feynman rules. The rules can be thought of as prescriptions to describe all manner of processes in quantum field theory and are embodied in a pictorial form, the famous *Feynman diagrams*. In this chapter we will develop the Feynman rules and show how to construct Feynman diagrams. The ultimate goal is to compute physical parameters for various particle interactions. We discuss these here.

In quantum theory we make experimental predictions by calculating the probability amplitude that a process will occur. This remains true in quantum field theory, where we calculate amplitudes for particle interactions such as decays and scattering events. The primary tool used to do such calculations is known as the *S matrix*. Any given physical process can be considered as a transition from an initial state $|i\rangle = |\alpha(t_0)\rangle$ to a final output state we denote by $|f\rangle = |\alpha(t)\rangle$, that is,

$$|i\rangle \rightarrow |f\rangle$$

This transition occurs via the action of a unitary operator, the S matrix, where S stands for scattering, in the following way.

$$|f\rangle = S|i\rangle$$

Since the S matrix is unitary, it satisfies

$$S^\dagger S = SS^\dagger = I$$

From ordinary quantum mechanics, we know that the time evolution of states is described using the unitary time evolution operator $U(t,t_0)$. Then the amplitude to evolve from $|\alpha_I(t_0)\rangle$ at time t_0 to a final state $|\alpha_F(t)\rangle$ at a later time t is

$$\langle \alpha_F(t)|U(t,t_0)|\alpha_I(t_0)\rangle \tag{7.1}$$

The initial and final states involve *free* particles that come in from $t=-\infty$, interact, and then move off as different free particle states at $t=+\infty$. An element of the S matrix is the limit of Eq. (7.1) where

$$S_{FI} = \lim_{\substack{t_0 \to -\infty \\ t \to +\infty}} \langle \alpha_F(t)|U(t,t_0)|\alpha_I(t_0)\rangle \tag{7.2}$$

In momentum space, the S matrix is proportional to the amplitude M_{FI} for a given process to occur, as follows:

$$S_{FI} \propto -i(2\pi)^4 \delta^4(p_F - p_I) M_{FI} \tag{7.3}$$

where p_F is the total four momentum of the outgoing states and similarly for the incoming momenta. The Dirac delta function enforces the conservation of momentum in the process. The *Feynman rules* allow us to calculate the amplitude M_{FI} for the process rather easily, using a graphical representation known as a Feynman diagram of each physical process that can occur. The amplitude M_{FI} is calculated using a perturbative process. Imagine taking the probability amplitude for a process, and expanding it in a series. There is one Feynman diagram for each term in the perturbative expansion, and we add them up to get the total amplitude. Suppose that M is the amplitude for a given event. The same initial and final particle states might result from a set of processes, each with amplitude M_i. The total amplitude is

$$M_{\text{total}} = \sum_{i=0}^{n} g^{k_i} M_i \tag{7.4}$$

where the g^{k_i} are coupling constants for each M_i. There will be a Feynman diagram for each term M_i in the sum. Each amplitude scales by a coupling constant denoted by g which describes the strength of the interaction, and k describes the order of the interaction. For a first-order process, $k = 1$; second-order $k = 2$ and so on. Higher order terms in Eq. (7.4) will have more factors of g. Therefore if g is small, as i gets larger, that is as we take more terms in the sum Eq. (7.4), the higher order terms will begin to become negligible and we can cut the sum off at some n to get a reasonable estimate of the amplitude, like

$$M = \sum_{i=0}^{n} g^{k_i} M_i$$

For example, in quantum electrodynamics (QED) the coupling constant is proportional to $\alpha = 1/137 \approx 0.0073$ which is a small number. Second-order probabilities are proportional to $\alpha^2 \approx 0.0000053$. So as factors of the coupling constant appear as products, the terms become small enough that we can ignore them in our calculations. This will make more sense as we do explicit derivations of amplitudes. In this chapter we will illustrate the procedure with simple and easy to understand scattering and decay events that are abstract yet instructive. In the next chapter we will begin to study physical processes when we examine QED.

In quantum field theory, it is helpful to examine the evolution of a system using the interaction picture which yields the amplitudes in Eq. (7.4). So we begin by reviewing interactions in quantum mechanics.

The Interaction Picture

In this section we will be considering quantum mechanics in three different pictures. The first two will be the Schrödinger picture, and the second is the interaction picture. In between is Heisenberg picture. To keep the states and operators in the two pictures separate, we will use an S subscript for the Schrödinger picture and an I subscript for the interaction picture.

We move to the interaction picture in quantum theory by splitting the Hamiltonian into two parts, a *free field Hamiltonian* H_0 which is time independent, and a time-dependent interaction Hamiltonian H_I.

$$H = H_0 + H_I \tag{7.5}$$

Let us denote the states and operators in the Schrödinger picture with a subscript S. For example, in the Schrödinger picture a state vector is written as $|\alpha\rangle_S$ and an operator is denoted by A_S. In this picture, operators are fixed and states evolve in time according to

$$i\frac{\partial}{\partial t}|\alpha(t)\rangle_S = H|\alpha(t)\rangle_S \qquad (7.6)$$

Here H is the full Hamiltonian. A state vector in the interaction picture $|\alpha(t)\rangle_I$ is related to the state vector in the Schrödinger picture via the action of the free part of the Hamiltonian.

$$|\alpha(t)\rangle_I = e^{iH_0 t}|\alpha(t)\rangle_S \qquad (7.7)$$

Now, using the interaction picture, we take an intermediate view between the Schrödinger picture and the Heisenberg picture, which moves the time evolution from the states to the operators. That is, in the interaction picture the operators also evolve in time. An interaction picture operator A_I is related to a Schrödinger picture operator A_S in the following way.

$$A_I = e^{iH_0 t}A_S e^{-iH_0 t} \qquad (7.8)$$

Now we differentiate Eq. (7.7) and use Eq. (7.6) to arrive at a dynamical equation for the state as shown here.

$$\frac{\partial}{\partial t}|\alpha(t)\rangle_I = \frac{\partial}{\partial t}\left(e^{iH_0 t}|\alpha(t)\rangle_S\right)$$

$$= iH_0 e^{iH_0 t}|\alpha(t)\rangle_S + e^{iH_0 t}\frac{\partial}{\partial t}|\alpha(t)\rangle_S$$

$$= iH_0 e^{iH_0 t}|\alpha(t)\rangle_S + e^{iH_0 t}\left(-iH|\alpha(t)\rangle_S\right)$$

$$= iH_0 e^{iH_0 t}|\alpha(t)\rangle_S + e^{iH_0 t}\left(-iH_0 - iH_I\right)|\alpha(t)\rangle_S$$

$$= -ie^{iH_0 t}H_I|\alpha(t)\rangle_S$$

$$= -iH_I|\alpha(t)\rangle_I$$

Therefore, we conclude that the time evolution of the states in the interaction picture is

$$i\frac{\partial}{\partial t}|\alpha(t)\rangle_I = H_I|\alpha(t)\rangle_I \qquad (7.9)$$

This tells us that the time evolution of the states is determined by the *interaction part of the Hamiltonian*. Now let's take a look at the interaction picture operators and see how they evolve with time. We do this by differentiating Eq. (7.8).

$$\frac{\partial}{\partial t} A_I = \frac{\partial}{\partial t}(e^{iH_0t} A_S e^{-iH_0t})$$

$$= iH_0 e^{iH_0t} A_S e^{-iH_0t} + e^{iH_0t}\left(\frac{\partial A_S}{\partial t}\right)e^{-iH_0t} - ie^{iH_0t} A_S H_0 e^{-iH_0t}$$

$$= iH_0 e^{iH_0t} A_S e^{-iH_0t} - ie^{iH_0t} A_S H_0 e^{-iH_0t}$$

$$= iH_0 A_I - iA_I H_0$$

$$= i[H_0, A_I]$$

That is, the time evolution of operators in the interaction picture is determined by *the free part of the Hamiltonian*.

$$\frac{\partial}{\partial t} A_I = i[H_0, A_I] \tag{7.10}$$

Let's summarize how the Hamiltonian affects time evolution in the interaction picture:

H_0 (free) affects time evolution of operators
H_I (interaction) affects time evolution of states

In quantum field theory, we have seen that the fields themselves are operators. Equation (7.10) implies that the time evolution of the fields will be characterized by the free field Hamiltonian.

Perturbation Theory

Now we know that the time evolution in quantum mechanics can also be described by a unitary operator $U(t,t_0)$.

$$|\alpha(t)\rangle_I = U(t,t_0)|\alpha(t_0)\rangle_I \tag{7.11}$$

Let's differentiate both sides of this equation. On the left, we have

$$\frac{\partial}{\partial t}|\alpha(t)\rangle_I = -iH_I|\alpha(t)\rangle_I = -iH_I U(t,t_0)|\alpha(t_0)\rangle_I$$

Quantum Field Theory Demystified

On the right hand side, we obtain

$$\frac{\partial}{\partial t}U(t,t_0)\big|\alpha(t_0)\big\rangle_I \tag{7.12}$$

Hence the time evolution of $U(t,t_0)$ is described by the equation

$$i\frac{\partial U}{\partial t}=H_I U \tag{7.13}$$

Now, if we set $t=t_0$ in Eq. (7.11) expect to see now change in the system. Let's compute the form of the U operator in this instant. We have

$$\big|\alpha(t_0)\big\rangle_I =U(t_0,t_0)\big|\alpha(t_0)\big\rangle_I$$

So it must be the case that

$$U(t_0,t_0)=1$$

The identity says there has been no change in the system in zero time. We take this as the initial condition in Eq. (7.13) and integrate to obtain

$$U(t,t_0)=1-i\int_{t_0}^{t} H_I(t')U(t',t_0)\,dt' \tag{7.14}$$

This is an integral equation which describes how the U operator changes as we go from time t_0 to time t. This immediately suggests an iterative refinement where we improve the computation of the U operator in a series of small steps.

As an example, let's start with a rough guess for which we will call U_0. Our first refinement will be U_1 and is given by

$$U_1(t,t_0)=1-i\int_{t_0}^{t} H_I(\tau)U_0(\tau,t_0)\,d\tau$$

However, we saw a large change as we went from U_0 to U_1 so we know we need to keep refining our calculation. So we will use the output U_1 to calculate U_2.

$$U_2(t,t_0)=1-i\int_{t_0}^{t} H_I(\tau)U_1(\tau,t_0)\,d\tau$$

These refinements produce a rather hideous expression as shown here.

$$U(t,t_0) = 1 - i \int_{t_0}^{t} H_I(t') \left(1 - i \int_{t_0}^{t'} H_I(t'') U(t'',t_0) dt'' \right) dt'$$

$$= 1 - i \int_{t_0}^{t} dt' \, H_I(t') + (-i)^2 \int_{t_0}^{t} dt' \int_{t_0}^{t'} dt'' H_I(t') H_I(t'') + \cdots$$

$$+ (-i)^n \int_{t_0}^{t} dt' \int_{t_0}^{t'} dt'' \cdots \int_{t_0}^{t^{(n-1)}} dt^{(n)} H_I(t') H_I(t'') \cdots H_I(t^{(n)})$$

We assume that the times satisfy $t > t' > t'' > \cdots > t^{(n)}$ so we can write the *nth* term as

$$U_n(t,t_0) = (-i)^n \int_{t_0}^{t} dt' \int_{t_0}^{t'} dt'' \cdots \int_{t_0}^{t^{(n-1)}} dt^{(n)} T\left\{ H_I(t') H_I(t'') \cdots H_I(t^{(n)}) \right\}$$

$$= \frac{(-i)^n}{n!} \int_{t_0}^{t} dt' \int_{t_0}^{t} dt'' \cdots \int_{t_0}^{t} dt^{(n)} T\left\{ H_I(t') H_I(t'') \cdots H_I(t^{(n)}) \right\}$$

where T is the time-ordering operator. Then we sum up these terms to obtain a *Dyson series*. We can find an approximate solution by cutting off the expansion at a suitable number of terms.

Now, returning to the states, we need to solve

$$i \frac{\partial}{\partial t} |\alpha(t)\rangle = H_I |\alpha(t)\rangle$$

where the Hamiltonian H_I is time dependent. Note the form of this equation. If the interaction H_I goes to 0, then the states are constant in time telling us that any *transitions in the state are due to interactions*. To simplify notation let's denote the initial state of the system at time t_0 as

$$|\alpha(t_0)\rangle = |i\rangle$$

The initial state of the system $|i\rangle$ is the state of the system before a scattering event, in other words the state of the system as $t \to -\infty$. This is the noninteracting state of the particles prior to the scattering event. The final state, long after the scattering event, is taken as $\lim_{t \to \infty} |\alpha(t)\rangle = S|i\rangle$. We wish to calculate the amplitude for a system in this state to end up in some specific final state $|f\rangle$

$$\langle f|S|i\rangle = S_{fi}$$

which is a component of the S matrix. Therefore the probability $\mathbf{P}_{fi}$ for a system that starts in the state $|i\rangle$ and ends up in the final state $|f\rangle$ after undergoing an interaction described by S is

$$\mathbf{P}_{fi}=\left|\langle f|S|i\rangle\right|^2 = S_{fi}S_{fi}^{*}$$

$$\left|\langle f|S|i\rangle\right|^2$$

The state of the system at time t can be written in an iterative expansion we used for the Dyson series. With an initial state $|i\rangle$, we have for the first two terms

$$|\alpha(t)\rangle = |i\rangle + (-i)\int_{-\infty}^{t} H_I(t')|\alpha(t')\rangle\, dt'$$

Since $S_{FI} = \lim_{\substack{t_0 \to -\infty \\ t \to +\infty}} \langle \alpha_F(t)|U(t,t_0)|\alpha_I(t_0)\rangle$, and we can write U in terms of a Dyson series, we can cast the S matrix as a series.

$$S = \sum_{n=0}^{\infty}(-i)^n \int_{-\infty}^{t} dt' \int_{-\infty}^{t'} dt'' \cdots \int_{-\infty}^{t^{(n-1)}} dt^n H_I(t_1)H_I(t_2)\cdots H_I(t_n) \qquad (7.15)$$

To summarize, to calculate the amplitude $\langle f|S|i\rangle$ we use perturbation theory and compute terms to a suitable order (to acceptable level of error). In quantum field theory we must describe processes like the creation of matter and antimatter:

$$_i e^- + e^- \to e^- + e^- + e^+ + e^-$$

You can see how we have a different set of particles in the before and after states (think special relativity). Because the sets of particles change, we use terms that annihilate particles in the initial state and create particles in the final state.

Confused? Who wouldn't be. Luckily Fenyman understood all this stuff well enough to distill it down to a simple recipe. We will now forget everything we've done so far and use the Feynman rules to calculate amplitudes.

Basics of the Feynman Rules

The crux of the perturbative expansion is this: we refine our calculation using corrections that are becoming smaller and smaller. The reduction in importance is quantified by the power of the perturbation parameter, which in this case is the coupling strength. At some point we are adding refinements that are too small to be measured and we know we can stop adding refinements.

In such a perturbation expansion, as we have seen earlier, the amplitude M for a given process can be computed using an expansion of the type

$$M = \sum_n g^{k_n} M_n$$

Each individual amplitude M_n is a specific particle reaction (scattering, decay) that can be drawn as a Feynman diagram. The higher the order of term M_n, the less likely it is to occur and the less it contributes to the overall amplitude. The terms M_n have the same incoming and outgoing particles, but represent different intermediate states. The intermediate states correspond to terms in the Dyson series. Since each term M_n is scaled by a coupling constant g, which represents the strength of a given interaction, when the relative strength of a given interaction is small then it can be analyzed using perturbation theory.

A Feynman diagram consists of one or more *external* lines that represent the incoming and outgoing particles, connected by a vertex. Time can be taken to flow from the bottom to the top of the diagram, or from the left to the right. For example, imagine a particle decay process involving particles A, B, and C that proceeds with A breaking up into particles B and C.

$$A \rightarrow B + C$$

If we draw time going from the bottom to the top, we obtain the Feynman diagram shown in Fig. 7.1.

If we draw a diagram such that time is moving from the left to the right, then we obtain the diagram shown in Fig. 7.2.

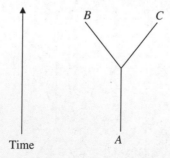

Figure 7.1   A Feynman diagram for the decay process $A \rightarrow B + C$. Time flows from the bottom of the diagram to the top.

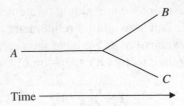

Figure 7.2 A Feynman diagram for the decay $A \rightarrow B + C$, with time flowing from left to right.

Feynman diagrams are a qualitative, symbolic representation of some particle interaction. So don't think of the flow of time as an actual time axis. Normally, the direction of time flow is not explicitly indicated on the diagram, rather it is understood from the context.

Scattering events will involve an intermediate state or particle that is drawn in the diagram as an *internal* line. Suppose that particles A and B scatter with particles C and D leaving the process, and let's suppose that the scattering process involves in intermediary I. The scattering event is

$$A + B \rightarrow C + D$$

It is represented by the Feynman diagram in Fig. 7.3, which includes the internal line with the intermediate state I. The correct way to interpret the intermediate state is that it is a force carrying particle that transmits the given force between the particles A and B. For example, if this were an electromagnetic interaction, say the scattering of an electron and a positron, the internal line would be a photon. The way the reaction is drawn in Fig. 7.3, particles A and

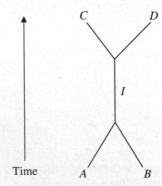

Figure 7.3 A Feynman diagram for the reaction $A + B \rightarrow C + D$.

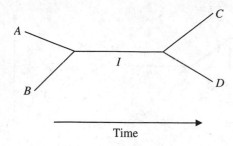

Figure 7.4   The process $A + B \rightarrow C + D$ with time flowing from left to right.

B meet and annihilate, producing the state I, which decays at a later time into the particles C and D.

Now let's draw the same reaction with time flowing from left to right. This is shown in Fig. 7.4.

Particles can scatter via the exchange of a boson (a force-carrying particle). Let's represent the scattering event

$$C + D \rightarrow C + D$$

where particles C and D exchange a boson B during a scattering event. This is shown in Fig. 7.5. Time flows from bottom to the top of the diagram. Particles C and D move in, scatter via the exchange of the boson B, and then move off.

Now we know that in quantum field theory, antiparticles, in addition to particles, take part in many processes. Let us indicate the labels for antiparticles by a tick mark so that A is the particle and A' is the antiparticle. The lines for particles in a Feynman diagram are indicated by an arrow that flows with the direction of time. The lines for antiparticles are indicated in a Feynman diagram

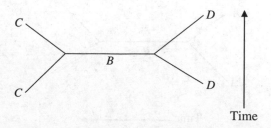

Figure 7.5   A scattering event $C + D \rightarrow C + D$ with the exchange of a boson B.

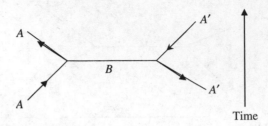

Figure 7.6   The scattering event $A + A' \rightarrow A + A'$, using arrows flowing with time to indicate a particle and arrows flowing against time to indicate an antiparticle.

by an arrow that flows in the opposite direction to the flow of time. Consider a reaction

$$A + A' \rightarrow A + A'$$

where A scatters with the antiparticle A' via the exchange of a boson B. The reaction is drawn in Fig. 7.6.

One of Feynman's brilliant observations was that a particle traveling *forward* in time is equivalent to an antiparticle traveling *backward* in time (refer to Chap. 3). This is why we arrows indicating a particle is traveling backward in time.

Now consider the annihilation reaction, where A and A' meet and annihilate, producing a boson B, which then decays into A and A'. This version of the reaction $A + A' \rightarrow A + A'$ is shown in Fig. 7.7.

Each line in a Feynman diagram is characterized by a four momentum. Suppose once again that $A + A' \rightarrow A + A'$ occurs with the exchange of a boson B as shown

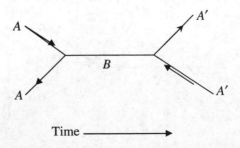

Figure 7.7   Another representation of the reaction $A + A' \rightarrow A + A'$.

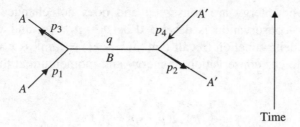

Figure 7.8   A Feynman diagram showing momenta.

in Fig. 7.6. We indicate the momenta of each incoming and outgoing particle (or external line) with a p. Internal momenta are indicated with a q. This is shown in Fig. 7.8.

Calculating Amplitudes

To actually calculate an amplitude M, we integrate over all of the internal momenta. Fortunately, all the integrals involve Dirac delta functions so they can be done by inspection. This is due to the sampling property of the delta function, that is,

$$\int_{-\infty}^{\infty} f(x)\,\delta(x-x')\,dx = f(x')$$

As mentioned earlier, the reason the delta functions are in the amplitudes is to enforce conservation of energy and momentum. At each vertex, we assign a positive sign for a momentum entering a vertex and a negative sign for a momentum leaving a vertex. For instance, consider the vertex shown in Fig. 7.9.

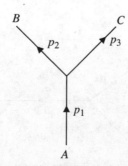

Figure 7.9   A particle decay showing momenta entering and leaving a vertex.

Momentum and energy are conserved and does not change because of the interaction. This conservation is described as $p_1 = p_2 + p_3$, and we include this using the Dirac delta function (recall that $\delta(x_1 - x_2) = 0$ unless $x_1 = x_2$). The delta function that enforces conservation of energy and momentum at the vertex shown in Fig. 7.9 is

$$(2\pi)^4 \delta(p_1 - p_2 - p_3)$$

The value of the delta function is 1 when $p_1 = p_2 + p_3$ and 0 otherwise.

The direction of the arrows indicates whether the given line is for a particle or antiparticle and does not have anything to do with the direction of momentum. If a line goes into a vertex, the momentum is entering the vertex. Following the direction of time (which can be up from the vertex or right from the vertex), lines leaving the vertex should be assigned momenta with minus signs.

Conservation of energy and momentum is enforced at a vertex involving an internal line as well. Consider the vertex shown in Fig. 7.10, where the boson B carries away momentum q.

The delta function that will enforce conservation of energy and momentum at the vertex shown in Fig. 7.10 is

$$(2\pi)^4 \delta(p_1 - p_3 - q)$$

We need to assign a direction to the internal momentum q. If we do so as shown in Fig. 7.11, then the delta function which must be of the form

$$\delta\left(\sum \text{incoming momenta} - \sum \text{outgoing momenta}\right)$$

is written as

$$(2\pi)^4 \delta(p_2 - p_4 + q)$$

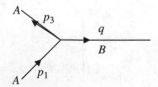

Figure 7.10   Conservation of energy and momentum is also enforced at a vertex with an internal line using a delta function.

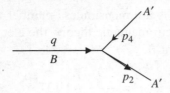

Figure 7.11   Conservation of momentum in this figure is enforced by
$$(2\pi)^4 \delta(p_2 - p_4 + q).$$

Steps to Construct an Amplitude

Constructing an amplitude from a Feynman diagram involves the following steps:

- Write down a delta function to conserve energy and momentum at each vertex. Multiply these terms together.
- Write down the one coupling constant for each vertex in the figure.
- Write down a propagator for each internal line.
- Multiply all the factors together.
- Integrate over internal momenta.

The total amplitude M is the sum of all the amplitudes M_i that can occur for a specific process with the same incoming and outgoing particles. Each M_i corresponds to a Feynman diagram. So the total amplitude for the process $A + A' \rightarrow A + A'$ is the sum of the two amplitudes represented by Figs. 7.6 and 7.7. In fact, higher-order diagrams for each of those processes can be drawn with different intermediate states. Now let's discuss the two remaining pieces of the process, the coupling constants and propagators.

COUPLING CONSTANTS

Every force has some fundamental strength, and the force manifests itself in the Feynman calculus as a coupling constant g. For quantum electrodynamics, for example, the coupling constant g_e is related to the fine structure constant α as

$$g_e = \sqrt{4\pi\alpha} \tag{7.16}$$

The fine structure constant is a dimensionless number that contains fundamental constants that appear in electromagnetic theory, the electric charge e, the speed of light c, and Planck's constant $\hbar$.

$$\alpha = \frac{e^2}{4\pi\hbar c} \approx 1/137$$

In a good theory like QED, or any interaction where the coupling constant is small ($\ll 1$), higher order diagrams contribute less and less because they include higher and higher powers of g. This means we can terminate the series at a point where we have the accuracy we need to describe a given process.

For each vertex in a Feynman diagram, we include one copy of the coupling constant as follows:

$$-ig$$

PROPAGATORS

We associate a *propagator* with each internal line in a Feynman diagram. A propagator is a factor that represents the transfer or propagation of momentum from one particle to another. Right now we will introduce propagators for three types of particles; you will come across others later in your studies.

The simplest case we can consider is an internal line for a spin-0 boson. In this case, the propagator is

$$\frac{i}{q^2 - m^2} \tag{7.17}$$

The mass in this term is the mass of the particle that corresponds to the internal line. In a Feynman diagram, an internal line for a spin-0 boson can be shown as a dashed line, as shown in Fig. 7.12.

For a spin-1/2 particle, we indicate the internal line the same way we would for an external line, as a solid line with an arrow pointing in the direction of momentum for a particle and against the direction of momentum for an antiparticle. In this case the propagator is

$$i\frac{q + m}{q^2 - m^2} = \frac{i}{q - m} \tag{7.18}$$

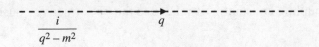

Figure 7.12 An internal line for a spin-0 boson.

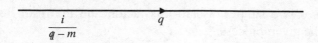

Figure 7.13   An internal line for a Fermion.

where

$$q\!\!\!/ = \gamma^{\mu} q_{\mu}$$

An internal line for a Fermion is shown in Fig. 7.13.

The photon propagator is

$$\frac{i}{k^2}\left(-g^{\mu\nu} + (1-\zeta)\frac{k^{\mu}k^{\nu}}{k^2}\right) \qquad (7.19)$$

In the Feynman gauge $\zeta = 1$, and the photon propagator is just

$$-\frac{i}{k^2}g^{\mu\nu} \qquad (7.20)$$

In the examples in this chapter, we will use spin-0 bosons as the force-carrying particles in all of our calculations because they are simpler to deal with.

Now let's review again the steps used to build up the amplitude from a Feynman diagram. We take each factor and multiply them together as a product.

- Write down one factor of $-ig$ for each vertex in the diagram where g is the coupling constant for the interaction depicted.

- Write down a delta function $(2\pi)^4 \delta\left(\sum \text{incoming momenta} - \sum \text{outgoing momenta}\right)$ for each vertex to conserve momentum.

- Add a propagator for each internal line.

The next step is to integrate over all internal momenta. For each internal momentum q we add an integration measure to enforce normalization in phase space:

$$\frac{1}{(2\pi)^4}d^4q \qquad (7.21)$$

Then we integrate for each internal momentum q. In the end, there will be a final delta function left over that enforces the conservation of energy and momentum for the *external lines*. We simply discard this factor, and the result left over is the amplitude for the given process.

EXAMPLE 7.1

A particle A annihilates with its antiparticle A', producing a spin-0 scalar boson B, which subsequently decays into A and A'. The mass of the scalar boson is m_B. Calculate the amplitude and probability for this process shown in Fig. 7.14.

SOLUTION

Following the rules, we start by writing down a factor of $-ig$ for each vertex. There are two vertices in Fig. 7.14, hence we get two factors of $-ig$.

$$(-ig)(-ig) = -g^2 \tag{7.22}$$

Next, we multiply this by the Dirac delta functions that will conserve 4-momentum at each vertex. For the first vertex at the bottom of Fig. 7.14, we have incoming momenta p_1 and p_2 and outgoing momentum q. This is represented by the delta function

$$(2\pi)^4 \delta(p_1 + p_2 - q)$$

We multiply this by Eq. (7.22), giving

$$-g^2 (2\pi)^4 \delta(p_1 + p_2 - q) \tag{7.23}$$

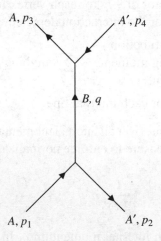

Figure 7.14 The annihilation-creation process with spin-0 boson.

At the top vertex, we have incoming momentum q, and outgoing momenta p_3 and p_4. We indicate this with the delta function

$$(2\pi)^4 \delta(q - p_3 + p_4)$$

Adding this to the product in Eq. (7.23) gives

$$-g^2 (2\pi)^4 \delta(p_1 + p_2 - q)(2\pi)^4 \delta(q - p_3 + p_4) \tag{7.24}$$

The next step is to add the propagator for the internal line. Since this is a spin-0 boson, we use Eq. (7.17), which we multiply by Eq. (7.24) to give

$$\frac{-ig^2}{q^2 - m_B^2} (2\pi)^4 \delta(p_1 + p_2 - q)(2\pi)^4 \delta(q - p_3 + p_4)$$

Now we integrate over q, using the measure

$$\frac{1}{(2\pi)^4} d^4 q$$

So we have

$$M = \int \frac{-ig^2}{q^2 - m_B^2} (2\pi)^4 \delta(p_1 + p_2 - q)\delta(q - p_3 + p_4)\, dq$$

This integral can be done by inspection. Using the second delta function, we set

$$q = p_3 + p_4$$

Therefore, the amplitude for this process is

$$M = \int \frac{-ig^2}{q^2 - m_B^2} (2\pi)^4 \delta(p_1 + p_2 - q)\delta(q - p_3 + p_4)\, dq$$

$$= \frac{-ig^2}{(p_3 + p_4)^2 - m_B^2}$$

where we have discarded the remaining delta function which is

$$(2\pi)^4 \delta(p_1 + p_2 - p_3 - p_4)$$

which enforces overall conservation of energy and momentum (at the external lines). The probability for the process to occur is the modulus squared of M, that is,

$$|M|^2 = \frac{g^4}{\left((p_3 + p_4)^2 - m_B{}^2\right)^2}$$

EXAMPLE 7.2
A particle decays as follows

$$u \rightarrow w + v$$

with a coupling strength g_w. The particle w then decays into

$$w \rightarrow v' + e$$

with a strength given by the same coupling constant. Draw the Feynman diagram for this process and calculate the amplitude for it to occur. The w particle is a spin-0 boson with mass m_w.

SOLUTION
The Feynman diagram for this process is shown in Fig. 7.15.

We include a factor of $-ig_w$ for each vertex. There are two vertices shown in Fig. 7.15, so we have

$$(-ig_w)^2 = -g_w^2$$

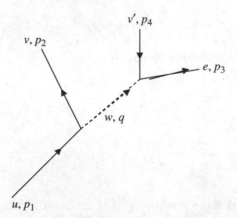

Figure 7.15 The process described in Example 7.2. Note that v' is an antiparticle, so the arrow for its external line is pointing in the opposite direction.

At vertex 1 (at the first vertex), there is incoming momentum p_1 and outgoing momenta p_2 and q. Therefore, we include a delta function giving

$$-g_w^2 (2\pi)^4 \delta(p_1 - p_2 - q)$$

The particle w decays into the final products, so it's represented by an internal line. The propagator for this particle is given by

$$\frac{i}{q^2 - m_w^2}$$

At the second vertex, we have incoming momentum q and outgoing momenta p_3 and p_4, so we multiply by a delta function

$$(2\pi)^4 \delta(q - p_3 - p_4)$$

Putting everything together, we get

$$-g_w^2 (2\pi)^4 \delta(p_1 - p_2 - q)\left(\frac{i}{q^2 - m_w^2}\right)(2\pi)^4 \delta(q - p_3 - p_4)$$

Now we integrate to obtain

$$\int -g_w^2 (2\pi)^4 \delta(p_1 - p_2 - q)\left(\frac{i}{q^2 - m_w^2}\right)(2\pi)^4 \delta(q - p_3 - p_4)\frac{d^4 q}{(2\pi)^4}$$

$$= -\frac{i g_w^2 (2\pi)^4 \delta(p_1 - p_2 - p_3 - p_4)}{(p_3 + p_4)^2 - m_w^2}$$

We have used $q = p_3 + p_4$ from the second delta function in the integral. To get the amplitude for the process to occur, we drop the $(2\pi)^4 \delta(p_1 - p_2 - p_3 - p_4)$ term

$$M = -\frac{i g_w^2 (2\pi)^4 \delta(p_1 - p_2 - p_3 - p_4)}{(p_3 + p_4)^2 - m_w^2}$$

Rates of Decay and Lifetimes

Decay processes are very important in nuclear and particle physics since many nuclei and particles are unstable—they will eventually decay into something else. In fact, very few particles are fundamental and immune to decay. So decay rates and lifetimes are an essential quantities of interest.

The rate of decay for a process is proportional to the squared amplitude.

$$\Gamma \propto |M|^2 \tag{7.25}$$

The lifetime of a particle that decays is the inverse of the amplitude squared.

$$\tau \propto \frac{1}{|M|^2} \tag{7.26}$$

Summary

Feynman diagrams allow us to represent the amplitude for a process to occur with a picture. External lines represent incoming and outgoing particle states. At each vertex, conservation of energy and momentum is enforced with a delta function, and the strength of the interaction is included with a coupling constant. Internal lines can represent force-carrying particles or particles that spontaneously decay into the end products. Each internal line is accompanied by a propagator that represents transfer of momentum to the final states.

Quiz

1. What is the amplitude for the process shown in Fig. 7.16?

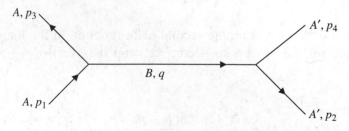

Figure 7.16   Feynman diagram for Question 1.

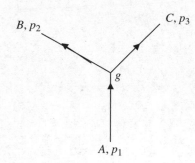

Figure 7.17 Feynman diagram for Question 1.

2. Find the lifetime for the decay as shown in Fig. 7.17.

3. An internal line corresponds to a spin-0 boson of mass m. The propagator is

   (a) $i\dfrac{\not{q} + m}{q^2 - m^2}$

   (b) $\dfrac{i}{q^2 - m^2}$

   (c) $\dfrac{i}{\not{q} - m}$

   (d) $\delta(q^2 - m^2)$

4. In the interaction picture,

   (a) The time evolution of states is governed by the free Hamiltonian

   (b) States are stationary, operators evolve according to the interaction part of the Lagrangian

   (c) States evolve according to the interaction part of the Hamiltonian, fields evolve according to the free part of the Hamiltonian

   (d) States obey the Heisenberg equation of motion

5. Each vertex in a Feynman diagram requires the addition of

   (a) One factor of the coupling constant $-ig$

   (b) One factor of the coupling constant $-g$

   (c) One factor of the coupling constant $-ig^2$

   (d) One factor of the coupling constant $-i\sqrt{g}$

6. What number is the coupling constant for quantum electrodynamics related to?

CHAPTER 8

Quantum Electrodynamics

Quantum electrodynamics or *QED* was the first true quantum field theory that was developed. As the name implies, it is a quantum field theory that describes electromagnetic interactions. It is sometimes called the prototype quantum field theory. In a sense, physicists would like all physical interactions described by a theory like quantum electrodynamics.

The development of quantum electrodynamics brought the notion of describing forces with particle exchange to the forefront. In quantum electrodynamics, electromagnetic forces are the result of the exchange of *virtual photons*. We say the photons are virtual because they are not observed directly, rather they are exchanged between two charged particles. The momentum carried by the photons causes a recoil between the two electrons giving rise to a repulsive force. We can illustrate a process like this with a Feynman diagram. The photon, which is the exchanged particle, is represented with a wavy line as shown in Fig. 8.1. A γ can also be used to indicate a photon.

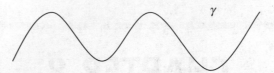

Figure 8.1 A schematic representation of a photon that we will use in a Feynman diagram for QED processes is a wavy line. We can include a γ for clarity if desired.

In Fig. 8.2, we show a basic QED process. This is the repulsion of two electrons mentioned in the introductory paragraph. In the diagram, time is moving in a direction from bottom to top. Two electrons enter and scatter with the exchange of a photon.

Recall that the strength of a given interaction is described by its coupling constant. The coupling constant for QED processes is the *fine structure constant* which is denoted by α. It has a precisely known numerical value given by

$$\alpha = 1/137$$

In terms of fundamental constants, it can be written as

$$\alpha = \frac{e^2}{\hbar c} \tag{8.1}$$

The fact that $\alpha \ll 1$ is very helpful. This means that if we expand some quantity in a series in terms of α, higher order terms will contribute less and less because $\alpha^n \to 0$ as n gets large. This fact makes QED calculations using perturbation theory, and in particular Feynman diagrams possible.

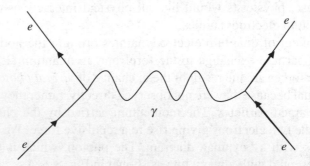

Figure 8.2 A basic QED process: repulsion between electrons. Two electrons enter from the bottom, exchange a photon, then move off.

Reviewing Classical Electrodynamics Again

In an earlier chapter we touched on a relativistic description of the electromagnetic field. We are going to review that here, and incorporate photon polarization into the picture. Once we have done that we can unify the electromagnetic field with the description of the electron using the Dirac equation to develop quantum electrodynamics.

We return to Maxwell's equations

$$\vec{\nabla}\cdot\vec{E}=\rho \qquad \vec{\nabla}\cdot\vec{B}=0$$
$$\vec{\nabla}\times\vec{E}+\frac{\partial\vec{B}}{\partial t}=0 \qquad \vec{\nabla}\times\vec{B}-\frac{\partial\vec{E}}{\partial t}=\vec{J} \qquad (8.2)$$

where $\vec{E}$ is the electric field, $\vec{B}$ is the magnetic field, ρ is the charge density, and $\vec{J}$ is the current density. In field theory, we work with the 4-vector potential

$$A^{\mu}=(\phi,\vec{A}) \qquad (8.3)$$

which allows us to define the electric and magnetic fields as

$$\vec{E}=-\vec{\nabla}\phi-\frac{\partial\vec{A}}{\partial t} \qquad \vec{B}=\vec{\nabla}\times\vec{A} \qquad (8.4)$$

The electromagnetic field tensor is defined as

$$F^{\mu\nu}=\partial^{\mu}A^{\nu}-\partial^{\nu}A^{\mu} \qquad (8.5)$$

which turns out to be the matrix given by

$$F^{\mu\nu}=\begin{pmatrix} 0 & -E_x & -E_y & -E_z \\ E_x & 0 & -B_z & B_y \\ E_y & B_z & 0 & -B_x \\ E_z & -B_y & B_x & 0 \end{pmatrix} \qquad (8.6)$$

Using this formulation, Maxwell's equations can be written in the compact form

$$\partial^{\alpha}F^{\mu\nu}+\partial^{\beta}F^{\alpha\mu}+\partial^{\mu}F^{\nu\alpha}=0$$
$$\partial_{\mu}F^{\mu\nu}=J^{\nu}$$

where $J^\mu = (\rho, \vec{J})$. The Lagrangian for the electromagnetic field is

$$L = -\frac{1}{4} F_{\mu\nu} F^{\mu\nu} - J^\mu A_\mu$$

From this Lagrangian, the field playing the role of canonical momentum to $F^{\mu\nu}(x)$ is

$$\pi^\mu(x) = \frac{\partial L}{\partial(\partial_0 A_\mu)} = -F^{\mu 0}(x) \tag{8.7}$$

The continuity equation expressing conservation of electric charge can be derived from this Lagrangian using the usual techniques and is

$$\partial_\mu J^\mu = 0 \tag{8.8}$$

A gauge transformation can be applied to the 4-vector potential A^μ by adding the derivative of a scalar field χ.

$$A'_\mu = A_\mu + \partial_\mu \chi \tag{8.9}$$

The field equations remain unchanged by such a mathematical transformation. This allows us to pick a form of A^μ that is convenient in some way. For example, we can require that the divergence of the four potential vanishes, something called the *Lorentz condition*.

$$\partial_\mu A^\mu = 0 \tag{8.10}$$

This equation allows us to approach the electromagnetic field in a similar way that the Klein-Gordon equation can be dealt with. While the vector potential plays an ancillary role as a mathematical tool in classical electrodynamics, in QED we treat A^μ itself as the photon field. In free space, the electromagnetic field will have plane wave solutions that we write as

$$A^\mu \propto e^{-ip \cdot x} \varepsilon^\mu(p)$$

As usual

$$p \cdot x = Et - \vec{p} \cdot \vec{x}$$

However, the photon is a massless particle with $E = |\vec{p}|$ and hence we have

$$p^{\mu} p_{\mu} = 0$$

The quantity $\varepsilon^{\mu}(p)$ is called the *polarization vector*. This vector plays the role of the spin part of the wave function for the photon. The Lorentz condition $\partial_{\mu} A^{\mu} = 0$ results in a constraint on the polarization vector.

$$0 = \partial_{\mu} A^{\mu} = \partial_{\mu} e^{-ip \cdot x} \varepsilon^{\mu}(p)$$

$$= \partial_{\mu} e^{-ip_{\mu}x^{\mu}} \varepsilon^{\mu}(p)$$

$$= -ip_{\mu} e^{-ip_{\mu}x^{\mu}} \varepsilon^{\mu}(p) + e^{-ip_{\mu}x^{\mu}} \partial_{\mu} \varepsilon^{\mu}(p)$$

$$= -ip_{\mu} e^{-ip_{\mu}x^{\mu}} \varepsilon^{\mu}(p)$$

$$\Rightarrow p_{\mu} \varepsilon^{\mu} = 0$$

Starting with the Lorentz gauge condition

$$\partial_{\mu} A^{\mu} = 0$$

we substitute the free-space form of the potential A and get the differential equation as shown here.

$$\partial_{\mu} \left[e^{-ip_{\mu}x^{\mu}} \varepsilon^{\mu}(p) \right] = 0$$

Now we apply the product rule for derivatives

$$\partial_{\mu} \left[e^{-ip_{\mu}x^{\mu}} \varepsilon^{\mu}(p) \right] = (-ip_{\mu}) e^{-ip_{\mu}x^{\mu}} \varepsilon^{\mu}(p) + e^{-ip_{\mu}x^{\mu}} \partial_{\mu} \varepsilon^{\mu}(p)$$

and make use of the fact that $\partial_{\mu} \varepsilon^{\mu}(p) = 0$ because $\varepsilon^{\mu}(p)$ is a function of momentum and does not depend on x^{μ}. This leads to

$$\partial_{\mu} \left[e^{-ip_{\mu}x^{\mu}} \varepsilon^{\mu}(p) \right] = (-ip_{\mu}) e^{-ip_{\mu}x^{\mu}} \varepsilon^{\mu}(p) = 0$$

Since the spacetime position x^{μ} is completely arbitrary, the exponential $e^{-ip_{\mu}x^{\mu}}$ can be nonzero and the only way to ensure the equality is to have

$$p_{\mu} \varepsilon^{\mu}(p) = 0$$

The form of the polarization vector is fixed by the choice of reference frame. It is common to take $\vec{p}$ in the z direction.

In the Coulomb gauge, where the gradient of the 3-vector potential is 0, that is,

$$\vec{\nabla} \cdot \vec{A} = 0$$

the polarization vector is perpendicular to the spatial component of momentum

$$\varepsilon \cdot \vec{p} = 0$$

This just says that the polarization is transverse—the polarization vector lies in a plane perpendicular to the direction of motion of the field. A massless spin-s particle has two possible spin states. Let's compute the spin states in the Coulomb gauge for the photon, which has $s = 1$. In this gauge, we take the time component of the polarization vector to be 0, that is,

$$\varepsilon^0 = 0$$

Then the two polarization states for a photon are

$$\varepsilon_1 = \begin{pmatrix} 0 \\ 1 \\ 0 \\ 0 \end{pmatrix} \qquad \varepsilon_2 = \begin{pmatrix} 0 \\ 0 \\ 1 \\ 0 \end{pmatrix}$$

Normalization of the polarization vector is expressed as

$$\varepsilon^\mu \cdot (\varepsilon^\nu)^* = g^{\mu\nu}$$

The Quantized Electromagnetic Field

The process of quantization takes us from classical electrodynamics to field theory. We quantize the electromagnetic field by imposing commutation relations and writing the field in terms of creation and annihilation operators. The canonical equal-time commutation rule is

$$\left[A_\mu(\vec{x}, t), \pi^\nu(\vec{y}, t) \right] = i g_\mu^\nu \delta^3(\vec{x} - \vec{y}) \tag{8.11}$$

Furthermore we have

$$\left[A_\mu(\vec{x},t), A^\nu(\vec{y},t)\right] = \left[\pi_\mu(\vec{x},t), \pi^\nu(\vec{y},t)\right] = 0 \tag{8.12}$$

We can quantize the electromagnetic field rather easily by looking at the classical free-space solution written in Fourier modes. The solution is summed over momentum $\vec{p}$ and polarization $\lambda = 1,2$ with complex expansion coefficients $a_{k,\lambda}$ as

$$\vec{A} = \frac{1}{\sqrt{V}} \sum_{\vec{p}} \sum_{\lambda} \frac{\varepsilon_\lambda(\vec{p})}{\sqrt{2\omega_p}} \left[a_{p,\lambda} e^{i(\vec{p}\cdot\vec{r}-\omega t)} + a^*_{p,\lambda} e^{-i(\vec{p}\cdot\vec{r}-\omega t)} \right]$$

To quantize the field, we promote the expansion coefficients to creation and annihilation operators as shown here.

$$a_{p,\lambda} \to \hat{a}_{p,\lambda}$$

$$a^*_{p,\lambda} \to a^\dagger_{p,\lambda}$$

The creation operator $a^\dagger_{p,\lambda}$ creates a photon of momentum $\vec{p}$ and polarization λ, while the annihilation operator $\hat{a}_{p,\lambda}$ destroys such a photon. These operators obey the usual commutation relations, where we now also take into account polarization so that

$$\left[a_{\lambda,p}, a^\dagger_{\lambda',p'}\right] = g^{\lambda\lambda'} 2p^0 (2\pi)^3 \delta^3(\vec{p}-\vec{p}') = \delta_{\lambda,\lambda'} 2p^0 (2\pi)^3 \delta^3(\vec{p}-\vec{p}')$$

and the field operator is

$$A_\mu = \int \frac{d^3p}{2p_0(2\pi)^3} \sum_\lambda \left[a_{\lambda,p} \varepsilon_\mu^{(\lambda)} e^{ip_\mu x^\mu} + a^\dagger_{\lambda,p} \left(\varepsilon_\mu^{(\lambda)}\right)^* e^{-ip_\mu x^\mu} \right]$$

$$= A_\mu^+ + A_\mu^-$$

where

$$A_\mu^+ = \int \frac{d^3p}{2p_0(2\pi)^3} \sum_\lambda \left[a_{\lambda,p} \varepsilon_\mu^{(\lambda)} e^{ip_\mu x^\mu} \right]$$

$$A_\mu^- = \int \frac{d^3p}{2p_0(2\pi)^3} \sum_\lambda \left[a^\dagger_{\lambda,p} \left(\varepsilon_\mu^{(\lambda)}\right)^* e^{-ip_\mu x^\mu} \right]$$

(where we have dropped the hats from the operators and we have broken up the field into positive and negative frequency components).

Gauge Invariance and QED

Since we are talking about QED, let's review gauge invariance for this theory. The gauge invariance we are going to have to satisfy is local and involves three terms: a Lagrangian term for the electromagnetic field, the Dirac Lagrangian, which will involve two terms, a kinetic energy term and a mass term; and an interaction term that couples the Dirac and electromagnetic fields. The kinetic energy part of the Lagrangian for the electromagnetic field is of the form

$$L_{EM} = -\frac{1}{4} F_{\mu\nu} F^{\mu\nu}$$

From the Dirac equation, we have the Lagrangian

$$L_{Dirac} = i\bar{\psi}\gamma^{\mu}\partial_{\mu}\psi - m\bar{\psi}\psi$$

The Lagrangian for interaction of a particle with charge q and the electromagnetic field is given by

$$L_{int} = -q\bar{\psi}\gamma^{\mu}\psi A_{\mu}$$

We can construct the total Lagrangian describing the electromagnetic field and interactions with a Dirac field like the electron by putting all of these terms together.

$$L = L_{EM} + L_{Dirac} + L_{int}$$
$$= -\frac{1}{4} F_{\mu\nu} F^{\mu\nu} + i\bar{\psi}\gamma^{\mu}\partial_{\mu}\psi - m\bar{\psi}\psi - q\bar{\psi}\gamma^{\mu}\psi A_{\mu}$$

Now, the Dirac portion of the Lagrangian is invariant under a global $U(1)$ symmetry; that is, the Lagrangian does not change when we change the field

$$\psi(x) \rightarrow e^{i\theta}\psi(x)$$

which of course implies

$$\bar{\psi}(x) \rightarrow e^{-i\theta}\bar{\psi}(x)$$

Recall that for a global symmetry, θ is just a parameter, a complex number—it is not a function of spacetime. This means that $\partial_{\mu}e^{i\theta} = 0$.

It is trivial to see that the mass term in the Dirac part of the Lagrangian is invariant under this transformation.

$$m\bar{\psi}\psi \rightarrow m[e^{-i\theta}\bar{\psi}(x)]\,[e^{i\theta}\psi(x)] = m\bar{\psi}\psi$$

Since the transformation is global, the kinetic energy term of the Dirac Lagrangian is invariant as well since the derivative ∂_μ doesn't affect it.

$$i\bar{\psi}\gamma^\mu\partial_\mu\psi \rightarrow ie^{-i\theta}\bar{\psi}(x)\gamma^\mu\partial_\mu[e^{i\theta}\psi(x)] = ie^{-i\theta}\bar{\psi}(x)\gamma^\mu e^{i\theta}\partial_\mu\psi(x)$$

$$= i\bar{\psi}\gamma^\mu\partial_\mu\psi$$

We have shown that the Dirac Lagrangian is invariant under a global $U(1)$ symmetry. This is all very nice, but what we are interested in for quantum field theory to maintain the spirit of relativity is invariance under a local transformation. Keep that in mind—the standard model of particle physics requires invariance under local transformations. Recall that this requires us to make the parameter θ space-time dependent

$$\theta \rightarrow \theta\,(x)$$

This would give us a local $U(1)$ symmetry. In this case the mass term of the Dirac Lagrangian is unchanged.

$$m\bar{\psi}\psi \rightarrow m[e^{-i\theta(x)}\bar{\psi}(x)][e^{i\theta(x)}\psi(x)] = m\bar{\psi}\psi$$

But for the kinetic energy term we have a problem since the derivative of the transformation term is no longer 0. That is, $\partial_\mu e^{i\theta(x)} = i[\partial_\mu\theta(x)]e^{i\theta(x)} \neq 0$. Here is the problem:

$$i\bar{\psi}\gamma^\mu\partial_\mu\psi \rightarrow ie^{-i\theta(x)}\bar{\psi}(x)\gamma^\mu\partial_\mu\left[e^{i\theta(x)}\psi(x)\right]$$

$$= ie^{-i\theta(x)}\bar{\psi}(x)\gamma^\mu e^{i\theta(x)}\partial_\mu\psi(x) - \bar{\psi}(x)\gamma^\mu\psi(x)\partial_\mu\theta$$

$$\neq i\bar{\psi}\gamma^\mu\partial_\mu\psi$$

Physically (and experimentally) we find invariance in nature and so we will insist our theory also has invariance. So the challenge becomes how do we recover invariance under a local gauge transformation? One way is to create a transformation of the electromagnetic field of the form, like

$$A^\mu \rightarrow A^\mu - \frac{1}{q}\partial^\mu\theta$$

This will cancel the errant terms in the kinetic energy term. To see this, examine the interaction portion of the Lagrangian as follows:

$$L_{int} = -q\bar{\psi}\gamma^{\mu}\psi A_{\mu} \rightarrow -q\bar{\psi}\gamma^{\mu}\psi\left(A_{\mu} - \frac{1}{q}\partial_{\mu}\theta\right)$$

$$= -q\bar{\psi}\gamma^{\mu}\psi A_{\mu} + \bar{\psi}\gamma^{\mu}\psi\partial_{\mu}\theta$$

$$= L_{int} + \bar{\psi}\gamma^{\mu}\psi\partial_{\mu}\theta$$

Now let's look at the results for both transformations together. We have the local $U(1)$ gauge transformation

$$\psi(x) \rightarrow e^{i\theta(x)}\psi(x)$$

and the new transformation which restores invariance

$$A^{\mu} \rightarrow A^{\mu} - \frac{1}{q}\partial^{\mu}\theta$$

The Dirac and interaction parts of the Lagrangian transform as

$$L_{Dirac} + L_{int} \rightarrow ie^{-i\theta(x)}\bar{\psi}\gamma^{\mu}e^{i\theta(x)}\partial_{\mu}\psi - \bar{\psi}\gamma^{\mu}\psi\partial_{\mu}\theta - m\bar{\psi}\psi - q\bar{\psi}\gamma^{\mu}\psi A_{\mu} + \bar{\psi}\gamma^{\mu}\psi\partial_{\mu}\theta$$

$$= L_{Dirac} + L_{int}$$

and we have restored the luster of the theory—the invariance.

The invariance is restored because we forced $A^{\mu} \rightarrow A^{\mu} - \frac{1}{q}\partial^{\mu}\theta$. That is, we have introduced the covariant derivative

$$D_{\mu} = \partial_{\mu} + iqA_{\mu}$$

This adjustment of the derivative operator is called the *minimal coupling prescription*. Hence the term

$$\bar{\psi}\gamma^{\mu}D_{\mu}\psi$$

is invariant under a local $U(1)$ transformation. We can understand the origin of the covariant derivative by considering how A^{μ} transforms under a Lorentz gauge transformation. It can be shown that the similarity transformation

$$U(\Lambda)A^{\mu}U^{-1}(\Lambda) = \Lambda^{\mu}_{\ \nu}A^{\nu} + \partial_{\mu}\theta(x)$$

This is why we require invariance under a transformation of the form $A^{\mu} \rightarrow A^{\mu} - \frac{1}{q}\partial^{\mu}\theta$. This gives us a Lagrangian that is invariant under a Lorentz gauge transformation.

Feynman Rules for QED

The Feynman rules for QED apply to any lepton–photon interactions, but for our purposes we will just discuss electrons and positrons. We use the Dirac states from Chap. 5 to represent electrons and positrons, so $u(p,s)$ is a particle state of momentum p and spin s and $v(p,s)$ is the antiparticle state. An incoming electron is a spinor state $u(p,s)$ with an arrow pointing in the direction of positive time flow. This is indicated schematically in Fig. 8.3.

An outgoing electron state has the arrow flowing with the direction of time, but we replace the spinor state $u(p,s)$ by its adjoint $\bar{u}(p,s)$. This is illustrated in Fig. 8.4.

Next we will need to represent incoming and outgoing positrons, which are the antiparticles of electrons. We use the spinors $v(p,s)$ for positrons. An incoming positron is represented by the adjoint spinor $\bar{v}(p,s)$. Since the positron is an antiparticle, the arrow used to represent it points *opposite* the direction of time flow. This is shown in Fig. 8.5.

Outgoing positron states are represented by the spinor $v(p,s)$. An outgoing positron state is shown in Fig. 8.6.

At each vertex of a Feynman diagram for a QED process we need to include a coupling constant, g_e. If we define it in terms of the fine structure constant as

$$g_e = \sqrt{4\pi\alpha} \tag{8.13}$$

then we need a term at each vertex of the form

$$ig_e\gamma^{\mu} \tag{8.14}$$

Next we consider an internal photon line. Since it is an internal line, it will be characterized by a propagator. The form that propagates a field from a spacetime point x

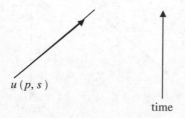

$u(p,s)$

time

Figure 8.3 An electron entering an interaction.

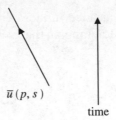

$$\bar{u}(p,s)$$

time

Figure 8.4 An electron leaving an interaction.

to a spacetime point y is determined by calculating the time ordered vacuum expectation value of the fields as shown here.

$$-i\Delta_{\mu\nu}(x-y) = \langle 0|T\{A_\mu(x)A_\nu(y)\}|0\rangle$$

This can be done by breaking up the field into positive and negative frequency components, like

$$-i\Delta_{\mu\nu}(x-y) = \langle 0|T\{A_\mu(x)A_\nu(y)\}|0\rangle$$

$$= \theta(x-y)\int \frac{d^3p\,d^3p'}{\sqrt{2p_0 2p^{0'}}(2\pi)^3}\varepsilon_\mu^{(\lambda)}(p)e^{ip_\mu x^\mu}\left[\varepsilon_\nu^{(\lambda')}(p')\right]^* e^{-ip'_\mu y^\mu}\delta_{\lambda,\lambda'}\delta^3(\vec{p}-\vec{p}')$$

$$+\theta(y-x)\int \frac{d^3p\,d^3p'}{\sqrt{2p_0 2p^{0'}}(2\pi)^3}\varepsilon_\nu^{(\lambda)}(p')e^{ip'_\mu y^\mu}\left[\varepsilon_\mu^{(\lambda')}(p\backslash)\right]^* e^{-ip_\mu x^\mu}\delta_{\lambda,\lambda'}\delta^3(\vec{p}-\vec{p}')$$

The Kronecker delta terms ensure that the polarization states are the same, that is, $\lambda' = \lambda$. We set $\lambda' \to \lambda$ and then readily do the integrals. The Dirac delta terms $\delta^3(\vec{p}-\vec{p}')$ enforce momentum conservation and the propagator simplifies to

$$-i\Delta_{\mu\nu}(x-y) = \int \frac{d^3p}{2|\vec{p}|(2\pi)^3}e^{ip\cdot(x-y)}\theta(x-y)\sum_\lambda \varepsilon_\mu^{(\lambda)}(p)\left[\varepsilon_\nu^{(\lambda')}(p)\right]^*$$

$$+\int \frac{d^3p}{2|\vec{p}|(2\pi)^3}e^{-ip\cdot(x-y)}\theta(y-x)\sum_\lambda \varepsilon_\nu^{(\lambda)}(p)\left[\varepsilon_\mu^{(\lambda')}(p)\right]^*$$

$$\bar{v}(p,s)$$

time

Figure 8.5 An incoming positron state.

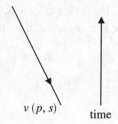

$v\,(p,s)$ time

Figure 8.6 An outgoing positron state.

But it can be shown that

$$\sum_{\lambda} \varepsilon_{\lambda}^{i}(p)\left[\varepsilon_{\lambda}^{j}(p)\right]^{*} = \delta_{ij} - \frac{p_{i}p_{j}}{|\vec{p}|^{2}} \qquad (8.15)$$

We are also taking the time component of the polarization vectors to be 0, that is, $\varepsilon_{\lambda}^{0}(p) = 0$. This just means that the polarization is fixed and does not vary with time. Then we define

$$P_{\mu\nu}(p) = \sum_{\lambda} \varepsilon_{\lambda}^{\mu}(p)\left[\varepsilon_{\lambda}^{\nu}(p)\right]^{*}$$

and use the following definition of the unit step or Heaviside function

$$\theta(x) = \frac{i}{2\pi} \int_{-\infty}^{\infty} \frac{e^{-isx}}{sx} ds \qquad (8.16)$$

Then we can write the photon propagator as (note that q is momentum in this context, and not charge)

$$\Delta_{\mu\nu}(x-y) = \int \frac{d^{4}q}{(2\pi)^{4}} P_{\mu\nu}(q) \frac{e^{iq\cdot(x-y)}}{q^{2} - i\varepsilon}$$

We have made the notational change $p \rightarrow q$ because this is momentum for an internal line in the Feynman diagram. It can be shown that this reduces the propagator to

$$\Delta_{\mu\nu} = -\frac{ig_{\mu\nu}}{q^{2}} \qquad (8.17)$$

In QED we can also have internal lines for electrons and positrons and the propagator is more complicated. For each internal line involving an electron or positron we include a factor

$$\frac{i(\gamma^{\mu}q_{\mu}+m)}{q^2-m^2} \tag{8.18}$$

The procedure is similar to that outlined in the last chapter, where conservation of momentum is enforced at each vertex using a Dirac delta function. However, in the case of QED spin must be taken into account. Let's see how to set up some basic calculations with a couple of examples.

Consider electron-electron scattering, sometimes known as Møller scattering, which is shown in Fig. 8.7.

To begin, we pick up a factor of $ig_e\gamma^{\mu}$ at each vertex. However, we need to be careful of the order of the factors because we are dealing with particles with spin and summations. We start at the left of the diagram, but move "backward in time" writing down factors from left to right. The exiting state on the left is the electron state $\bar{u}(p_3,s_3)$ and the input state is $u(p_1,s_1)$. So we get

$$\bar{u}(p_3,s_3)ig_e\gamma^{\mu}u(p_1,s_1)$$

Now we stir in a Dirac delta function to enforce conservation of momentum at the left vertex. We have an incoming momentum p_1 associated with $u(p_1,s_1)$ and an outgoing momentum p_3 associated with $\bar{u}(p_3,s_3)$. We also have a momentum q that is carried away from this vertex by the photon. The appropriate delta function is

$$(2\pi)^4\delta(p_1-p_3-q)$$

Figure 8.7 The simplest representation of Møller scattering.

So far we have

$$\bar{u}(p_3,s_3)ig_e\gamma^\mu u(p_1,s_1)(2\pi)^4\delta(p_1-p_3-q)$$

The next step is to add the photon propagator Eq. (8.17). Since there is one internal photon line, we only need one factor. Our expression for the Feynman diagram becomes

$$-\bar{u}(p_3,s_3)ig_e\gamma^\mu u(p_1,s_1)(2\pi)^4\delta(p_1-p_3-q)\left(\frac{ig_{\mu\nu}}{q^2}\right)$$

Now we multiply another factor of $ig_e\gamma^\mu$ for the vertex on the right side, along with terms for the electron states entering and leaving at that vertex, which is $\bar{u}(p_4,s_4)ig_e\gamma^\mu u(p_2,s_2)$. All together we have

$$-\bar{u}(p_3,s_3)ig_e\gamma^\mu u(p_1,s_1)(2\pi)^4\delta(p_1-p_3-q)\left(\frac{ig_{\mu\nu}}{q^2}\right)\bar{u}(p_4,s_4)ig_e\gamma^\mu u(p_2,s_2)$$

which we can rearrange as

$$-i(2\pi)^4 g_e^2\bar{u}(p_3,s_3)\gamma^\mu u(p_1,s_1)\delta(p_1-p_3-q)\left(\frac{g_{\mu\nu}}{q^2}\right)\bar{u}(p_4,s_4)\gamma^\mu u(p_2,s_2)$$

Integrating overall internal momenta q gives

$$-iM=\int -i(2\pi)^4 g_e^2\bar{u}(p_3,s_3)\gamma^\mu u(p_1,s_1)\delta(p_1-p_3-q)\left(\frac{g_{\mu\nu}}{q^2}\right)\bar{u}(p_4,s_4)\gamma^\mu u(p_2,s_2)d^4q$$

$$=-ig_e^2\bar{u}(p_3,s_3)\gamma^\mu u(p_1,s_1)\left(\frac{g_{\mu\nu}}{(p_1-p_3)^2}\right)\bar{u}(p_4,s_4)\gamma^\mu u(p_2,s_2)$$

$$\Rightarrow M=-g_e^2\bar{u}(p_3,s_3)\gamma^\mu u(p_1,s_1)\left(\frac{g_{\mu\nu}}{(p_1-p_3)^2}\right)\bar{u}(p_4,s_4)\gamma^\mu u(p_2,s_2)$$

Overall conservation of momentum is enforced by adding a delta function of the form $(2\pi)^4\delta(p_1+p_2-p_3-p_4)$ which can be ignored when writing down the amplitude.

It turns out we aren't done. Møller scattering includes one more lowest-order diagram, which is shown in Fig. 8.8.

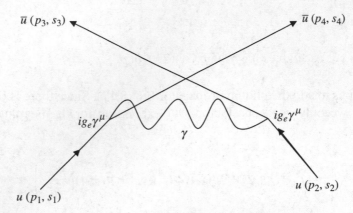

Figure 8.8 To complete the calculation for Møller scattering we need to include this diagram.

This process isn't exactly identical to the one in Fig. 8.7. This time $u(p_1,s_1)$ enters but then crosses over and exits as $\bar{u}(p_4,s_4)$, while $u(p_2,s_2)$ enters on the right and exits on the left as $\bar{u}(p_3,s_3)$. Conservation of momentum is enforced by the Dirac delta function on the left side of the diagram where the photon is emitted with momentum q.

$$(2\pi)^4 \delta(p_1 - p_4 - q)$$

Aside from this, the form of amplitude for this process is similar to the last one we wrote down, but we are swapping output states as shown here.

$$\bar{u}(p_3,s_3) \rightleftarrows \bar{u}(p_4,s_4)$$

Since we are exchanging two identical fermions, we must make a sign change. In total, the amplitude for the process illustrated in Fig. 8.8 is

$$M' = g_e^2 \bar{u}(p_4,s_4)\gamma^\mu u(p_1,s_1)\left(\frac{g_{\mu\nu}}{(p_1-p_4)^2}\right)\bar{u}(p_3,s_3)\gamma^\mu u(p_2,s_2)$$

The total amplitude for the process is the sum of the two amplitudes we have written down, that is,

$$M_{\text{Møller}} = M + M'$$

$$= -g_e^2 \bar{u}(p_3,s_3)\gamma^\mu u(p_1,s_1)\left(\frac{g_{\mu\nu}}{(p_1-p_3)^2}\right)\bar{u}(p_4,s_4)\gamma^\mu u(p_2,s_2)$$

$$+ g_e^2 \bar{u}(p_4,s_4)\gamma^\mu u(p_1,s_1)\left(\frac{g_{\mu\nu}}{(p_1-p_4)^2}\right)\bar{u}(p_3,s_3)\gamma^\mu u(p_2,s_2)$$

This is a useful theoretical result. To calculate a measurable quantity we have to

- Pick a reference frame, either the lab frame or the center of mass frame
- Assign helicities to the particles or average/sum over all possible spin states

Let's consider the process in Fig. 8.7. It is easy to choose the center of mass frame. Let's suppose that all particles have a helicity of +1. This means that if the particle is moving along the positive z axis, its wave function will be of the form

$$u = \sqrt{E+m} \begin{pmatrix} 1 \\ 0 \\ \dfrac{p}{E+m} \\ 0 \end{pmatrix}$$

modulo a normalization factor. If the particle is moving toward negative z, the wave function will be of the form

$$u = \sqrt{E+m} \begin{pmatrix} 0 \\ 1 \\ 0 \\ \dfrac{p}{E+m} \end{pmatrix}$$

where m is the mass of the electron. When the particles are incoming, we have the situation illustrated in Fig. 8.9.

Now, since $u(p_1, s_1)$ moves in the direction of positive z, the state is

$$u(p_1, s_1) = \sqrt{E+m} \begin{pmatrix} 1 \\ 0 \\ \dfrac{p}{E+m} \\ 0 \end{pmatrix}$$

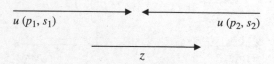

$u(p_1, s_1)$ $u(p_2, s_2)$

z

Figure 8.9   The incoming particle states. The state $u(p_1, s_1)$ moves in the direction of positive z.

All states have momentum p since we are using the center of mass frame. Now, $u(p_2, s_2)$ is moving in the direction of negative z so the state is

$$u(p_2, s_2) = \sqrt{E+m} \begin{pmatrix} 0 \\ 1 \\ 0 \\ \dfrac{p}{E+m} \end{pmatrix}$$

Now let's look at the outgoing states, shown in Fig. 8.10. The direction of motion is reversed for each state, so the form of the states is different. Also recall that $\overline{\psi} = \psi^\dagger \gamma^0$. We have

$$\overline{u}(p_3, s_3) = u^\dagger(p_3, s_3)\gamma^0 = \sqrt{E+m}\begin{pmatrix} 0 & 1 & 0 & \dfrac{p}{E+m} \end{pmatrix}\begin{pmatrix} 1 & 0 & 0 & 0 \\ 0 & 1 & 0 & 0 \\ 0 & 0 & -1 & 0 \\ 0 & 0 & 0 & -1 \end{pmatrix}$$

$$= \sqrt{E+m}\begin{pmatrix} 0 & 1 & 0 & -\dfrac{p}{E+m} \end{pmatrix}$$

In the case of $\overline{u}(p_4, s_4)$, the state is moving in the positive z direction and so can be written as

$$\overline{u}(p_4, s_4) = u^\dagger(p_4, s_4)\gamma^0 = \sqrt{E+m}\begin{pmatrix} 1 & 0 & \dfrac{p}{E+m} & 0 \end{pmatrix}\begin{pmatrix} 1 & 0 & 0 & 0 \\ 0 & 1 & 0 & 0 \\ 0 & 0 & -1 & 0 \\ 0 & 0 & 0 & -1 \end{pmatrix}$$

$$= \sqrt{E+m}\begin{pmatrix} 1 & 0 & -\dfrac{p}{E+m} & 0 \end{pmatrix}$$

Now we can use these results to do some explicit calculations. Recall that the amplitude we found for the process shown in Fig. 8.7 was given by

$$M = -g_e^2 \overline{u}(p_3, s_3)\gamma^\mu u(p_1, s_1)\left(\frac{g_{\mu\nu}}{(p_1 - p_3)^2}\right)\overline{u}(p_4, s_4)\gamma^\mu u(p_2, s_2) \qquad (8.19)$$

$$\xleftarrow{\hspace{2cm}} \overline{u}(p_3, s_3) \qquad\qquad \overline{u}(p_4, s_4) \xrightarrow{\hspace{2cm}}$$

$$\xrightarrow{\hspace{4cm}} z$$

Figure 8.10   The outgoing states reverse direction.

So we need to calculate

$$\bar{u}(p_4,s_4)\gamma^0 u(p_2,s_2), \bar{u}(p_4,s_4)\gamma^1 u(p_2,s_2), \bar{u}(p_4,s_4)\gamma^2 u(p_2,s_2)$$

and

$$\bar{u}(p_4,s_4)\gamma^3 u(p_2,s_2)$$

There is only one way to do it, using brute force. The first term is

$$\bar{u}(p_4,s_4)\gamma^0 u(p_2,s_2) =$$

$$\sqrt{E+m}\begin{pmatrix} 1 & 0 & -\dfrac{p}{E+m} & 0 \end{pmatrix}\begin{pmatrix} 1 & 0 & 0 & 0 \\ 0 & 1 & 0 & 0 \\ 0 & 0 & -1 & 0 \\ 0 & 0 & 0 & -1 \end{pmatrix}\sqrt{E+m}\begin{pmatrix} 0 \\ 1 \\ 0 \\ \dfrac{p}{E+m} \end{pmatrix}$$

$$= (E+m)\begin{pmatrix} 1 & 0 & -\dfrac{p}{E+m} & 0 \end{pmatrix}\begin{pmatrix} 0 \\ 1 \\ 0 \\ -\dfrac{p}{E+m} \end{pmatrix}$$

$$= 0$$

Next we find

$$\bar{u}(p_4,s_4)\gamma^1 u(p_2,s_2) =$$

$$\sqrt{E+m}\begin{pmatrix} 1 & 0 & -\dfrac{p}{E+m} & 0 \end{pmatrix}\begin{pmatrix} 0 & 0 & 0 & 1 \\ 0 & 0 & 1 & 0 \\ 0 & -1 & 0 & 0 \\ -1 & 0 & 0 & 0 \end{pmatrix}\sqrt{E+m}\begin{pmatrix} 0 \\ 1 \\ 0 \\ \dfrac{p}{E+m} \end{pmatrix}$$

$$= (E+m)\begin{pmatrix} 1 & 0 & -\dfrac{p}{E+m} & 0 \end{pmatrix}\begin{pmatrix} \dfrac{p}{E+m} \\ 0 \\ -1 \\ 0 \end{pmatrix}$$

$$= 2p$$

Continuing we find

$$\bar{u}(p_4,s_4)\gamma^2 u(p_2,s_2) =$$

$$\sqrt{E+m}\begin{pmatrix} 1 & 0 & -\dfrac{p}{E+m} & 0 \end{pmatrix}\begin{pmatrix} 0 & 0 & 0 & -i \\ 0 & 0 & i & 0 \\ 0 & i & 0 & 0 \\ -i & 0 & 0 & 0 \end{pmatrix}\sqrt{E+m}\begin{pmatrix} 0 \\ 1 \\ 0 \\ \dfrac{p}{E+m} \end{pmatrix}$$

$$= (E+m)\begin{pmatrix} 1 & 0 & -\dfrac{p}{E+m} & 0 \end{pmatrix}\begin{pmatrix} \dfrac{-ip}{E+m} \\ 0 \\ i \\ 0 \end{pmatrix}$$

$$= -2ip$$

and finally

$$\bar{u}(p_4,s_4)\gamma^3 u(p_2,s_2) =$$

$$\sqrt{E+m}\begin{pmatrix} 1 & 0 & -\dfrac{p}{E+m} & 0 \end{pmatrix}\begin{pmatrix} 0 & 0 & 1 & 0 \\ 0 & 0 & 0 & -1 \\ -1 & 0 & 0 & 0 \\ 0 & 1 & 0 & 0 \end{pmatrix}\sqrt{E+m}\begin{pmatrix} 0 \\ 1 \\ 0 \\ \dfrac{p}{E+m} \end{pmatrix}$$

$$= (E+m)\begin{pmatrix} 1 & 0 & -\dfrac{p}{E+m} & 0 \end{pmatrix}\begin{pmatrix} 0 \\ -\dfrac{p}{E+m} \\ 0 \\ 1 \end{pmatrix}$$

$$= 0$$

Using

$$u(p_1,s_1) = \sqrt{E+m}\begin{pmatrix} 1 \\ 0 \\ \dfrac{p}{E+m} \\ 0 \end{pmatrix}, \quad \bar{u}(p_3,s_3) = \sqrt{E+m}\begin{pmatrix} 0 & 1 & 0 & -\dfrac{p}{E+m} \end{pmatrix}$$

We also obtain

$$\bar{u}(p_3,s_3)\gamma^0 u(p_1,s_1) = 0$$

$$\bar{u}(p_3,s_3)\gamma^1 u(p_1,s_1) = 2p$$

$$\bar{u}(p_3,s_3)\gamma^2 u(p_1,s_1) = 2ip$$

$$\bar{u}(p_3,s_3)\gamma^3 u(p_1,s_1) = 0$$

which the reader should verify.

Now we apply the summation convention to each term in the amplitude [Eq. (8.19)]. For the left term we have

$$
\begin{aligned}
g_{\mu\nu}\bar{u}(p_4,s_4)\gamma^\mu u(p_2,s_2) &= g_{00}\bar{u}(p_4,s_4)\gamma^0 u(p_2,s_2) + g_{11}\bar{u}(p_4,s_4)\gamma^1 u(p_2,s_2) \\
&\quad + g_{22}\bar{u}(p_4,s_4)\gamma^2 u(p_2,s_2) + g_{33}\bar{u}(p_4,s_4)\gamma^3 u(p_2,s_2) \\
&= \bar{u}(p_4,s_4)\gamma^1 u(p_2,s_2) + \bar{u}(p_4,s_4)\gamma^2 u(p_2,s_2) \\
&= 2p(1-i)
\end{aligned}
$$

For $\bar{u}(p_3,s_3)\gamma^\mu u(p_1,s_1)$ we get $2p(1+i)$ and so Eq. (8.19) becomes

$$M = -g_e^{\,2}\bar{u}(p_3,s_3)\gamma^\mu u(p_1,s_1)\left(\frac{g_{\mu\nu}}{(p_1-p_3)^2}\right)\bar{u}(p_4,s_4)\gamma^\mu u(p_2,s_2)$$

$$= -g_e^{\,2}\frac{2p(1+i)2p(1-i)}{(p_1-p_3)^2} = -g_e^{\,2}\frac{8p^2}{(p_1-p_3)^2}$$

Using $E^2 = m^2 + p^2$ you can show that $(p_1-p_3)^2 = 2m^2 - 2(E^2 + p^2) = -4p^2$ and so the amplitude becomes

$$M = -g_e^{\,2}\frac{8p^2}{-4p^2} = 2g_e^{\,2}$$

Often, the helicities of the particles are not known. When this is the case, we say that the cross section is *unpolarized*. In that case it is necessary to average and sum over the spins. We compute the average over all spins for incoming particles and then sum over all possible spin states for outgoing particles. A useful tool for doing so is

$$\sum_s u(p,s)\bar{u}(p,s) = \frac{p+m}{2m} \qquad (8.20)$$

(consider deriving this relation). For example, let's look at the lowest order Feynman diagram for electron–muon scattering. This is Fig. 8.7, but we replace the incoming and outgoing particle on the right with a muon. The amplitude is

$$M = -g_e^2 \bar{u}(p_3)\gamma^\mu u(p_1)\frac{g_{\mu\nu}}{q^2}\bar{u}(p_4)\gamma^\mu u(p_2)$$

where $\bar{u}(p_3)\gamma^\mu u(p_1)$ is due to the electron states and $\bar{u}(p_4)\gamma^\mu u(p_2)$ is due to the muon states. The final amplitude, when summing and averaging over all outgoing and incoming spins is

$$\left|\bar{M}\right|^2 = \frac{g_e^4}{q^4}L_e^{\mu\nu}L_{\mu\nu}{}^{\text{muon}}$$

We will focus on the electron term only, the muon term is similar. The electron term is

$$L_e^{\mu\nu} = \frac{1}{2}\sum_s \left[\bar{u}(p_3)\gamma^\mu u(p_1)\right]\left[\bar{u}(p_3)\gamma^\mu u(p_1)\right]^*$$

Using Eq. (8.20) we find

$$L_e^{\mu\nu} = \frac{1}{2}Tr(\not{p}_3 + m)\gamma^\mu(\not{p}_1 + m)\gamma^\nu \tag{8.21}$$

These terms can be evaluated using the so-called trace theorems. These include

$$Tr(I) = 4$$
$$Tr(\not{a}\not{b}) = 4a \cdot b$$
$$Tr(\not{a}\not{b}\not{c}\not{d}) = 4[(a \cdot b)(c \cdot d) - (a \cdot c)(b \cdot d) + (a \cdot d)(b \cdot c)] \tag{8.22}$$
$$\gamma_\mu \not{a}\gamma^\mu = -2\not{a}$$
$$Tr(\gamma^\mu\gamma^\nu) = 4g^{\mu\nu}$$

These theorems are based only on basic linear algebra and the properties of the Dirac matrices.

Multiplying out the terms in Eq. (8.21) and applying trace theorems gives the result

$$L_e^{\mu\nu} = \frac{1}{2}Tr(\not{p}_3 + m)\gamma^\mu(\not{p}_1 + m)\gamma^\nu$$
$$= \frac{1}{2}Tr(\not{p}_3\gamma^\mu \not{p}_1\gamma^\nu + m\not{p}_3\gamma^\mu\gamma^\nu + m\gamma^\mu \not{p}_1\gamma^\nu + m^2\gamma^\mu\gamma^\nu)$$
$$= 2p_3^\mu p_1^\nu + 2p_3^\nu p_1^\mu - 2(p_3 \cdot p_1 - m^2)g^{\mu\nu}$$

At last we are done. The amplitude for the electron–muon scattering is

$$\left|\bar{M}\right|^2 = \frac{g_e^4}{q^4} L_e^{\mu\nu} L_{\mu\nu}^{\text{muon}} = \frac{g_e^4}{q^4} \left[Tr(p_3 + m)\gamma^\mu (p_1 + m)\gamma^\nu \right]$$

$$\times \left[p_3^\mu p_1^\nu + p_3^\nu p_1^\mu - (p_3 \cdot p_1 - m^2)g^{\mu\nu} \right]$$

Summary

In this first treatment of quantum electrodynamics, we have introduced the basic concept of electromagnetic forces as being due to the exchange of photons. Quantum electrodynamics ties together the Dirac theory of the electron (and other leptons, the principles are the same) with electromagnetics, which is a description of the photon field. It does this by considering interactions which are mediated by the photon, the force-carrying particle for the electromagnetic interaction. Electrons and other charged particles interact electromagnetically by exchanging photons. In this initial treatment, we have ignored higher order processes which include internal "loops" that lead to divergences.

The gauge symmetry for quantum electrodynamics is a local $U(1)$ symmetry. The requirement that the Lagrangian be invariant under this symmetry led to the minimal coupling prescription of the covariant derivative. We then extended this concept to compute measurable quantities like scattering amplitudes.

Quiz

1. Compute $[D_\mu, D_\nu]$.

2. The Lagrangian of quantum electrodynamics can be best described as

    (a) Admitting a local $U(1)$ symmetry

    (b) Admitting a global $U(1)$ symmetry

    (c) Admitting a local $SU(2)$ symmetry

    (d) Admitting a local $SU(1)$ symmetry

3. Write down the amplitude for electron-positron scattering as shown in Fig. 8.11.

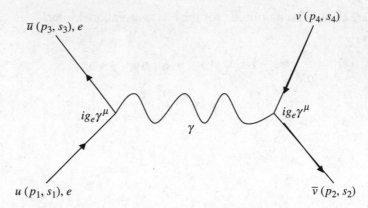

Figure 8.11 Electron-muon scattering to lowest order.

4. The minimal coupling prescription for the QED Lagrangian is

    (a) $D_\mu = \partial_\mu + ig_e A_\mu$

    (b) $D_\mu = \partial_\mu - ig_e A_\mu$

    (c) $D_\mu = \partial_\mu + iqA_\mu$

    (d) $D_\mu = \partial_\mu + iq\gamma^\mu A_\mu$

5. In a QED process an incoming antiparticle state is written as

    (a) $\bar{v}(p,s)$

    (b) $\bar{u}(p,s)$

    (c) $u(p,s)$

    (d) $v(p,s)$

CHAPTER 9

Spontaneous Symmetry Breaking and the Higgs Mechanism

Let's review some important concepts. Noether's theorem relates conservation laws to symmetries in the Lagrangian. When quantum theory is invoked, these symmetries can take the form of invariance under a unitary transformation. For example, a $U(1)$ symmetry means that a Lagrangian $L = L(\varphi, \partial_\mu \varphi)$ is invariant under a transformation of the form

$$\varphi(x) \to \varphi'(x) = e^{-i\theta} \varphi(x) \qquad (9.1)$$

When θ does not depend on the spacetime coordinate x, then we say that Eq. (9.1) is a global symmetry. In quantum field theory global symmetries represent something that cannot be measured, like the phase of a wave function in quantum mechanics. The wave function $\psi(x,t)$ and $e^{-i\alpha}\psi(x,t)$ give the same physical predictions. On the other hand, if θ does depend on the spacetime coordinate so that $\theta = \theta(x)$, then Eq. (9.1) depends on where you are and hence represents a *local* symmetry. Local symmetries are very important in relativistic physics because they represent the physical fact that quantities that are conserved like charge and lepton number are conserved *locally*. Charge would not be conserved locally if you could have a current on earth disappear and suddenly reappear on the moon. The charge must travel across the intervening space to appear on the moon, and the way it moves from the earth to the moon is dictated by the fact that nothing travels faster than the speed of light. Said another way, a local symmetry preserves *causality* as required by special relativity.

Two pictures help distinguish between local and global $U(1)$ transformations. We know from complex variables that the exponential $e^{i\theta}$ represents a point on the unit circle. We can think of this point as a vector—an arrow from the origin to the point on the unit circle. In a global transformation, let θ assume a fixed value, so that $e^{i\theta}$ has a constant value throughout spacetime. Hence the name global. On the other hand, $e^{i\theta(x)}$ where x is a point in spacetime, has a different value depending upon location because $\theta(x)$ is now a function on spacetime. So we talk of a value at a specific point. The figure below shows these two cases. On the left, notice that each vector is the same, but on the right, the direction of the vector depends on it's location. So the left side illustrates a global transformation, while the right side illustrates a local transformation.

A global transformation: The value of $e^{i\theta}$ is the same everywhere.	A local transformation: The value of $e^{i\theta}$ depends upon the space-time location.

We saw in Chap. 3 that more complicated unitary transformations such as $SU(2)$ appear in quantum field theory. It is also possible to have Lagrangians that are invariant under other types of transformations, such as $\varphi \to -\varphi$. We will use this type of symmetry to introduce the concept of *spontaneous symmetry breaking*. An example of a Lagrangian that is invariant under $\varphi \to -\varphi$ is the Lagrangian of the so-called (φ^4) theory.

$$L = \partial_\mu \varphi \, \partial^\mu \varphi - m^2 \varphi^2 + \lambda \varphi^4 \tag{9.2}$$

Clearly, since the field φ only appears in the Lagrangian in terms of even powers and even derivatives, the Lagrangian is unchanged under the transformation $\varphi \to -\varphi$. We will see, however, this Lagrangian is even more interesting than it appears at first sight.

It turns out that in many cases, a system that has some symmetry that exists in the Lagrangian may have a ground state (i.e., a vacuum state) that *does not* satisfy the same symmetry. This is the case for the Lagrangian given in Eq. (9.2). When a situation like this exists, we say that the system has undergone *spontaneous symmetry breaking*. Before jumping into the mathematics of the situation, let's describe the concept with a simple physical example. Imagine an upside down steel bowl placed on flat ground. We place a marble on top of the bowl right in the center. This system is symmetric—from the point of view of the marble every direction from the top of the bowl to the ground is equivalent. However, the system is *unstable*. The marble starts out at rest, but the slightest perturbation will send it rolling down the bowl to the ground. In analogy with quantum field theory, think of the marble sitting on top of the bowl as a ground state that is unstable.

Now suppose that the marble is perturbed and rolls off the bowl. It will roll in one particular direction and come to rest below on the flat ground. In short, the perturbation has *spontaneously broken* the symmetry that existed before. Moreover, the marble has now arrived at a state of minimum potential energy. In short, the marble was not really in the ground state when it was resting on top of the bowl— the true ground state of potential energy exists when the symmetry is broken and the marble finds itself resting on the ground below.

Symmetry Breaking in Field Theory

In quantum field theory, we often have Lagrangians that exhibit similar properties to an upside down bowl. We will see a vacuum state that is an apparent ground state, but in fact there will be a *true* ground state or vacuum state of lower energy that leads to symmetry breaking. What is vacuum? Vacuum is the state with no fields, that is, $\varphi = 0$. In our calculations by applying perturbation theory, we expand about $\varphi = 0$, the fields are then viewed as fluctuations about the ground state. You can think of $\varphi = 0$ as the minimum of potential energy.

However, when considering different Lagrangians, it turns out that the state with $\varphi = 0$ is not always the minimum. Remember that the Lagrangian is the difference between kinetic energy T and potential energy V.

$$L = T - V$$

In field theory, remember that kinetic energy terms will be of the form $\partial_\mu \varphi \partial^\mu \varphi$. The potential V will be some function of the fields φ so that $V = V(\varphi)$. Therefore, to find the minimum we use ordinary calculus, that is, we seek to find the minimum of the potential by computing its derivative. That is, we will find a φ that allows us to satisfy

$$\frac{\partial V}{\partial \varphi} = 0 \tag{9.3}$$

This procedure will give us the true ground state of the system, which may not be $\varphi = 0$.

EXAMPLE 9.1

Consider the Lagrangian for φ^4 theory when

$$L = \frac{1}{2}(\partial_\mu \varphi)^2 - \frac{1}{2}m^2 \varphi^2 - \frac{1}{4}\lambda \varphi^4 \tag{9.4}$$

where φ is a real scalar field. Describe the minimum of potential energy when $m^2 > 0$ and $m^2 < 0$.

SOLUTION

The kinetic energy term in the Lagrangian is

$$\frac{1}{2}(\partial_\mu \varphi)^2$$

The potential is

$$V(\varphi) = \frac{1}{2}m^2 \varphi^2 + \frac{1}{4}\lambda \varphi^4$$

What force does this potential create? We compute the derivative of V with respect to the field φ:

$$\frac{\partial V}{\partial \varphi} = m^2 \varphi + \lambda \varphi^3$$

$$= \varphi(m^2 + \lambda \varphi^2)$$

We obtain the minima by setting this expression equal to 0. One extremum jumps out immediately, it is the one we naively expect to find

$$\varphi = 0$$

This case corresponds to the case $m^2 > 0$, which represents a scalar field of mass m. The φ^4 term represents self-interactions of the field with a coupling strength given by λ. The e potential in this case is shown in Fig. 9.1.

When the ground state is at $\varphi = 0$, it obviously satisfies the symmetry present in the Lagrangian, $\varphi \to -\varphi$, and does so trivially.

Now let's consider the other alternative minimum that result from our calculation of $\partial V / \partial \varphi = 0$. In this case we have

$$m^2 + \lambda \varphi^2 = 0$$

Since φ appears as a square, this leads to two possible minima given by

$$\varphi = \pm \sqrt{\frac{-m^2}{\lambda}} = \pm v \tag{9.5}$$

In this case, in order for the field φ to be real, it must be the case that $m^2 < 0$. This is a situation which corresponds to the ball sitting on top of the bowl. The potential in this case is shown in Fig. 9.2. Notice that $\varphi = 0$ corresponds to the unstable point where the marble is resting on top of the bowl. We can go to one or the other minimum, where $\varphi = +v$ or $\varphi = -v$, giving the true ground state. But choosing one of the other breaks the symmetry. This is analogous to the marble rolling off the bowl and coming to rest at some particular point on the ground.

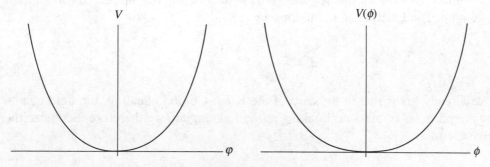

Figure 9.1   The potential for the Lagrangian given in Eq. (9.4), for the case of $m^2 > 0$. The minimum is at $\varphi = 0$.

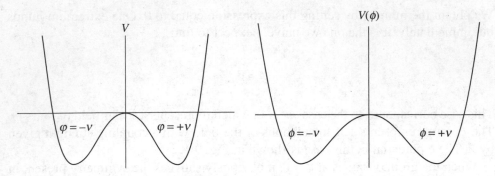

Figure 9.2   The potential for the Lagrangian given in Eq. (9.4) when $m^2 < 0$.

In the case of the Lagrangian, we have found the true minimum at $\varphi = \pm v$. The point $\varphi = 0$ is an unstable point, therefore a perturbative expansion about this point will not converge. In contrast, a perturbative expansion about one of $\varphi = \pm v$ will converge, allowing us to do calculations using the Feynman rules.

However, the symmetry has been broken. There are two ground states, the minima at $\varphi = +v$ and $\varphi = -v$. In the next section we will see that the Lagrangian is no longer invariant under $\varphi \to -\varphi$. While we loose this benefit, we will gain knowledge, specifically the true mass of the particle associated with the field φ.

Mass Terms in the Lagrangian

A key task in exercises involving spontaneous symmetry breaking is the ability to recognize mass terms in the Lagrangian. Doing so is usually pretty simple; to see this let's go back to square one—which means digging up the Klein-Gordon equation. The Lagrangian in this case is

$$L = \frac{1}{2}(\partial_\mu \varphi)^2 - \frac{1}{2}m^2\varphi^2$$

We already know that in the case of the Klein-Gordon equation, the field quanta φ are particles of mass m. Looking at the Lagrangian, we therefore recognize the mass term as

$$-\frac{1}{2}m^2\varphi^2$$

where $m^2 > 0$ and m is the mass of the associated particle. So this is a straightforward exercise. We conclude that

> A mass term in the Lagrangian is one that is quadratic in the fields, which is a term of the form $\alpha^2 \varphi^2$ for some α.

However, it turns out that identifying the mass terms in a Lagrangian by inspection is not always possible. Many Lagrangians have mass terms that are hidden in one way or another. To see this, consider a fictitious Lagrangian given by

$$L = \frac{1}{2}(\partial_\mu \varphi)^2 + \ln(1 - \alpha \varphi) \qquad (9.6)$$

Is there a mass term in this Lagrangian? By inspection, we don't see any terms that are quadratic in the field, so we might jump to the conclusion that $m = 0$ in this case. The Lagrangian appears to describe a massless field like a photon, say. But a closer look at Eq. (9.6) will reveal otherwise. Once again we call upon our skills we learned in freshman calculus. To expand the Lagrangian in a series we need to expand the logarithmic term. The trick is to start with the geometric series as shown here.

$$\frac{1}{1 - \alpha x} = 1 + \alpha x + (\alpha x)^2 + (\alpha x)^3 + \cdots \text{ for } |\alpha x| < 1$$

(If you have forgotten this expansion, just do the division manually.) To introduce a logarithm, we need to integrate this expression to get

$$\ln(1 - \alpha x) = -\alpha \left[\int 1 + \alpha x + (\alpha x)^2 + \cdots dx \right]$$

which is just

$$\ln(1 - \alpha x) = -\alpha x - \frac{1}{2}\alpha^2 x^2 - \frac{1}{3}\alpha^3 x^3 - O(x^4)$$

Writing Eq. (9.6) using this expansion, we see that the Lagrangian does in fact contain a term that is quadratic in the field.

$$L = \frac{1}{2}(\partial_\mu \varphi)^2 + \ln(1 - \alpha \varphi)$$

$$= \frac{1}{2}(\partial_\mu \varphi)^2 - \alpha \varphi - \frac{1}{2}\alpha^2 \varphi^2 - \frac{1}{3}\alpha^3 \varphi^3 - O(\varphi^4)$$

Provided that, $\alpha^2 > 0$, this Lagrangian describes particles with mass $\alpha = m$. The mass term has been disguised by the original representation of the Lagrangian given in Eq. (9.6). So, when it's not obvious as to whether or not a given Lagrangian contains a mass term

- Expand the potential in a series.
- Look for terms that are quadratic in the fields.

EXAMPLE 9.2

Do the Lagrangians

$$L_1 = \frac{1}{2}(\partial_\mu \varphi)^2 - e^{\alpha^3 \varphi^3}$$

$$L_2 = \frac{1}{2}(\partial_\mu \varphi)^2 - e^{\alpha \varphi}$$

represent massive or massless fields?

SOLUTION

We apply our little recipe and follow the guidance provided by the Klein-Gordon equation. That is, we use the expansion of the exponential function and look for terms that are quadratic in the field. Recall that

$$e^{\alpha x} = 1 + \alpha x + \frac{1}{2!}(\alpha x)^2 + \frac{1}{3!}(\alpha x)^3 + O(x^4)$$

In the case of L_1, we have

$$L_1 = \frac{1}{2}(\partial_\mu \varphi)^2 - e^{\alpha^3 \varphi^3}$$

$$= \frac{1}{2}(\partial_\mu \varphi)^2 - 1 - \alpha^3 \varphi^3 - \frac{1}{2}\alpha^6 \varphi^6$$

plus higher order terms. There is not a term involving φ^2, hence we conclude that L_1 is a Lagrangian for a massless field.

Now let's consider L_2. Expanding the exponential in this case gives us

$$L_2 = \frac{1}{2}(\partial_\mu \varphi)^2 - e^{\alpha \varphi}$$

$$= \frac{1}{2}(\partial_\mu \varphi)^2 - 1 - \alpha \varphi - \frac{1}{2}\alpha^2 \varphi^2 - \frac{1}{6}\alpha^3 \varphi^3 - \cdots$$

The presence of the term

$$\frac{1}{2}\alpha^2\varphi^2$$

tells us that L_2 is a Lagrangian for a massive field. The mass of the particle is given by comparison with the Klein-Gordon equation, hence the mass is $m = \alpha$.

Aside on Units

Now a short note on units and scaling with mass terms. When we put all the $\hbar$'s and c's back in the Klein-Gordon equation, it is written as

$$\frac{1}{c^2}\frac{\partial^2\varphi}{\partial t^2} - \nabla^2\varphi - \frac{m^2 c^2}{\hbar^2}\varphi = 0$$

So, if we have a term in the Lagrangian we are considering that looks like

$$\frac{1}{2}\alpha^2\varphi^2$$

Then the mass of the particle is related to the constant α in the following way.

$$\alpha = \frac{mc}{\hbar} \qquad\qquad (9.7)$$

That is, the mass of the particle is

$$m = \frac{\hbar\alpha}{c} \qquad\qquad (9.8)$$

where α is a unitless number and m will have dimensions of mass inherited from the values used for $\hbar$ and c.

If the quadratic term in the Lagrangian is missing the ½ scale factor, that is, if it contains a term

$$\alpha^2\varphi^2$$

you need to account for the missing ½ when comparing to the Klein-Gordon equation to get the mass. In this case we have a relationship given by

$$\alpha^2 = \frac{1}{2}\frac{m^2 c^2}{\hbar^2}$$

Hence, the mass of the particle is

$$m = \sqrt{2}\,\frac{\alpha\hbar}{c}$$

Spontaneous Symmetry Breaking and Mass

Now that we know how to recognize mass terms in the Lagrangian, let's go back to the φ^4 theory and reconsider the situation. To keep you from having to flip back and forth through the pages, remember that the Lagrangian we used in Example 9.1 was given by

$$L = \frac{1}{2}(\partial_\mu\varphi)^2 - \frac{1}{2}m^2\varphi^2 - \frac{1}{4}\lambda\varphi^4$$

We found that the true ground state or minima of the potential was spontaneously broken and given by

$$\varphi = \pm\sqrt{\frac{-m^2}{\lambda}} = \pm v$$

Now what to do with this information? The minimum is not at $\varphi = 0$, instead it's located at $\varphi = \pm v$. We consider the case where $\varphi = v$ and rescale the field to represent this fact

$$\varphi(x) = v + \eta(x) \tag{9.9}$$

We've written the field as fluctuations described by $\eta(x)$ about the right-hand minimum v. Next we will rewrite the Lagrangian using the new form [Eq. (9.9)]. The kinetic energy terms are easy to write down because v is just a number, so

$$\partial_\mu\varphi(x) = \partial_\mu[v + \eta(x)] = \partial_\mu\eta(x)$$

Now we square Eq. (9.9) to give

$$\varphi^2 = (v+\eta)^2 = v^2 + 2v\eta + \eta^2$$

and the fourth power of the field becomes

$$\varphi^4 = (v+\eta)^4 = v^4 + 4v^3\eta + 6v^2\eta^2 + 4v\eta^3 + \eta^4$$

Putting these terms together, the Lagrangian becomes

$$L = \frac{1}{2}(\partial_\mu \varphi)^2 - \frac{1}{2}m^2\varphi^2 - \frac{1}{4}\lambda\varphi^4$$

$$= \frac{1}{2}(\partial_\mu \eta)^2 - \frac{1}{2}m^2(v^2 + 2v\eta + \eta^2) - \frac{1}{4}\lambda(v^4 + 4v^3\eta + 6v^2\eta^2 + 4v\eta^3 + \eta^4)$$

Again, we remember that v is just a number. We can drop all terms that are constant from the Lagrangian since a constant does not contribute to the field equations for the system. The first term in the potential can then be written as

$$\frac{1}{2}m^2(v^2 + 2v\eta + \eta^2) = -\frac{1}{2}\lambda v^2(v^2 + 2v\eta + \eta^2)$$

$$= -\lambda v^3\eta - \frac{1}{2}\lambda v^2\eta^2$$

where we used Eq. (9.5) to write $m^2 = -\lambda v^2$ and we dropped constant terms. Dropping constant terms from the last term of the potential gives

$$L = \frac{1}{2}(\partial_\mu \eta)^2 - \frac{1}{2}m^2(v^2 + 2v\eta + \eta^2) - \frac{1}{4}\lambda(v^4 + 4v^3\eta + 6v^2\eta^2 + 4v\eta^3 + \eta^4)$$

$$= \frac{1}{2}(\partial_\mu \eta)^2 + \lambda v^3\eta + \frac{1}{2}\lambda v^2\eta^2 - \lambda v^3\eta - \frac{3}{2}\lambda v^2\eta^2 - \lambda v\eta^3 - \frac{1}{4}\lambda\eta^4$$

Finally, we arrive at the new Lagrangian

$$L = \frac{1}{2}(\partial_\mu \eta)^2 - \lambda v^2\eta^2 - \lambda v\eta^3 - \frac{1}{4}\lambda\eta^4 \qquad (9.10)$$

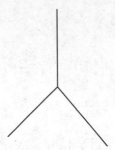

Figure 9.3   A Feynman diagram representation of the self-interaction term in the Lagrangian $\lambda v\eta^3$.

Now we apply our rule. Look for terms that are quadratic in the fields η; they should have negative signs in front of them. The mass term for Eq. (9.10) is

$$\lambda v^2 \eta^2$$

Comparing to a mass term in the Klein-Gordon–type Lagrangian

$$\frac{1}{2}m^2\phi^2$$

we see that the mass of the particle in the case of Eq. (9.10) is

$$m = \sqrt{2\lambda v^2} = \sqrt{2\lambda}\, v$$

Notice that we have taken into account the missing ½ factor. What about the other terms in the Lagrangian? These represent self-interaction terms of the field $\eta(x)$. In particular, the cubic term η^3 is a vertex in a Feynman diagram with three legs and a coupling given by λv. This is illustrated in Fig. 9.3.

The last term, $\frac{1}{4}\lambda\eta^4$, is another self-interaction term that will have four legs in a Feynman diagram. This is illustrated in Fig. 9.4.

Figure 9.4   The $\frac{1}{4}\lambda\eta^4$ term in the Lagrangian given in Eq. (9.10) is represented by a four legged vertex in a Feynman diagram.

We have now accounted for every term in the Lagrangian:

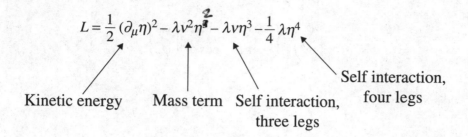

$$L = \frac{1}{2}(\partial_\mu \eta)^2 - \lambda v^2 \eta^2 - \lambda v \eta^3 - \frac{1}{4}\lambda \eta^4$$

Kinetic energy · Mass term · Self interaction, three legs · Self interaction, four legs

Lagrangians with Multiple Particles

In most, if not all, real cases of physical interest, the spontaneous symmetry breaking of a given Lagrangian will result in the appearance of more than one particle. It may be that these particles have different masses; perhaps some will have mass and some will not. Let's illustrate this with a complex field and a Lagrangian that gives rise to one massive and one massless particle. First let's define the field in terms of two real fields φ_1 and φ_2.

$$\varphi = \frac{\varphi_1 + i\varphi_2}{\sqrt{2}} \tag{9.11}$$

The Lagrangian we consider is

$$L = \partial_\mu \varphi^\dagger \partial^\mu \varphi - m^2 \varphi^\dagger \varphi + \lambda(\varphi^\dagger \varphi)^2 \tag{9.12}$$

Now

$$\varphi^\dagger \varphi = \left(\frac{\varphi_1 - i\varphi_2}{\sqrt{2}}\right)\left(\frac{\varphi_1 + i\varphi_2}{\sqrt{2}}\right) = \frac{1}{2}(\varphi_1^2 + \varphi_2^2)$$

Using this the Lagrangian in Eq. (9.12) becomes

$$L = \frac{1}{2}(\partial_\mu \varphi_1)^2 + \frac{1}{2}(\partial_\mu \varphi_2)^2 - \frac{1}{2}m^2(\varphi_1^2 + \varphi_2^2) + \frac{1}{4}\lambda(\varphi_1^4 + \varphi_2^4) + \frac{1}{2}\lambda\varphi_1^2\varphi_2^2 \tag{9.13}$$

The potential is

$$V = \frac{1}{2}m^2(\varphi_1^2 + \varphi_2^2) - \frac{1}{4}\lambda(\varphi_1^4 + \varphi_2^4) - \frac{1}{2}\lambda\varphi_1^2\varphi_2^2$$

The Lagrangian has a (φ_1, φ_2) symmetry that can be described by rotations in (φ_1, φ_2) space. These can be written in the matrix form

$$\begin{pmatrix} \varphi_1' \\ \varphi_2' \end{pmatrix} = \begin{pmatrix} \cos\alpha & \sin\alpha \\ -\sin\alpha & \cos\alpha \end{pmatrix} \begin{pmatrix} \varphi_1 \\ \varphi_2 \end{pmatrix}$$

That is,

$$\varphi_1' = \cos\alpha\, \varphi_1 + \sin\alpha\, \varphi_2$$
$$\varphi_2' = -\sin\alpha\, \varphi_1 + \cos\alpha\, \varphi_2$$

The minima of the potential lie on a circle that is described by

$$\varphi_1^2 + \varphi_2^2 = \frac{m^2}{\lambda}$$

To break the $U(1)$ symmetry, we think back to the original example of a marble sitting on top of the bowl. We pick out a specific direction. Following the notation of the last example, we denote the minima by v using a subscript to indicate the minima of φ_1 and φ_2. In this case, we pick the minimum at

$$v_1 = \frac{m}{\sqrt{\lambda}} \qquad v_2 = 0 \tag{9.14}$$

Now we rewrite the field. This time we need two fields χ and ψ that fluctuate about the minimum given by Eq. (9.14). We have

$$\varphi = \frac{m}{\sqrt{\lambda}} + \frac{\chi + i\psi}{\sqrt{2}}$$

We have taken

$$\chi = \varphi_1 - \frac{m}{\sqrt{\lambda}} \qquad \psi = \varphi_2 \tag{9.15}$$

The change in the coordinate systems in going from $\begin{pmatrix} \varphi_1 \\ \varphi_2 \end{pmatrix}$ to $\begin{pmatrix} \chi \\ \psi \end{pmatrix}$ amounts to shifting the coordinate system to the right by an amount $\frac{m}{\sqrt{\lambda}}$. In other words, we have shifted the origin to the actual minimum of the potential. Then,

$$\varphi_1^2 = \left(\chi + \frac{m}{\sqrt{\lambda}} \right)^2 = \chi^2 + 2\frac{m}{\sqrt{\lambda}}\chi + \frac{m^2}{\lambda}$$

$$\varphi_2^2 = \psi^2$$

and for the quadratic terms

$$\varphi_1^4 = \left(\chi^2 + 2\frac{m}{\sqrt{\lambda}}\chi + \frac{m^2}{\lambda} \right)^2$$

$$= \chi^4 + 4\frac{m}{\sqrt{\lambda}}\chi^3 + 6\frac{m^2}{\lambda}\chi^2 + 4\frac{m^3}{\sqrt{\lambda^3}}\chi + \frac{m^4}{\lambda^2}$$

and

$$\varphi_2^4 = \psi^4$$

Finally,

$$\frac{1}{2}\lambda\varphi_1^2\varphi_2^2 = \frac{1}{2}\lambda\left(\chi^2\psi^2 + \frac{2m}{\sqrt{\lambda}}\chi\psi^2 + \frac{m^2}{\lambda}\psi^2 \right)$$

Remember that terms of the form φ^n where $n > 2$ represent interaction terms. To get the mass terms, we need to ignore those and look at the *free Lagrangian*. Also remember we can drop constants because they do not contribute to the field equations that would be derived from the Lagrangian. Now, dropping everything except the quadratic terms, the free parts of the potential are

$$\frac{1}{2}m^2\left(\varphi_1^2 + \varphi_2^2 \right)_{\text{free}} = \frac{1}{2}m^2(\chi^2 + \psi^2)$$

$$-\frac{1}{4}\lambda\left(\varphi_1^4 + \varphi_2^4 \right)_{\text{free}} = -\frac{1}{4}\lambda\left(6\frac{m^2}{\lambda}\chi^2 \right)$$

$$-\frac{1}{2}\lambda\left(\varphi_1^2\varphi_2^2 \right)_{\text{free}} = -\frac{1}{2}\lambda\left(\frac{m^2}{\lambda}\psi^2 \right)$$

Notice that

$$\frac{1}{2}m^2\left(\varphi_1^2+\varphi_2^2\right)_{\text{free}}-\frac{1}{2}\lambda\left(\varphi_1^2\varphi_2^2\right)_{\text{free}}=\frac{1}{2}m^2(\chi^2+\psi^2)-\frac{1}{2}\lambda\left(\frac{m^2}{\lambda}\psi^2\right)$$

$$=\frac{1}{2}m^2\chi^2$$

Putting all this together, the free or noninteraction part of the Lagrangian is

$$L_{\text{free}}=\frac{1}{2}(\partial_\mu\chi)^2+\frac{1}{2}(\partial_\mu\psi)^2-\frac{1}{2}m^2\chi^2$$

Hence, spontaneous symmetry breaking of the $U(1)$ symmetry given by a rotation in (φ_1,φ_2) space for the complex field in Eq. (9.11) together with the Lagrangian in Eq. (9.12) gives us a field χ with mass m and a field χ that is *massless*. The mass m is defined by the minima on the circle that we choose to be

$$v_1=\frac{m}{\sqrt{\lambda}}$$

Hence $m=\sqrt{\lambda}v_1$. This example involved the use of scalar fields, so the particles associated with these fields are spin-0 particles. When a massless spin-0 particle appears in a theory due to symmetry breaking, it is called a *Goldstone boson*.

The Higgs Mechanism

In the previous section, we examined spontaneous symmetry breaking by considering the case of a complex field with two real components, and a $U(1)$ symmetry. For simplicity, we've considered global gauge invariance. Now we wish to extend this idea to a more complicated situation in which we include a gauge field A_μ and require *local* gauge invariance under a $U(1)$ transformation. We will start with a massless gauge field A_μ, and show that symmetry breaking results in a massive vector field. We'll see later that in electroweak theory this type of procedure gives rise to the massive vector bosons, the $W^\pm$ and the Z^0. The mechanism involving spontaneous symmetry breaking involving a gauge field and local $U(1)$ invariance is known as the *Higgs mechanism*, named for its founder Peter Higgs who discovered the effect in 1964. A major task ahead for experimentalists when the Large Hadron Collider begins operation in the summer of 2008 will be to find the quanta of the Higgs field $h(x)$.

Once again, recall that $U(1)$ invariance implies that the Lagrangian is invariant under a transformation of the form

$$\varphi \rightarrow \varphi' = e^{-i\theta}\varphi$$

Previously, we considered the case of a global gauge transformation, wherein θ is a scalar, just a number, and not a function of spacetime. Now, however, we extend this concept and let $\theta \rightarrow \theta(x)$, giving us a local gauge transformation. We want the Lagrangian to be invariant under the transformation that is,

$$\varphi(x) \rightarrow \varphi'(x) = e^{-iq\theta(x)}\varphi(x) \tag{9.16}$$

Now q is a number, but $\theta \rightarrow \theta(x)$ is a function of spacetime, meaning that the transformation varies from point to point. Invariance with a local gauge transformation will require the introduction of a *gauge field*. In analogy with electrodynamics, we will see that the requirement of local gauge invariance forces us to use a covariant derivative in order to restore the invariance of the Lagrangian. Define the gauge field as a vector potential A_μ in analogy to electromagnetics. The gauge field must also be invariant under a $U(1)$ gauge transformation, with the transformation assuming the form

$$A_\mu \rightarrow A_\mu' = A_\mu + \partial_\mu \theta \tag{9.17}$$

This is the same θ present in Eq. (9.16), and we note that A_μ is a function of spacetime as well, that is, $A_\mu = A_\mu(x)$.

Up until this point, in this chapter we have been using ordinary derivatives in the Lagrangian. That is, the kinetic terms in the Lagrangian have been of the form

$$\partial_\mu \varphi^\dagger \partial^\mu \varphi$$

In order to have a gauge invariant Lagrangian, considering Eqs. (9.16) and (9.17), we will need to use a covariant derivative. A suitable covariant derivative in this case is

$$D_\mu = \partial_\mu + iqA_\mu \tag{9.18}$$

With this definition, the Lagrangian assumes the form

$$L = D_\mu \varphi^\dagger D^\mu \varphi - V(\varphi^\dagger \varphi) - \frac{1}{4}F_{\mu\nu}F^{\mu\nu} \tag{9.19}$$

Summarizing, this Lagrangian describes a theory that includes a complex scalar field φ and a *massless* gauge field A_μ. With a gauge field in the theory, terms like $F_{\mu\nu}F^{\mu\nu}$ represent kinetic energy terms. In analogy with electrodynamics, the following definition is used.

$$F_{\mu\nu} = \partial_\mu A_\nu - \partial_\nu A_\mu \qquad (9.20)$$

If a mass term was present, we would see a contraction on the gauge field $A_\mu A^\mu$ that is analogous to the quadratic mass terms of the scalar field. Since the Lagrangian given in Eq. (9.19) has no terms of this form, the gauge field is massless. What we will see in a moment is that a certain type of symmetry breaking will cause the gauge field A_μ to acquire mass. This is the essence of the Higgs mechanism.

The potential in Eq. (9.19) is given by

$$V(\varphi^\dagger \varphi) = \frac{m^2}{2v^2}(\varphi^\dagger \varphi - v^2)^2 \qquad (9.21)$$

where we have included a constant v in anticipation of the search for a minimum. In this case, v is the minimum for the theory when the symmetry is *unbroken*.

Now we proceed as in the previous examples. This time, we seek the minimum potential energy when both the gauge field A_μ and potential V vanish. When the symmetry is unbroken, the minimum of the potential, the vacuum state, is at

$$|\varphi|^2 = v$$

where v is some real number. But we can do a gauge transformation since

$$\varphi'^\dagger \varphi' = (\varphi^\dagger e^{iq\theta(x)})(e^{-iq\theta(x)}\varphi) = \varphi^\dagger \varphi = |\varphi|^2 = v$$

So a local gauge transformation gives us the same minimum. And if the field is complex, there are infinity vacuum states. Since the gauge transformation gives the same minimum, we have a symmetry.

How can we break the symmetry? There is a hint in the fact that the minimum is obtained with the squared amplitude of a complex field. We can break the symmetry by requiring that the field be *real*. So the situation we have is

- The value v is the minimum of the potential when the symmetry is unbroken.
- We seek a gauge transformation that gives the field in terms of fluctuations about v.

This can be done when the vacuum v is perturbed by a real field $h(x)$.

$$\varphi \to \varphi' = v + \frac{h(x)}{\sqrt{2}} \qquad (9.22)$$

The field $h(x)$ is the Higgs field. Since the field is real, $\varphi^\dagger = \varphi$ and the potential becomes

$$V = \frac{m^2}{2v^2} \left[\sqrt{2}vh(x) + \frac{h^2(x)}{2} \right]^2$$

$$= m^2 h^2 + \frac{m^2 h^2}{2v^2} \left(\sqrt{2}vh + \frac{h^2}{4} \right) \qquad (9.23)$$

Using Eq. (9.22) together with the definition of the covariant derivative in Eq. (9.18), we get

$$D^\mu \varphi' = (\partial^\mu + iqA'^\mu)\left(v + \frac{h}{\sqrt{2}} \right)$$

$$= \frac{1}{\sqrt{2}} \partial^\mu h + iqvA'^\mu + \frac{iqh}{\sqrt{2}} A'^\mu$$

Similarly, we find that

$$D_\mu \varphi' = (\partial_\mu - iqA')\left(v + \frac{h}{\sqrt{2}} \right)$$

$$= \frac{1}{\sqrt{2}} \partial_\mu h - iqvA'_\mu - \frac{iqh}{\sqrt{2}} A'_\mu$$

Therefore,

$$D_\mu \varphi' D^\mu \phi' = \left(\frac{1}{\sqrt{2}} \partial_\mu h - iqvA'_\mu - \frac{iqh}{\sqrt{2}} A'_\mu \right)\left(\frac{1}{\sqrt{2}} \partial^\mu h + iqvA'^\mu + \frac{iqh}{\sqrt{2}} A'^\mu \right)$$

$$= \frac{1}{2} \partial_\mu h \partial^\mu h + q^2 v^2 A'_\mu A'^\mu + \sqrt{2} q^2 vhA'_\mu A'^\mu + \frac{q^2 h}{2} A'_\mu A'^\mu$$

We can put this together with the expression we obtained for the potential to obtain the full Lagrangian in the case of the gauge transformation in Eq. (9.22) where we chose the field to be real. Let's drop the primes on the vector potential terms to simplify writing. We find that

$$L = D_\mu \varphi^\dagger D^\mu \varphi - V(\varphi^\dagger \varphi) - \frac{1}{4} F_{\mu\nu} F^{\mu\nu}$$

$$= \frac{1}{2} \partial_\mu h \partial^\mu h + q^2 v^2 A_\mu' A'^\mu + \sqrt{2} q^2 v h A_\mu' A'^\mu + \frac{q^2 h}{2} A_\mu' A'^\mu$$

$$- m^2 h^2 - \frac{m^2 h^2}{2 v^2} \left(\sqrt{2} v h + \frac{h^2}{4} \right)$$

This Lagrangian has several components. The first part we will take a look at is the free part of the Lagrangian involving the Higgs field $h(x)$. This is

$$L_{\text{free}}^h = \frac{1}{2} \partial_\mu h \partial^\mu h - m^2 h^2 \tag{9.24}$$

By now, this should look very familiar. It's a Klein-Gordon equation type Lagrangian for a scalar field $h(x)$ with mass

$$\sqrt{2}\, m \tag{9.25}$$

So in the example we've done here, the Higgs field is a scalar field, a spin-0 boson with mass $\sqrt{2}m$. Now let's look at some of the other terms. Next we have a free Lagrangian for the gauge field. This is given by

$$L_{\text{free}}^B = -\frac{1}{4} F_{\mu\nu} F^{\mu\nu} + q^2 v^2 A_\mu' A'^\mu$$

This is a remarkable result. The kinetic term $-\frac{1}{4} F_{\mu\nu} F^{\mu\nu}$ was present in the original Lagrangian in Eq. (9.19), but before symmetry breaking the gauge field was massless. By choosing a real field, that is, a perturbation about the unbroken vacuum v, we have picked up a mass term which is given by

$$q^2 v^2 A_\mu' A'^\mu$$

We can determine the mass by comparison with a mass term that would appear in a Klein-Gordon type Lagrangian

$$\frac{1}{2} M^2 \phi^2$$

for some field ϕ. Hence, comparing the two terms, we see that symmetry breaking has given rise to a vector boson with mass

$$M = \sqrt{2}\, qv \qquad (9.26)$$

The remaining terms in the Lagrangian are interaction terms. The first term we can write down includes the self-interaction terms for the Higgs field as shown here.

$$L_{\text{int}}^{h} = -\frac{m^2 h^2}{2v^2}\left(\sqrt{2}vh + \frac{h^2}{4}\right)$$

And finally, there is an interaction Lagrangian representing coupling between the Higgs field h and the gauge field A_{μ}. This is

$$L_{\text{int}}^{\text{coup}} = q^2 A'_{\mu} A'^{\mu}\left(\sqrt{2}vh + \frac{1}{2}h^2\right)$$

Summary

In this chapter we learned about spontaneous symmetry breaking, a process which leads to the appearance of massive particles in the Lagrangian. The procedure works by considering a Lagrangian with some vacuum state. The system is then reconsidered by breaking the symmetry, leading us to a new vacuum state. Gauge invariance leads to the appearance of new particles. For a scalar theory, a massive Goldstone boson appears. When we combine a complex scalar theory with a massless gauge field, breaking the symmetry by forcing the field to be a real fluctuation about the unbroken vacuum v leads to the appearance of a massive scalar field called the Higgs field, and the gauge field acquires mass. The Higgs field and the gauge field are coupled through an interaction Lagrangian.

Quiz

1. Suppose that $L = \frac{1}{2}(\partial_{\mu}\varphi)^2 + \cosh(b\varphi)$. Does this describe a massive or massless particle?

2. A mass term appears in the Lagrangian

   (a) As a squared scalar multiplying the field

   (b) As a scalar term multiplying the field squared

(c) As a scalar term multiplying the field to the fourth power

(d) Must be put in by hand

Consider a Lagrangian with a potential given by

$$V = \frac{\lambda}{4}\left(\varphi^*\varphi - \frac{\mu^2}{\lambda}\right)^2$$

3. Let $\varphi \to \psi(x)e^{i\theta(x)}$. Write down the form of the Lagrangian.

4. Identify the mass term in the resulting Lagrangian.

5. Is there a self-interaction term?

CHAPTER 10

Electroweak Theory

In this chapter we will explore the electroweak part of the standard model of particle physics, which unifies the electromagnetic and weak interactions. The gauge group that does this is

$$SU(2) \otimes U(1)$$

The weak interactions are mediated by the $SU(2)$ gauge bosons, which includes the charged $W^{\pm}$ and the neutral Z^0. The $U(1)$ sector of the interaction is the electromagnetic interaction, which is mediated by the massless photon. The theory that describes the electroweak interaction is known as the *Weinberg-Salam* model, after the two codiscoverers of the theory. They shared the Nobel Prize with Sheldon Glashow in 1979 for the development of this theory and their prediction of the $W^{\pm}$ and Z^0 masses.

The Higgs field is introduced into the model causing spontaneous symmetry breaking. This leads the electron and its heavy partners, the muon (m) and the tau (τ), to acquire mass. In addition, the gauge bosons $W^{\pm}$ and Z^0 acquire mass, but the photon remains massless. So far so good—the results are in good agreement with experiment. However, the Weinberg-Salam model also predicts that neutrinos are massless. Recent experimental evidence indicates that while their mass may be small (<1 eV), neutrinos probably *do* have mass. This problem is currently one of the great outstanding problems[1] in particle physics, and solving the neutrino mass problem may lead to new physics beyond the standard model.

In this chapter, we will focus on the electroweak interactions of leptons, and will leave out hadron interactions.

Right- and Left-Handed Spinors

Let's briefly review the concept of left- and right-handed spinors. We write a Dirac spinor as a two-component object, with the top component being the right-handed spinor and the lower component being the left-handed spinor.

$$\psi = \begin{pmatrix} \psi_R \\ \psi_L \end{pmatrix} \tag{10.1}$$

Each component, ψ_R and ψ_L, is itself a two-component object. We can pick out the left- and right-handed components of a Dirac field ψ by using an operator composed of the identity and the γ_5 matrix. Refreshing our memory, the γ_5 matrix is a 4×4 matrix given by

$$\gamma_5 = \begin{pmatrix} 1 & 0 \\ 0 & -1 \end{pmatrix} = \begin{pmatrix} 1 & 0 & 0 & 0 \\ 0 & 1 & 0 & 0 \\ 0 & 0 & -1 & 0 \\ 0 & 0 & 0 & -1 \end{pmatrix} \tag{10.2}$$

Now let's see how we can pick out the left- and right-handed components of ψ. First we write

$$\frac{1}{2}(1 - \gamma_5) = \frac{1}{2}\left[\begin{pmatrix} 1 & 0 \\ 0 & 1 \end{pmatrix} - \begin{pmatrix} 1 & 0 \\ 0 & -1 \end{pmatrix} \right] = \begin{pmatrix} 0 & 0 \\ 0 & 1 \end{pmatrix}$$

---

[1]Neutrinos with mass could solve the long-standing solar neutrino deficit. The nuclear physics of our sun is well-understood and in agreement with measurement except for the number of neutrinos. Terrestrial measurements by different laboratories agree that the solar neutrino flux is one-third of what is expected, a problem that has been solved by neutrino oscillations.

Hence,

$$\frac{1}{2}(1-\gamma_5)\psi = \begin{pmatrix} 0 & 0 \\ 0 & 1 \end{pmatrix}\begin{pmatrix} \psi_R \\ \psi_L \end{pmatrix} = \begin{pmatrix} 0 \\ \psi_L \end{pmatrix} = \psi_L$$

Similarly, we have

$$\frac{1}{2}(1+\gamma_5)\psi = \begin{pmatrix} 1 & 0 \\ 0 & 0 \end{pmatrix}\begin{pmatrix} \psi_R \\ \psi_L \end{pmatrix} = \begin{pmatrix} \psi_R \\ 0 \end{pmatrix} = \psi_R$$

Also notice that we can write the Dirac field as

$$\psi = \begin{pmatrix} \psi_R \\ \psi_L \end{pmatrix} = \begin{pmatrix} 0 \\ \psi_L \end{pmatrix} + \begin{pmatrix} \psi_R \\ 0 \end{pmatrix} = \psi_L + \psi_R$$

A Massless Dirac Lagrangian

We begin with the standard Dirac Lagrangian, setting the mass term to 0. This gives

$$L = i\bar{\psi}\gamma^\mu\partial_\mu\psi \tag{10.3}$$

where as usual

$$\bar{\psi} = \psi^\dagger\gamma^0$$

We wish to split up the Lagrangian into two parts, one for the left-handed spinor and one for the right-handed spinor. This is actually very straightforward. Proceeding we have

$$\begin{aligned} L &= i\bar{\psi}\gamma^\mu\partial_\mu\psi \\ &= i(\bar{\psi}_L + \bar{\psi}_R)\gamma^\mu\partial_\mu(\psi_L + \psi_R) \\ &= i\bar{\psi}_L\gamma^\mu\partial_\mu\psi_L + i\bar{\psi}_R\gamma^\mu\partial_\mu\psi_R + i(\bar{\psi}_L\gamma^\mu\partial_\mu\psi_R + \bar{\psi}_R\gamma^\mu\partial_\mu\psi_L) \end{aligned}$$

The last term actually vanishes. This is because

$$\bar{\psi}_L \gamma^\mu \partial_\mu \psi_R = \frac{1}{2}\left(\frac{1-\gamma_5}{2}\right)\bar{\psi}\gamma^\mu \partial_\mu \left(\frac{1+\gamma_5}{2}\right)\psi$$

$$= \frac{1}{4}\left(1-\gamma_5+\gamma_5-\gamma_5^2\right)\bar{\psi}\gamma^\mu \partial_\mu \psi$$

$$= \frac{1}{4}\left(1-\gamma_5^2\right)\bar{\psi}\gamma^\mu \partial_\mu \psi$$

But since

$$\gamma_5^5 = \begin{pmatrix} 1 & 0 \\ 0 & -1 \end{pmatrix}\begin{pmatrix} 1 & 0 \\ 0 & -1 \end{pmatrix} = I$$

the mixed terms vanish. So we are left with

$$L = i\bar{\psi}_L \gamma^\mu \partial_\mu \psi_L + i\bar{\psi}_R \gamma^\mu \partial_\mu \psi_R \tag{10.4}$$

And the Lagrangian separates nicely into left- and right-handed parts. In the case of electroweak theory, there is an asymmetry between left- and right-handed weak interactions. As a result, the actual Lagrangian used will reflect this.

Leptonic Fields of the Electroweak Interactions

For reasons as of yet unclear, fundamental particles belong to one of three families. The distinction is one of mass; otherwise particles within a family behave in a similar fashion (have the same charge and spin, for example). When considering just leptons, the fields of the electroweak interaction consists of the electron (e), muon (μ), and tau (τ), together with their corresponding neutrinos. These are the three families of leptons. In short we can write

$$L = L_e + L_\mu + L_\tau$$

However, the muon and tau are just heavier duplicates of the electron, so we can learn everything we need to about electroweak theory by just focusing on the electron.[2]

---

[2]So what are the two other families? Mysteries like this are driving physicists to look at options like string theory.

The electron field and its associated neutrino field are combined together into a two-component object. Considering left-handed components only, we have the left-handed spinor

$$\psi_L = \begin{pmatrix} v_e \\ e_L \end{pmatrix}$$

(10.5)

where v_e is the electron neutrino and e_L is a left-handed electron field. The left- and right-handed electron fields are related to the electron e in the usual way.

$$e_L = \left(\frac{1-\gamma_5}{2} \right) e \qquad e_R = \left(\frac{1+\gamma_5}{2} \right) e$$

(10.6)

If we take the mass of the neutrino to be 0, that is, $m_{v_e} = 0$, then there is only a left-handed component of the neutrino field. Since the field is entirely left-handed, it satisfies the equation

$$\left(\frac{1-\gamma_5}{2} \right) v_e = v_e$$

(10.7)

With no right-handed component of the neutrino field, we can define

$$\psi_R = \begin{pmatrix} 0 \\ e_R \end{pmatrix}$$

(10.8)

When considering only electrons, the Lagrangian describing the Dirac fields of the electroweak interaction can therefore be written in the form stated above, that is,

$$L = i\bar{\psi}_L \gamma^\mu \partial_\mu \psi_L + i\bar{\psi}_R \gamma^\mu \partial_\mu \psi_R$$

If we wanted to consider the full theory for leptons, we would simply add terms for the muon and tau, which would be identical in form to that of the electron.

Charges of the Electroweak Interaction

Charged current interactions work as follows. Charged currents couple to left-handed particles and to right-handed antiparticles. In electroweak theory, there exist three types of charge.

Description	Label
Electric Charge	Q
Weak Isospin	I
Weak Hypercharge	Y

These charges are related by the *Gell-Mann-Nishijima relation*

$$Q = I^3 + \frac{Y}{2} \tag{10.9}$$

where I^3 is the third component of weak isospin. The neutrino is assigned an isospin of

$$I_\nu^3 = +\frac{1}{2}$$

while a left-handed electron has

$$I_e^3 = -\frac{1}{2}$$

The right-handed electron field has

$$I_R = 0$$

For a left-handed spinor, $Y = -1$, while for a right-handed spinor, $Y = -2$. Hence overall the charges for left- and right-handed spinors in electroweak theory area are

$$I_L^3 = +\frac{1}{2} \qquad Y_L = -1 \qquad Q_L = 0 \text{ (neutrino)}$$

$$I_L^3 = -\frac{1}{2} \qquad Y_L = -1 \qquad Q_L = -1 \text{ (electron—left handed)} \tag{10.10}$$

$$I_R^3 = 0 \qquad Y_L = -2 \qquad Q_R = -1 \text{ (electron—right handed)}$$

The weak hypercharge Y and weak isospin charge I are independent, hence

$$[Y, \vec{I}] = 0$$

In the next section we will see that there are four gauge fields denoted by W_μ^1, W_μ^2, W_μ^3, and B_μ that correspond to the weak isospin charge I and weak hypercharge Y, respectively. If a particle has a given type of charge, then it can interact with the field associated with that charge, and the value of the charge determines the strength of that interaction. Since the neutrino participates in weak interactions but does not interact with the photon, it has quantum numbers $I_L^3 = +\frac{1}{2}, Y_L = -1$, and $Q_L = 0$. A left-handed electron participates in the weak interaction and interacts with the photon and accordingly has all nonzero charges.

A right-handed electron is a little different. It has $I_R^3 = 0$, $Y_R = -2$, and $Q_R = -1$. As we will see below, the isospin charge allows interactions with the gauge bosons W_μ. A right-handed electron does not interact with the gauge field. It will, however, interact with the B_μ field, and in fact does so with twice the strength of a left-handed electron. The right-handed electron does interact with the photon since $Q_R = -1$.

Unitary Transformations and the Gauge Fields of the Theory

Next we consider the possible symmetries of the theory and proceed to introduce the gauge bosons. As mentioned above, electroweak theory has two independent symmetries $SU(2)$ and $U(1)$; we call the combination

$$SU(2) \otimes U(1)$$

The $SU(2)$ symmetry leads to three gauge bosons as mentioned here.

$$SU(2): W_\mu^1, W_\mu^2, W_\mu^3$$

The conservation of the weak hypercharge Y actually arises from invariance under a $U(1)$ transformation. Hence, there is an additional gauge field associated with $U(1)$ invariance. We denote this field as B_μ. Summarizing:

$$U(1): B_\mu$$

After we introduce the gauge fields, the Lagrangian will be expanded as

$$L = L_{\text{leptons}} + L_{\text{gauge}}$$

Let's take a look at the $U(1)$ transformation first. It is clear by looking at the Lagrangian that it is invariant under a standard $U(1)$ transformation on the right-handed spinor

$$\psi_R \rightarrow \psi_R' = e^{i\beta} \psi_R$$

where β is a scalar. This transformation does not change the Lagrangian

$$L \to L' = \bar{\psi}_L \gamma^\mu \partial_\mu \psi_L + i\bar{\psi}_R e^{-i\beta} \gamma^\mu \partial_\mu e^{i\beta} \psi_R + i\bar{L} \gamma^\mu \partial_\mu L$$
$$= i\bar{\psi}_L \gamma^\mu \partial_\mu \psi_L + i\bar{\psi}_R \gamma^\mu \partial_\mu \psi_R = L$$

So clearly the Lagrangian is invariant under $\psi_R \to \psi'_R = e^{i\beta} \psi_R$. Naively, one would expect the $U(1)$ transformation to be the same for the left-handed field, $\psi_L \to \psi'_L = e^{i\beta} \psi_L$, but it is not. We already know that the left- and right-handed fields have different weak hypercharge, so we expect them to transform differently. The correct transformation for the left-handed field is of the form

$$\psi_L \to \psi'_L = e^{in\beta} \psi_L$$

Since $Y_R = -2$, but $Y_L = -1$, the left-handed field interacts at half the strength, so $n = \frac{1}{2}$ and the correct transformation is

$$\psi_L \to \psi'_L = e^{i\beta/2} \psi_L$$

We can arrange the neutrino, left- and right-handed electrons into a single object as

$$\begin{pmatrix} v_e \\ e_L \\ e_R \end{pmatrix}$$

Then the $U(1)$ transformation can be written in nice matrix form as

$$\begin{pmatrix} v_e \\ e_L \\ e_R \end{pmatrix} \to \begin{pmatrix} v_e' \\ e_L' \\ e_R' \end{pmatrix} = \begin{pmatrix} e^{i\beta/2} & 0 & 0 \\ 0 & e^{i\beta/2} & 0 \\ 0 & 0 & e^{i\beta} \end{pmatrix} \begin{pmatrix} v_e \\ e_L \\ e_R \end{pmatrix} \tag{10.11}$$

Now, given a gauge field B_μ, we define the field strength tensor $f_{\mu\nu}$ where

$$f_{\mu\nu} = \partial_\mu B_\nu - \partial_\nu B_\mu \tag{10.12}$$

Hence, the gauge field B_μ is included in the Lagrangian with the addition of the term

$$L_B = -\frac{1}{4} f_{\mu\nu} f^{\mu\nu} \tag{10.13}$$

We want to introduce this term but preserve of the action under variation $\delta S = 0$. Once more the gauge field B_μ forces us to introduce extra terms into the derivative in order to maintain covariance. This is possible if we take

$$\partial_\mu \to \partial_\mu + \frac{i g_B}{2} B_\mu$$

where g_B is a coupling constant associated with the gauge field B_μ. In a similar fashion, we will define a field strength tensor associated with the gauge fields W_μ^1, W_μ^2, and W_μ^3. First let's consider $SU(2)$ transformations.

We now reintroduce the Pauli matrices in anticipation of including $SU(2)$ as

$$\tau_1 = \begin{pmatrix} 0 & 1 \\ 1 & 0 \end{pmatrix} \qquad \tau_2 = \begin{pmatrix} 0 & -i \\ i & 0 \end{pmatrix} \qquad \tau_3 = \begin{pmatrix} 1 & 0 \\ 0 & -1 \end{pmatrix} \tag{10.14}$$

Since these are just the Pauli matrices, we have the $SU(2)$ algebra.

$$[\tau_i, \tau_j] = 2i\varepsilon_{ijk}\tau_k \tag{10.15}$$

The τ_i generators define weak isospin space. We now use them to consider a $SU(2)$ transformation of the form

$$U(\alpha) = \exp(i\alpha_j \tau_j/2) \tag{10.16}$$

The right-handed electron spinor e_R is invariant under the $SU(2)$ transformation.

$$\psi_R \to \psi_R' = U(\alpha)\psi_R = \psi_R$$

However, the left-handed spinor transforms in the usual way as

$$\psi_L \to \psi_L' = e^{-i(\tau \cdot \alpha)/2}\psi_L$$

The reason for these transformation properties is that the right-handed electron e_R does not carry any weak isospin charge ($\vec{I}_R = I_R^3 = 0$) which is associated with the $SU(2)$ transformation—hence it does not couple to the W_μ^1, W_μ^2, and W_μ^3 fields.

The electron neutrino and left-handed electron state do carry isospin charge, so we need to apply the $SU(2)$ transformation in that case. In matrix form, the $SU(2)$ transformation can be written as

$$
\begin{pmatrix} v_e \\ e_L \\ e_R \end{pmatrix} \rightarrow \begin{pmatrix} v_e' \\ e_L' \\ e_R' \end{pmatrix} = \begin{pmatrix} e^{-i(\tau \cdot \alpha)/2} & 0 & 0 \\ 0 & e^{-i(\tau \cdot \alpha)/2} & 0 \\ 0 & 0 & 1 \end{pmatrix} \begin{pmatrix} v_e \\ e_L \\ e_R \end{pmatrix} \tag{10.17}
$$

Now let's consider the field strength tensor corresponding to the gauge fields W_μ^1, W_μ^2, and W_μ^3. It assumes the form

$$
F_{\mu\nu}^\ell = \partial_\mu W_\nu^\ell - \partial_\nu W_\mu^\ell - g_W \varepsilon^{\ell mn} W_\mu^m W_\nu^n \tag{10.18}
$$

We add the field tensor to the Lagrangian by summing up $F_{\mu\nu}^\ell F^{\ell,\mu\nu}$ over $\ell = 1,2,3$ to include each of the gauge fields W_μ^1, W_μ^2, and W_μ^3. That is, we take the trace and can write the contribution to the Lagrangian as

$$
L_W = -\frac{1}{8} Tr(F_{\mu\nu} F^{\mu\nu}) \tag{10.19}
$$

To keep the derivative covariant, now we need to add an extra term to account for the presence of Eq. (10.18) in the Lagrangian. This is done by adding the following term to the derivative:

$$
ig_W \frac{\vec{\tau}}{2} \cdot \vec{W}_\mu = \frac{ig_W}{2} \left(\tau_1 W_\mu^1 + \tau_2 W_\mu^2 + \tau_3 W_\mu^3 \right)
$$

Since the right-handed lepton field does not participate in the interaction involving weak isospin, this term is not added to the derivative in that case. We only add the term to account for the presence of the gauge field B_μ. Hence,

$$
i\bar{\psi}_R \gamma^\mu \partial_\mu \psi_R \rightarrow i\bar{\psi}_R \gamma^\mu \left(\partial_\mu + \frac{ig_B}{2} B_\mu \right) \psi_R
$$

For the left-handed field, we have

$$
i\bar{\psi}_L \gamma^\mu \partial_\mu \psi_L \rightarrow i\bar{\psi}_L \gamma^\mu \left(\partial_\mu + \frac{ig_B}{2} B_\mu + ig_W \frac{\vec{\tau}}{2} \cdot \vec{W}_\mu \right) \psi_L
$$

So the total leptonic portion of the Lagrangian is

$$L_{\text{Lepton}} = i\bar{\psi}_R \gamma^\mu \left(\partial_\mu + \frac{ig_B}{2} B_\mu \right) \psi_R + i\bar{\psi}_L \gamma^\mu \left(\partial_\mu + \frac{ig_B}{2} B_\mu + ig_W \frac{\vec{\tau}}{2} \cdot \vec{W}_\mu \right) \psi_L$$

The total Lagrangian includes the gauge field Lagrangians as

$$\begin{aligned} L &= L_{\text{lepton}} + L_{\text{gauge}} \\ &= i\bar{\psi}_R \gamma^\mu \left(\partial_\mu + \frac{ig_B}{2} B_\mu \right) \psi_R + i\bar{\psi}_L \gamma^\mu \left(\partial_\mu + \frac{ig_B}{2} B_\mu + ig_W \frac{\vec{\tau}}{2} \cdot \vec{W}_\mu \right) \psi_L \\ &\quad - \frac{1}{4} f_{\mu\nu} f^{\mu\nu} - \frac{1}{8} Tr(F_{\mu\nu} F^{\mu\nu}) \end{aligned} \qquad (10.20)$$

When you step back and look at the Lagrangian, the asymmetry between the chiral fields strikes a note of discord. While the Standard Model has been a great success, it now leaves theorists with many questions that will need to be solved with physics beyond the Standard Model.

Weak Mixing or Weinberg Angle

In the following sections we will see that it is convenient to relate the coupling constants g_B and g_W in the following way:

$$\tan\theta_W = \frac{g_W}{g_B} \qquad (10.21)$$

The angle θ_W is often called the *Weinberg angle*. It is also sometimes called the *weak mixing angle*. The reason is that it mixes the gauge fields to give

$$\begin{aligned} A_\mu &= B_\mu \cos\theta_W + W_\mu^3 \sin\theta_W \\ Z_\mu &= -B_\mu \sin\theta_W + W_\mu^3 \cos\theta_W \end{aligned} \qquad (10.22)$$

We can view this as a rotation, writing the relationship in matrix form:

$$\begin{pmatrix} A_\mu \\ Z_\mu \end{pmatrix} = R(\theta_W) \begin{pmatrix} B_\mu \\ W_\mu^3 \end{pmatrix} = \begin{pmatrix} \cos\theta_W & \sin\theta_W \\ -\sin\theta_W & \cos\theta_W \end{pmatrix} \begin{pmatrix} B_\mu \\ W_\mu^3 \end{pmatrix}$$

The gauge field A_μ is nothing other than the electromagnetic vector potential which couples the photon to the theory. We will explore this in more detail after we introduce the Higgs mechanism.

Symmetry Breaking

At this point we have put together a theory describing two leptons, the electron and its corresponding neutrino, and four gauge bosons. All the particles described so far are massless. It's time to introduce mass into the theory and this will be done using the symmetry breaking methods of Chap. 9. We will introduce a Higgs field that will force the gauge bosons to acquire mass. In this section we will also introduce the photon field explicitly (besides talking about charge) so that we can present a unified picture of the electroweak interaction.

Following the example of Chap. 9, we introduce the Higgs field as a scalar (spin-0) field. However this time it is a two-component object as shown here.

$$\varphi = \begin{pmatrix} \varphi^A \\ \varphi^B \end{pmatrix} \tag{10.23}$$

Each component is a complex scalar field and is written as

$$\varphi^A = \frac{\varphi_3 + i\varphi_4}{\sqrt{2}}$$
$$\varphi^B = \frac{\varphi_1 + i\varphi_2}{\sqrt{2}} \tag{10.24}$$

So the Higgs field is really composed of four real, scalar fields. We see that

$$\varphi^\dagger \varphi = (\varphi^A)^\dagger \varphi^A + (\varphi^B)^\dagger \varphi^B$$
$$= \frac{1}{2}\left(\varphi_1^2 + \varphi_2^2 + \varphi_3^2 + \varphi_4^2\right)$$

The Higgs field carries charges of the weak interaction, specifically

$$Y_\varphi = +1 \qquad I_\varphi = 1/2 \tag{10.25}$$

The $SU(2)$ transformation is given in Eq. (10.16), which we restate here.

$$U(\alpha) = \exp(i\alpha_j \tau_j /2)$$

We have a gauge freedom that can be exploited to simplify the form of the Higgs field. This can be done by insisting that each of the "angles" are functions of space-time $\alpha_j = \alpha_j(x)$; this is more than whimsy. It is because we want a local symmetry.

Here is how the gauge freedom helps us. We can choose the $\alpha_j = \alpha_j(x)$ in such a way that

$$\varphi^A = 0$$

and

$$\varphi^B = \varphi_0 + \frac{h(x)}{\sqrt{2}}$$

This leaves us with a particularly simple and useful form of the Higgs field as

$$\varphi = \begin{pmatrix} 0 \\ \varphi_0 + \dfrac{h(x)}{\sqrt{2}} \end{pmatrix} \qquad (10.26)$$

The parameter φ_0 and field $h(x)$ are both real. The φ_0 parameter allows us to break the symmetry. Instead of taking the ground state to be $\varphi \to 0$, we take it to be

$$\varphi_G = \begin{pmatrix} 0 \\ \varphi_0 \end{pmatrix}$$

The Higgs field will make its appearance in the Lagrangian Eq. (10.20) in several ways. It will do so through a potential $V = V(\varphi^\dagger \varphi)$, kinetic energy terms $D_\mu \varphi D^\mu \varphi$, and via interaction terms that couple the Higgs field to the electron and gauge bosons, giving them mass. Let's use Eq. (10.26) and write the form of the potential. First

$$\varphi^\dagger \varphi = \begin{pmatrix} 0 & \varphi_0 + \dfrac{h}{\sqrt{2}} \end{pmatrix} \begin{pmatrix} 0 \\ \dfrac{h}{\sqrt{2}} \end{pmatrix}$$

$$= \varphi_0^2 + \sqrt{2}\varphi_0 h + \frac{1}{2}h^2$$

The potential is

$$V(\varphi^\dagger \varphi) = \mu^2 \varphi^\dagger \varphi + \lambda(\varphi^\dagger \varphi)^2 \qquad (10.27)$$

If $\mu^2 > 0$, then the minimum of the potential is located at $\varphi = 0$. Following the procedure of Chap. 9, we take $\mu^2 < 0$ to break the symmetry, since this causes the Higgs field to attain a vacuum expectation value.

What is the new minimum? As usual, we apply elementary calculus to get

$$\frac{\partial V}{\partial \varphi} = \mu^2 \varphi^\dagger + 2\lambda(\varphi^\dagger \varphi) = 0$$

That is,

$$\varphi_{min} = -\frac{\mu^2}{2\lambda}$$

Defining

$$\varphi_0^{\;2} = -\frac{\mu^2}{2\lambda}$$

We arrive at the result $\varphi_G = \begin{pmatrix} 0 \\ \varphi_0 \end{pmatrix}$. The Higgs field has a mass given by

$$m_h = \sqrt{-2\mu^2} \tag{10.28}$$

A review of Chap. 9 will help clarify this result.

Giving Mass to the Lepton Fields

A Dirac Lagrangian consisting of left- and right-handed fields with mass m is written as

$$L = i\bar{\psi}_L \gamma^\mu \partial_\mu \psi_L + i\bar{\psi}_R \gamma^\mu \partial_\mu \psi_R - m(\bar{\psi}_L \psi_R + \bar{\psi}_R \psi_L) \tag{10.29}$$

Hence a mass term in the Lagrangian is of the form

$$-m(\bar{\psi}_L \psi_R + \bar{\psi}_R \psi_L) \tag{10.30}$$

where m is a scalar (a number, not a function dependent on spacetime). In Weinberg-Salam theory, an interaction term (known as the *Yukawa* term) is introduced that couples the matter fields to the Higgs field. The Yukawa coupling G_e

which defines the strength of the interaction between the Higgs field and the electron-lepton fields. The interaction Lagrangian is

$$L_{int} = -G_e(\bar{\psi}_L \varphi \psi_R + \bar{\psi}_R \varphi^\dagger \psi_L) \qquad (10.31)$$

Let's look at each term.

$$\bar{\psi}_L \varphi = (\bar{v}_e \quad \bar{e}_L)\begin{pmatrix} \varphi^A \\ \varphi^B \end{pmatrix} = \bar{v}_e \varphi^A + \bar{e}_L \varphi^B$$

Now, using the gauge choice that led to the form of the Higgs field in Eq. (10.26), the neutrino term drops out—a key step—and we have

$$\bar{\psi}_L \varphi = (\bar{v}_e \quad \bar{e}_L)\begin{pmatrix} \varphi^A \\ \varphi^B \end{pmatrix} = \bar{v}_e \varphi^A + \bar{e}_L \varphi^B = \bar{e}_L\left(\varphi_0 + \frac{h(x)}{\sqrt{2}}\right)$$

Using $\psi_R = \begin{pmatrix} 0 \\ e_R \end{pmatrix}$ we have

$$\bar{\psi}_L \varphi \psi_R = \bar{e}_L\left(\varphi_0 + \frac{h(x)}{\sqrt{2}}\right)e_R \qquad (10.32)$$

For the next term in Eq. (10.31), we have

$$\bar{\psi}_R \varphi^\dagger \psi_L = (0 \quad \bar{e}_R)\begin{pmatrix} 0 & \varphi_0 + \frac{h(x)}{\sqrt{2}} \end{pmatrix}\begin{pmatrix} v_e \\ e_L \end{pmatrix}$$

$$= (0 \quad \bar{e}_R)\left(\varphi_0 + \frac{h(x)}{\sqrt{2}}\right)e_L$$

$$= \bar{e}_R\left(\varphi_0 + \frac{h(x)}{\sqrt{2}}\right)e_L$$

Once again, the neutrino term drops out. Using these results the interaction Lagrangian becomes

$$L_{int} = -G_e(\bar{\psi}_L \varphi \psi_R + \bar{\psi}_R \varphi^\dagger \psi_L)$$

$$= -G_e\left[\bar{e}_L\left(\varphi_0 + \frac{h(x)}{\sqrt{2}}\right)e_R + \bar{e}_R\left(\varphi_0 + \frac{h(x)}{\sqrt{2}}\right)e_L\right]$$

$$= -G_e\varphi_0(\bar{e}_L e_R + \bar{e}_R e_L) - G_e\frac{h(x)}{\sqrt{2}}(\bar{e}_L e_R + \bar{e}_R e_L)$$

Looking for a mass, we ignore the second term because a mass term in the Lagrangian is multiplied by an overall scalar (a number). The second term is multiplied by the Higgs field $h(x)$. So the mass term in the Lagrangian is

$$L_{\text{mass}} = -G_e \varphi_0 (\overline{e}_L e_R + \overline{e}_R e_L) \tag{10.33}$$

Comparison with Eq. (10.30) leads us to the mass of the electron

$$m = G_e \varphi_0 \tag{10.34}$$

The conclusion is that the interaction of the Higgs field with the electron field leaves the neutrino massless and gives the electron the mass defined in Eq. (10.34). So while these are instructive results, they do not seem to completely reflect what we see in nature as the neutrinos described here have no mass. As noted earlier experimental results indicate that neutrinos have a small, but nonzero mass. This indicates that the theory is incomplete.

Gauge Masses

Now we will see how the Higgs mechanism gives mass to the gauge bosons. The gauge bosons will acquire a mass through the action of the covariant derivative on the Higgs field. First let's set up some notation that is frequently used. The fields W_μ^1 and W_μ^2 are electrically charged, and can be combined into the physical fields as shown here.

$$W_\mu^+ = \frac{W_\mu^1 - iW_\mu^2}{\sqrt{2}} \tag{10.35}$$

$$W_\mu^- = \frac{W_\mu^1 + iW_\mu^2}{\sqrt{2}} \tag{10.36}$$

We have a covariant derivative given by

$$D_\mu = \partial_\mu + i\frac{g_B}{2} + i\frac{g_W}{2}\vec{\tau}\cdot\vec{W}_\mu \tag{10.37}$$

Now we apply the covariant derivative to the Higgs field

$$D_\mu\varphi = \partial_\mu\varphi + i\frac{g_B}{2}\varphi + i\frac{g_W}{2}\vec{\tau}\cdot\vec{W}_\mu\varphi \tag{10.38}$$

Notice that

$$\vec{\tau} \cdot \vec{W}_\mu = \tau_1 W_\mu^1 + \tau_2 W_\mu^2 + \tau_3 W_\mu^3$$

$$= \begin{pmatrix} 0 & 1 \\ 1 & 0 \end{pmatrix} W_\mu^1 + \begin{pmatrix} 0 & -i \\ i & 0 \end{pmatrix} W_\mu^2 + \begin{pmatrix} 1 & 0 \\ 0 & -1 \end{pmatrix} W_\mu^3$$

$$= \begin{pmatrix} W_\mu^3 & W_\mu^1 - iW_\mu^2 \\ W_\mu^1 + iW_\mu^2 & -W_\mu^3 \end{pmatrix}$$

So

$$i\frac{g_W}{2} \vec{\tau} \cdot \vec{W}_\mu \varphi = i\frac{g_W}{2} \begin{pmatrix} W_\mu^3 & W_\mu^1 - iW_\mu^2 \\ W_\mu^1 + iW_\mu^2 & -W_\mu^3 \end{pmatrix} \begin{pmatrix} 0 \\ \varphi_0 + \dfrac{h(x)}{\sqrt{2}} \end{pmatrix}$$

$$= i\frac{g_W}{2} \begin{pmatrix} \sqrt{2}W_\mu^+ \left(\varphi_0 + \dfrac{h(x)}{\sqrt{2}} \right) \\ -W_\mu^3 \left(\varphi_0 + \dfrac{h(x)}{\sqrt{2}} \right) \end{pmatrix}$$

The first term in Eq. (10.37) just gives the ordinary derivative of the Higgs field.

$$\partial_\mu \left(\begin{pmatrix} 0 \\ \varphi_0 + \dfrac{h(x)}{\sqrt{2}} \end{pmatrix} \right) = \begin{pmatrix} 0 \\ \dfrac{1}{\sqrt{2}} \partial_\mu h \end{pmatrix}$$

This term is the kinetic energy term and does not contribute to the generation of the boson masses, so we won't worry about it. But putting everything together we have

$$D_\mu \varphi = \begin{pmatrix} 0 \\ \dfrac{1}{\sqrt{2}} \partial_\mu h \end{pmatrix} + i\frac{g_W}{2} \begin{pmatrix} \sqrt{2}W_\mu^+ \left(\varphi_0 + \dfrac{h(x)}{\sqrt{2}} \right) \\ -W_\mu^3 \left(\varphi_0 + \dfrac{h(x)}{\sqrt{2}} \right) \end{pmatrix} + i\frac{g_B}{2} B_\mu \begin{pmatrix} 0 \\ \varphi_0 + \dfrac{h(x)}{\sqrt{2}} \end{pmatrix}$$

Quantum Field Theory Demystified

To find the mass terms, we calculate $(D_\mu\varphi)^\dagger D_\mu\varphi$ and only keep terms of quadratic order. Now

$$(D_\mu\varphi)^\dagger = \left(0 \quad \frac{1}{\sqrt{2}}\partial_\mu h\right) - i\frac{g_W}{2}\left(\sqrt{2}W_\mu^-\left(\varphi_0 + \frac{h(x)}{\sqrt{2}}\right) \quad - W_\mu^3\left(\varphi_0 + \frac{h(x)}{\sqrt{2}}\right)\right)$$
$$- i\frac{g_B}{2}B_\mu\left(0 \quad \varphi_0 + \frac{h(x)}{\sqrt{2}}\right)$$

The first term in the product $(D_\mu\varphi)^\dagger D_\mu\varphi$ is

$$\frac{1}{2}(\partial_\mu h)^2$$

The next term we get is

$$-i\frac{g_W}{2}\left(\sqrt{2}W_\mu^-\left(\varphi_0 + \frac{h(x)}{\sqrt{2}}\right) \quad - W_\mu^3\left(\varphi_0 + \frac{h(x)}{\sqrt{2}}\right)\right) i\frac{g_W}{2}\begin{pmatrix} \sqrt{2}W^{+,\mu}\left(\varphi_0 + \frac{h(x)}{\sqrt{2}}\right) \\ -W^{3,\mu}\left(\varphi_0 + \frac{h(x)}{\sqrt{2}}\right) \end{pmatrix}$$

$$= \frac{g_W^2}{2}W_\mu^- W^{+,\mu}\left(\varphi_0 + \frac{h(x)}{\sqrt{2}}\right)^2 + \frac{g_W^2}{4}W_\mu^3 W^{3,\mu}\left(\varphi_0 + \frac{h(x)}{\sqrt{2}}\right)^2$$

And then we have a cross term

$$-i\frac{g_W}{2}\left(\sqrt{2}W_\mu^-\left(\varphi_0 + \frac{h(x)}{\sqrt{2}}\right) \quad - W_\mu^3\left(\varphi_0 + \frac{h(x)}{\sqrt{2}}\right)\right) i\frac{g_B}{2}B^\mu\begin{pmatrix} 0 \\ \varphi_0 + \frac{h(x)}{\sqrt{2}} \end{pmatrix}$$

$$= -\frac{g_W g_B}{4}W_\mu^3 B^\mu\left(\varphi_0 + \frac{h(x)}{\sqrt{2}}\right)^2$$

The second cross term is

$$-i\frac{g_B}{2}B_\mu\left(0 \quad \varphi_0 + \frac{h(x)}{\sqrt{2}}\right) i\frac{g_W}{2}\begin{pmatrix} \sqrt{2}W^{+,\mu}\left(\varphi_0 + \frac{h(x)}{\sqrt{2}}\right) \\ -W^{3,\mu}\left(\varphi_0 + \frac{h(x)}{\sqrt{2}}\right) \end{pmatrix}$$

$$= -\frac{g_B g_W}{4}B_\mu W^{3,\mu}\left(\varphi_0 + \frac{h(x)}{\sqrt{2}}\right)^2$$

The final quadratic term of interest is

$$-i\frac{g_B}{2}B_\mu\begin{pmatrix} 0 & \varphi_0+\dfrac{h(x)}{\sqrt{2}} \end{pmatrix}i\frac{g_B}{2}B^\mu\begin{pmatrix} 0 \\ \varphi_0+\dfrac{h(x)}{\sqrt{2}} \end{pmatrix}$$

$$=\frac{g_B^{\,2}}{4}B_\mu B^\mu\left(\varphi_0+\frac{h(x)}{\sqrt{2}}\right)^2$$

In the last section, we noted that mass terms will be multiplied by numbers. So we can ignore any terms that include $h(x)$. For the boson fields we are looking for terms of the form $m^2 A_\mu A^\mu$.

We wish to write these expressions in terms of the physical fields. Using the Weinberg angle, we can invert Eq. (10.22) to give

$$\begin{pmatrix} B_\mu \\ W_\mu^3 \end{pmatrix}=R(-\theta_w)\begin{pmatrix} A_\mu \\ Z_\mu \end{pmatrix}=\begin{pmatrix} \cos\theta_w & -\sin\theta_w \\ \sin\theta_w & \cos\theta_w \end{pmatrix}\begin{pmatrix} A_\mu \\ Z_\mu \end{pmatrix}$$

$$B_\mu = A_\mu \cos\theta_w - Z_\mu \sin\theta_w \qquad\qquad (10.39)$$

$$W_\mu^3 = A_\mu \sin\theta_w + Z_\mu \cos\theta_w \qquad\qquad (10.40)$$

Keeping only the leading term in $\left(\varphi_0+\dfrac{h(x)}{\sqrt{2}}\right)^2$, which is the only term that gives a scalar (number), we find

$$\frac{g_w^{\,2}}{2}W_\mu^- W^{+,\mu}\varphi_0^{\,2}+\frac{g_w^{\,2}}{4}W_\mu^3 W^{3,\mu}\varphi_0^{\,2}$$

$$=\frac{g_w^{\,2}}{2}W_\mu^- W^{+,\mu}\varphi_0^{\,2}+\frac{g_w^{\,2}}{4}(A_\mu \sin\theta_w + Z_\mu \cos\theta_w)$$

$$\times(A^\mu \sin\theta_w + Z^\mu \cos\theta_w)\,\varphi_0^{\,2}$$

$$=\frac{g_w^{\,2}}{2}W_\mu^- W^{+,\mu}\varphi_0^{\,2}+\frac{g_w^{\,2}}{4}(A_\mu A^\mu \sin^2\theta_w + A_\mu Z^\mu \sin\theta_w \cos\theta_w$$

$$+ Z_\mu A^\mu \sin\theta_w \cos\theta_w + Z_\mu Z^\mu \cos^2\theta_w)\,\varphi_0^{\,2} \qquad (10.41)$$

The next term becomes

$$-\frac{g_W g_B}{4} W_\mu^3 B^\mu \varphi_0^2$$

$$= -\frac{g_W g_B}{4}(A_\mu \sin\theta_W + Z_\mu \cos\theta_W)(A^\mu \cos\theta_W - Z^\mu \sin\theta_W)\varphi_0^2$$

$$= -\frac{g_W g_B}{4}(A_\mu A^\mu \sin\theta_W \cos\theta_W - A_\mu Z^\mu \sin^2\theta_W$$
$$+ Z_\mu A^\mu \cos^2\theta_W - Z_\mu Z^\mu \sin\theta_W \cos\theta_W)\varphi_0^2 \qquad (10.42)$$

The second cross term becomes

$$-\frac{g_B g_W}{4} B_\mu W^{3,\mu} \varphi_0^2$$

$$= -\frac{g_B g_W}{4}(A_\mu \cos\theta_W - Z_\mu \sin\theta_W)(A^\mu \sin\theta_W + Z^\mu \cos\theta_W)\varphi_0^2$$

$$= -\frac{g_B g_W}{4}(A_\mu A^\mu \sin\theta_W \cos\theta_W + A_\mu Z^\mu \cos^2\theta_W$$
$$- Z_\mu A^\mu \sin^2\theta_W - Z_\mu Z^\mu \sin\theta_W \cos\theta_W)\varphi_0^2 \qquad (10.43)$$

And the final term is

$$\frac{g_B^2}{4} B_\mu B^\mu \varphi_0^2$$

$$= \frac{g_B^2}{4}(A_\mu \cos\theta_W - Z_\mu \sin\theta_W)(A^\mu \cos\theta_W - Z^\mu \sin\theta_W)\varphi_0^2$$

$$= \frac{g_B^2}{4}(A_\mu A^\mu \cos^2\theta_W - A_\mu Z^\mu \sin\theta_W \cos\theta_W$$
$$- Z_\mu A^\mu \sin\theta_W \cos\theta_W + Z_\mu Z^\mu \sin^2\theta_W)\varphi_0^2 \qquad (10.44)$$

Now we add up Eqs. (10.41) through (10.44). We are looking for mass terms in the field, so we will ignore mixed terms that describe interactions like $A_\mu Z^\mu$. Let's look at each combination of the fields. We have

$$A_\mu A^\mu \left[-\frac{g_B^2}{4}\cos^2\theta_W - \frac{g_B g_W}{4}\sin\theta_W \cos\theta_W - \frac{g_W g_B}{4}\sin\theta_W \cos\theta_W + \frac{g_W^2}{4}\sin^2\theta_W \right]$$

$$= A_\mu A^\mu \left[\frac{g_B^2}{4}\cos^2\theta_W + \frac{g_W^2}{4}\sin^2\theta_W - \frac{g_B g_W}{2}\sin\theta_W \cos\theta_W \right]$$

The mixing angle describes how the two forces mix. As you can see from the figure below, if $\theta_w = 0$, then we have pure coupling to the W bosons, and no coupling to the Z boson. A nonzero θ_w less than 90° indicates coupling to both fields (thus the term weak mixing angle) as shown in the following illustration.

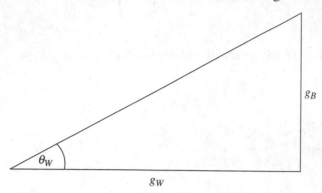

From the diagram we see that

$$\cos\theta_w = \frac{g_W}{\sqrt{g_B^2 + g_W^2}} \qquad \sin\theta_w = \frac{g_B}{\sqrt{g_B^2 + g_W^2}} \tag{10.45}$$

and this becomes

$$A_\mu A^\mu \left[\frac{g_B^2}{4}\cos^2\theta_w + \frac{g_W^2}{4}\sin^2\theta_w - \frac{g_B g_W}{2}\sin\theta_w \cos\theta_w \right]$$

$$= A_\mu A^\mu \left[\frac{g_B^2}{4}\frac{g_W^2}{\left(g_B^2 + g_W^2\right)} + \frac{g_W^2}{4}\frac{g_B^2}{\left(g_B^2 + g_W^2\right)} - \frac{g_B g_W}{2}\frac{g_B}{\sqrt{g_B^2 + g_W^2}}\frac{g_W}{\sqrt{g_B^2 + g_W^2}} \right]$$

$$= 0$$

This tells us that A_μ is a massless field. In fact it couples to the electric charge, so we know this is our photon field. Now, for the Z field we get

$$Z_\mu Z^\mu \left[\frac{g_B^2}{4}\cos^2\theta_w + \frac{g_W^2}{4}\sin^2\theta_w + \frac{g_B g_W}{2}\sin\theta_w \cos\theta_w \right]\varphi_0^2$$

$$= Z_\mu Z^\mu \left[\frac{g_B^2}{4}\frac{g_W^2}{\left(g_B^2 + g_W^2\right)} + \frac{g_W^2}{4}\frac{g_B^2}{\left(g_B^2 + g_W^2\right)} + \frac{g_B g_W}{2}\frac{g_B}{\sqrt{g_B^2 + g_W^2}}\frac{g_W}{\sqrt{g_B^2 + g_W^2}} \right]\varphi_0^2$$

$$= Z_\mu Z^\mu \left(\frac{g_B^2 + g_W^2}{4} \right)\varphi_0^2$$

Therefore, the mass of the Z particle is

$$M_Z = \left(\frac{\sqrt{g_B^2 + g_W^2}}{2} \right) \varphi_0 \qquad (10.46)$$

A similar exercise shows that the mass of the $W_\mu^\pm$ is

$$M_W = \frac{\varphi_0 g_W}{2} \qquad (10.47)$$

We can also write

$$\sin^2 \theta_W = 1 - \left(\frac{M_W}{M_Z} \right)^2 \qquad (10.48)$$

Theoretical bounds on the masses are

$$M_W = \frac{\varphi_0 g_W}{2} = \frac{38}{\sin \theta_W} \text{ GeV } \geq 38 \text{ GeV}$$

$$M_Z = \frac{\varphi_0 g_W}{2 \cos \theta_W} = \frac{76}{\sin \theta_W} \text{ GeV } \geq 76 \text{ GeV} \qquad (10.49)$$

Hence $M_W < M_Z$, which is born out in experiment. Measured masses indicate that

$$\sin^2 \theta_W = 0.222 \qquad (10.50)$$

which tells us that θ_w is approximately 28.1 (which is the angle used to draw the figure).

The ability of the Weinberg-Salam model to predict the masses in this fashion is what earned its discoverers the Nobel Prize. Moreover, it gives us confidence in the Higgs mechanism as a valid means for adding mass to the standard model despite the fact that experimental observation as yet eludes us. However, at the time of writing solid experimental observation of the Higgs still eludes us. Researchers are confident it will be seen when the Large Hadron Collider (LHC) begins operation in Switzerland sometime in 2008—but we will have to wait and see.

Summary

The Weinberg-Salam model combines leptons and gauge bosons into a single Lagrangian that has a $SU(2) \otimes U(1)$ symmetry. The original formulation of the Lagrangian is for massless particles, including a massless electron neutrino, massless electron, and four massless gauge fields. Spontaneous symmetry breaking using the Higgs mechanism breaks the $SU(2)$ symmetry and gives rise to masses for the electron and the gauge bosons $W^{\pm}$ and Z^0 that mediate the weak force. The symmetry breaking also introduces the photon field in the model, unifying the electromagnetic and weak interactions into a single Lagrangian.

Quiz

1. In electroweak theory, the neutrino has
   (a) No charge
   (b) $I_L^3 = +\frac{1}{2}, Y_L = -1, Q_L = 0$
   (c) $I_R^3 = 0, Y_L = -2, Q_R = -1$
   (d) $I_L^3 = -\frac{1}{2}, Y_L = -1, Q_L = -1$

2. In electroweak theory, the charges on the electron for a left-handed spinor are
   (a) $I_L^3 = -\frac{1}{2}, Y_L = -1, Q_L = -1$
   (b) $I_L^3 = \frac{1}{2}, Y_L = -1, Q_L = -1$
   (c) $I_R^3 = 0, Y_L = -2, Q_R = -1$
   (d) $I_L^3 = +\frac{1}{2}, Y_L = -1, Q_L = +1$

3. The introduction of the gauge field B_μ forces us to change the derivative operator as
   (a) $\partial_\mu \to \partial_\mu - ig_B \frac{Y}{2} B_\mu$
   (b) $\partial_\mu \to \partial_\mu + ig_B \frac{\vec{\tau}}{2} \cdot \vec{B}_\mu$
   (c) The derivative is unchanged
   (d) $\partial_\mu \to \partial_\mu + ig_B \frac{Y}{2} B_\mu$

4. In the weak interaction, invariance under a $U(1)$ transformation is associated with

   (a) The supercharge S_Y

   (b) The hypercharge Y

   (c) The weak isospin charge I

   (d) Coupling of the weak isospin charge to the weak hypercharge Y

5. Under a weak interaction $SU(2)$ transformation, the left-handed field transforms as

   (a) $\psi_L \to \psi'_L = e^{-i(\tau \cdot \alpha)/2}\psi_L = \psi_L$

   (b) $\psi_L \to \psi'_L = e^{-i(\tau \cdot \alpha)/2}\psi_L$

   (c) $\psi_L \to \psi_R$

   (d) There is no $SU(2)$ transformation in weak theory

6. The Weinberg angle

   (a) Describes scattering angles between leptons undergoing weak mediated collisions

   (b) Describes the ratio of the weak hypercharge to the weak isospin charge

   (c) Is related to the mixing of the coupling constants associated with the gauge fields of the weak theory

   (d) Describes the ratio of the weak isospin charge to the weak hypercharge

7. The mass of the W and Z particles are related as

   (a) $M_Z = \frac{M_W}{\cos\theta_W}$

   (b) $M_Z = \frac{M_W}{\sin\theta_W}$

   (c) Cannot be related directly

   (d) $M_Z = \frac{M_W}{\tan\theta_W}$

8. If we define the standard electron charge as measured by experiment as q, the charge on the electron predicted by electroweak theory

   (a) Is greater than or equal to the electron charge q

   (b) Is the same as the electron charge q

   (c) The charge cannot be predicted from the theory

CHAPTER 11

Path Integrals

One approach to quantum field theory, which is helpful in advanced contexts like string theory, is to use what is known as the *path integral* approach. A path integral is really a method for calculating amplitude for a quantum transition from one state to another. Our treatment will be very brief and introductory; for a detailed description see the text *Quantum Field Theory* by Lowell Brown.

Gaussian Integrals

It turns out that most calculations involving path integrals boil down to a simple looking integral known as a *Gaussian integral*. So before jumping into path integration directly, we will quickly summarize what a Gaussian integral is and how it is calculated.

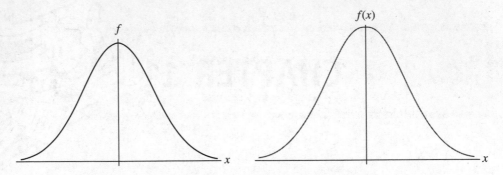

Figure 11.1 A plot of $f(x) = e^{-x^2}$.

The simplest Gaussian integral is an integral over all space in one dimension of the Gaussian function e^{-x^2}. A plot of this function, shown in Fig. 11.1, shows that it is localized in a small region about the origin.

As a result, we expect that the integral converges to a small finite value. Unfortunately, there is no elementary way to calculate

$$I = \int_{-\infty}^{\infty} e^{-x^2} dx$$

We must instead use a trick. The motivation is this: we can't evaluate this integral on the real line, but looking at the x^2 term, we want to see if we will have better luck in the plane using polar coordinates. First we square the integral. Now, since x is just a dummy variable when it appears in the integrand, we can call it something else—and we will opt for y for good reason. So we have

$$I^2 = \int_{-\infty}^{\infty} e^{-x^2} dx \int_{-\infty}^{\infty} e^{-y^2} dy = \int_{-\infty}^{\infty} e^{-(x^2+y^2)} dxdy$$

Next, we will change to polar coordinates. Recall that

$$x = r\cos\theta$$
$$y = r\sin\theta$$

and so

$$x^2 + y^2 = r^2 \cos^2\theta + r^2 \sin^2\theta = r^2$$

In transforming to polar coordinates, the element of area changes as

$$dxdy \rightarrow rdrd\theta$$

This leads to

$$I^2 = \int_{-\infty}^{\infty} e^{-(x^2+y^2)} dx dy = \int_0^{2\pi} d\theta \int_0^{\infty} e^{-r^2} r dr$$

and we have tamed the exponential term. The integral over r can be handled using a basic substitution $u = r^2$:

$$\int_0^{\infty} e^{-r^2} r dr = \frac{1}{2} \int_0^{\infty} e^{-u} du = \frac{1}{2}$$

so,

$$I^2 = \int_0^{2\pi} d\theta \int_0^{\infty} e^{-r^2} r dr = \frac{1}{2} \int_0^{2\pi} d\theta = \frac{1}{2} 2\pi = \pi$$

Taking the square root we obtain the result

$$I = \int_{-\infty}^{\infty} e^{-x^2} dx = \sqrt{\pi} \tag{11.1}$$

The integral in Eq. (11.1) can be extended into more complicated situations. Start by considering multiplication by constant a

$$I' = \int_{-\infty}^{\infty} e^{-ax^2} dx$$

Knowing the result of Eq. (11.1), this can be done using a substitution technique. We set $y = \sqrt{a}x$ and then

$$I' = \int_{-\infty}^{\infty} e^{-ax^2} dx = \frac{1}{\sqrt{a}} \int_{-\infty}^{\infty} e^{-y^2} dy = \sqrt{\frac{\pi}{a}} \tag{11.2}$$

The next extension of the basic Gaussian integral in Eq. (11.1) is to add powers of x. For example,

$$I = \int_{-\infty}^{\infty} x^n e^{-ax^2} dx$$

If n is odd, it is easy to see that this is just 0. Let $n = 1$. A plot of $f(x) = xe^{-x^2}$ in Fig. 11.2 shows this is an odd function, so it must integrate to zero since we are integrating over the entire real line. The integration over $(-\infty, 0)$ exactly cancels the integration over $(0, \infty)$.

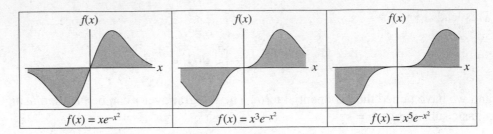

Figure 11.2   A plot of the first few odd powers on the exponential, all of which obviously integrate to zero.

Hence,

$$I = \int_{-\infty}^{\infty} x^n e^{-ax^2} dx = 0 \text{ for } n \text{ odd} \tag{11.3}$$

To obtain the result for even powers of n, we resort to more trickery. Start with Eq. (11.2), and take the derivative of both sides with respect to a. On the left we get

$$\frac{d}{da} \int_{-\infty}^{\infty} e^{-ax^2} dx = -\int_{-\infty}^{\infty} x^2 e^{-ax^2} dx$$

On the right we get

$$\frac{d}{da} \sqrt{\frac{\pi}{a}} = -\frac{\sqrt{\pi}}{2a^{3/2}}$$

Equating the two, we've found that

$$\int_{-\infty}^{\infty} x^2 e^{-ax^2} dx = \frac{\sqrt{\pi}}{2a^{3/2}}$$

A few plots of even powers of x against the exponential are illustrated in Figure 11.3.

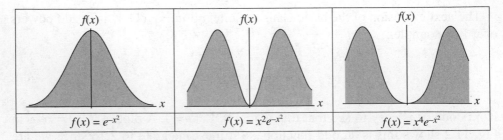

Figure 11.3   A plot of the first few even powers on the exponential. These integrals require some tinkering to compute.

This iteration procedure can be carried out ad infinitum to obtain more and more results. It turns out that for some general even n

$$\int_{-\infty}^{\infty} x^n e^{-ax^2} dx = \frac{1 \cdot 3 \cdot 5 \cdots (n+1)\sqrt{\pi}}{2^{n/2} a^{(n+1)/2}} \tag{11.4}$$

We can continue playing this game of introducing new terms into the Gaussian integral. One important Gaussian integral is

$$\int_{-\infty}^{\infty} e^{-ax^2 + bx} dx$$

This integral can be transformed into one of the form in Eq. (11.1) by completing the square. We write

$$-a\frac{x^2}{2} + bx = -\frac{a}{2}\left(x^2 - \frac{2b}{a}x\right) = -\frac{a}{2}\left(x - \frac{b}{a}\right)^2 + \frac{b^2}{2a}$$

Hence

$$\int_{-\infty}^{\infty} e^{-ax^2 + bx} dx = \int_{-\infty}^{\infty} e^{-\frac{a}{2}\left(x - \frac{b}{a}\right)^2} e^{b^2/2a} dx$$

$$= e^{b^2/2a} \int_{-\infty}^{\infty} e^{-\frac{a}{2}\left(x - \frac{b}{a}\right)^2} dx$$

$$= e^{b^2/2a} \sqrt{\frac{2\pi}{a}} \tag{11.5}$$

In n dimensions, a Gaussian integral takes the form

$$\int e^{-x^T A x} d^n x$$

where A is an $n \times n$ matrix of coefficients, x is an n dimensional column vector, and $d^n x = dx_1 dx_2 \ldots dx_n$. Let's consider a two-dimensional case where

$$A = \begin{pmatrix} a & b \\ b & c \end{pmatrix} \qquad x = \begin{pmatrix} x \\ y \end{pmatrix}$$

The argument of the exponential reduces to a scalar

$$x^T A x = \begin{pmatrix} x & y \end{pmatrix} \begin{pmatrix} a & b \\ b & c \end{pmatrix} \begin{pmatrix} x \\ y \end{pmatrix} = x(ax + by) + y(bx + cy)$$

Then we have

$$\int e^{-x^T A x} d^n x = \int_{-\infty}^{\infty} \int_{-\infty}^{\infty} e^{-(ax^2 + 2bxy + cy^2)} dx dy$$

$$= \int_{-\infty}^{\infty} \int_{-\infty}^{\infty} e^{-a(x + b/ay)^2 - \left(\frac{ac - b^2}{a}\right) y^2} dx dy$$

$$= \sqrt{\frac{\pi}{a}} \int_{-\infty}^{\infty} e^{-\left(\frac{ac - b^2}{a}\right) y^2} dy = \frac{\pi}{\sqrt{ac - b^2}}$$

But $ac - b^2$ is just the determinant of the matrix A. In general, we can write

$$\int e^{-x^T A x} d^n x = \frac{\pi^{n/2}}{\sqrt{\det A}} \tag{11.6}$$

This can be extended to the results

$$\int e^{-\frac{1}{2} x^T A x + J \cdot x} d^n x = \frac{(2\pi)^{n/2}}{\sqrt{\det A}} e^{\frac{J}{2} \cdot A^{-1} \cdot J} \tag{11.7}$$

$$\int e^{-\frac{i}{2} x^T A x + i J \cdot x} d^n x = \frac{(2\pi i)^{n/2}}{\sqrt{\det A}} e^{-i \frac{J}{2} \cdot A^{-1} \cdot J} \tag{11.8}$$

Basic Path Integrals

Now that we know how to do some Gaussian integrals, we are ready to take a look at path integration. Our development follows closely the clear expositions by Zee and Hatfield, but it can be found in many quantum field theory books. The basic trick is to reduce the problem to a Riemann sum.

A path integral is a way to calculate the amplitude for a system that starts off in some state $|i\rangle$ to end up in a state $|f\rangle$ by adding up the amplitudes for the system to pass through all possible paths from $|i\rangle$ to $|f\rangle$. As a specific example, the path integral can be constructed by considering the amplitude for a particle to pass from a point x_0 to a point x_N. If the dynamics of the system are described by a Hamiltonian H, then the amplitude in question is

$$\langle x_N | e^{-i\hat{H}t} | x_0 \rangle \tag{11.9}$$

We can consider the simplest possible case, where

$$\hat{H} = \frac{\hat{p}^2}{2m}$$

for a free particle of mass m. If we want to rewrite the amplitude in Eq. (11.9) so that we consider every possible path from x_0 to x_N, we start by splitting up the path into smaller pieces and then use a limiting procedure. We begin by dividing up the time interval into N equally spaced intervals Δt.

$$\Delta t = \frac{t}{N} \tag{11.10}$$

Then

$$e^{-i\hat{H}t} = e^{-i\hat{H}(N\Delta t)}$$

$$= e^{-i\hat{H}(\Delta t + \Delta t + \cdots + \Delta t)} = e^{-i\hat{H}\Delta t} e^{-i\hat{H}\Delta t} e^{-i\hat{H}\Delta t} \cdots e^{-i\hat{H}\Delta t}$$

So the amplitude can be rewritten as

$$\langle x_N | e^{-i\hat{H}t} | x_0 \rangle = \langle x_N | e^{-i\hat{H}\Delta t} e^{-i\hat{H}\Delta t} e^{-i\hat{H}\Delta t} \cdots e^{-i\hat{H}\Delta t} | x_0 \rangle$$

Now we use the fact that the position eigenstates form a complete set of states. That is,

$$\int dx |x\rangle\langle x| = I \tag{11.11}$$

We can split up the interval into N parts and each little part will be a complete set of states, that is,

$$\int dx_j |x_j\rangle\langle x_j| = I$$

We stick these in between the exponential factors to give

$$\langle x_N | e^{-i\hat{H}t} | x_0 \rangle = \langle x_N | e^{-i\hat{H}\Delta t} e^{-i\hat{H}\Delta t} e^{-i\hat{H}\Delta t} \cdots e^{-i\hat{H}\Delta t} | x_0 \rangle$$

$$= \langle x_N | e^{-i\hat{H}\Delta t} I e^{-i\hat{H}\Delta t} I e^{-i\hat{H}\Delta t} I \cdots I e^{-i\hat{H}\Delta t} | x_0 \rangle$$

$$= \langle x_N | e^{-i\hat{H}\Delta t} \int dx_{N-1} | x_{N-1}\rangle\langle x_{N-1} | e^{-i\hat{H}\Delta t} \int dx_{N-2} | x_{N-2}\rangle$$

$$\times \langle x_{N-2} | e^{-i\hat{H}\Delta t} \cdots \int dx_1 | x_1\rangle\langle x_1 | e^{-i\hat{H}\Delta t} | x_0 \rangle$$

This expression can be written more compactly as a product, like

$$\langle x_N | e^{-i\hat{H}t} | x_0 \rangle = \prod_{j=1}^{N-1} \int dx_j \langle x_N | e^{-i\hat{H}\Delta t} | x_{N-1} \rangle \langle x_{N-1} | e^{-i\hat{H}\Delta t} | x_{N-2} \rangle \langle x_{N-2} | e^{-i\hat{H}\Delta t} \cdots | x_1 \rangle \langle x_1 | e^{-i\hat{H}\Delta t} | x_0 \rangle$$

Right now the situation probably looks pretty hopeless. But we can calculate each individual term by bringing up the Gaussian integrals in the last section. Getting there will take some additional complications. We begin by recalling that the momentum eigenstates also form a complete set.

$$\frac{1}{2\pi} \int dp |p\rangle\langle p| \tag{11.12}$$

Noting that

$$\langle x | p \rangle = e^{ipx} \qquad \langle p | x \rangle = e^{-ipx} \tag{11.13}$$

and that for a free particle

$$\hat{H} | p \rangle = \frac{\hat{p}^2}{2m} | p \rangle = \frac{p^2}{2m} | p \rangle$$

$$\Rightarrow e^{-i\hat{H}\Delta t} | p \rangle = e^{-ip^2/2m\Delta t} | p \rangle$$

we can get expressions like $\langle x_j | e^{-i\hat{H}\Delta t} | x_{j-1} \rangle$ into a form that can be readily integrated. We have

$$\langle x_j | e^{-i\hat{H}\Delta t} | x_{j-1} \rangle = \langle x_j | e^{-i\hat{H}\Delta t} \left(\frac{1}{2\pi} \int dp |p\rangle\langle p| \right) | x_{j-1} \rangle$$

$$= \frac{1}{2\pi} \langle x_j | \int dp \left(e^{-i\hat{H}\Delta t} | p \rangle \right) \langle p | x_{j-1} \rangle$$

$$= \frac{1}{2\pi} \langle x_j | \int dp \left(e^{-i\hat{H}\Delta t} | p \rangle \right) e^{-ipx_{j-1}}$$

$$= \frac{1}{2\pi} \langle x_j | \int dp \left(e^{-ip^2/2m\Delta t} | p \rangle \right) e^{-ipx_{j-1}}$$

$$= \frac{1}{2\pi} \int dp \left(e^{-ip^2/2m\Delta t} \langle x_j | p \rangle \right) e^{-ipx_{j-1}}$$

$$= \frac{1}{2\pi} \int dp \, e^{-ip^2/2m\Delta t} e^{ip(x_j - x_{j-1})}$$

This integral is nothing other than Eq. (11.5). Taking

$$a = \frac{i}{m} \Delta t \qquad J = x_j - x_{j-1}$$

we get

$$\left\langle x_j \left| e^{-i\hat{H}\Delta t} \right| x_{j-1} \right\rangle = \frac{1}{2\pi} \int dp \, e^{-ip^2/2m\Delta t} e^{ip(x_j - x_{j-1})}$$

$$= \frac{1}{2\pi} \left(-i \frac{2\pi m}{\Delta t} \right)^{1/2} e^{i(x_j - x_{j-1})^2 m/2\Delta t}$$

Using this expression for each term in the original product allows us to write the compact expression as follows:

$$\left\langle x_N \left| e^{-i\hat{H}t} \right| x_0 \right\rangle = \left(-i \frac{m}{2\pi \Delta t} \right)^{N/2} \prod_{j=1}^{N-1} \int dx_j \, e^{\sum_{j=0}^{N-1} im/2\Delta t \left[(x_j - x_{j-1})/\Delta t \right]^2} \qquad (11.14)$$

Now we let $\Delta t \to 0$. In the limit, the term $(x_j - x_{j-1})/\Delta t$ just becomes a derivative

$$\lim_{\Delta t \to 0} \frac{x_j - x_{j-1}}{\Delta t} = \frac{dx}{dt} = \dot{x}$$

and the summation $\sum_{j=0}^{N-1} \to \int_0^t dt'$, so Eq. (11.14) becomes

$$\left\langle x_N \left| e^{-i\hat{H}t} \right| x_0 \right\rangle = \lim_{N \to \infty} \left(-i \frac{m}{2\pi \Delta t} \right)^{N/2} \prod_{j=1}^{N-1} \int dx_j \, e^{i\int_0^t dt' \frac{m}{2} \dot{x}^2}$$

The path integral measure is then

$$Dx = \lim_{N \to \infty} \prod_{j=1}^{N-1} \int dx_j \left(-i \frac{m}{2\pi \Delta t} \right)^{N/2} \qquad (11.15)$$

where $\Delta t = \frac{t}{N}$. We can then write the path integral as the compact expression

$$\left\langle x_N \left| e^{-i\hat{H}t} \right| x_0 \right\rangle = \int Dx \, e^{i\int_0^t dt' \frac{m}{2}\dot{x}^2} \qquad (11.16)$$

Quantum Field Theory Demystified

242

This tells us that the amplitude for a particle to take the path from x_0 to x_N is proportional to the exponential of the action S, which you recall as

$$S = \int L\, dt$$

In this example, we have considered a free particle, so $L = \frac{1}{2} m \dot{x}^2$. This is a nice result which ties quantum theory to classical mechanics. We can write the general amplitude for a system with Lagrangian L as

$$\langle F | e^{-i\hat{H}t} | I \rangle = \int Dx\, e^{i \int_0^t dt' L(q, \dot{q})} \tag{11.17}$$

where $|I\rangle$ and $|F\rangle$ are the initial and final states of the system. In a quantum field theory, we can use a path integral to compute the amplitude for a transition from a state $\varphi_1(t_1)$ to a state $\varphi_2(t_2)$.

$$\langle \varphi_2(t_2) | \varphi_1(t_1) \rangle = \langle \varphi_2 | e^{-i\hat{H}(t_2 - t_1)} | \varphi_1 \rangle$$
$$= \int_{\varphi_1}^{\varphi_2} D\varphi \exp\left(i \int L\, d^4 x \right) \tag{11.18}$$

For example, if we let $\varphi_1 = \varphi_2 = \varphi_0$ be the ground state, we can calculate the energy of the vacuum using path integral methods. When an external source $J(x)$ is present, the path integral becomes

$$\langle \varphi_2(t_2) | \varphi_1(t_1) \rangle = \int_{\varphi_1}^{\varphi_2} D\varphi \exp\left(i \int [L + \varphi(x) J(x)] d^4 x \right) \tag{11.19}$$

We define the vacuum to vacuum amplitude by

$$Z[J] = e^{iW[J]} = \langle 0 | 0 \rangle_J = \int D\varphi \exp\left[\int d^4 x \left(L + \varphi(x) J(x) \right) \right] \tag{11.20}$$

Summary

In this chapter we have succinctly introduced the notion of a path integral. This is a method that can be used to calculate the amplitude for a system to transition from one state to another by considering all possible paths between the two states. Path integrals can be used to calculate any quantity in field theory such as the vacuum expectation value for a field. The use of Gaussian integrals is important in the calculation of path integrals.

Quiz

1. The integral $\int_{-\infty}^{\infty} x^3 e^{-x^2} dx$ is

   (a) $\frac{4\sqrt{\pi}}{2^{3/2} a^2}$

   (b) 0

   (c) Indeterminate

2. The integral $\int_0^{\infty} x^3 e^{-x^2} dx$ is

   (a) ½

   (b) 0

   (c) $\int_0^{\infty} x^3 e^{-x^2} dx \to \infty$ unbounded

3. Compute $\int e^{-x^T A x} d^n x$ when $x = \begin{pmatrix} x \\ y \\ z \end{pmatrix}$ and $A = \begin{pmatrix} 1 & 0 & 3 \\ 0 & 2 & 4 \\ 1 & 0 & 2 \end{pmatrix}$.

   (a) $i \frac{\pi^{3/2}}{\sqrt{2}}$

   (b) $\frac{\pi^{3/2}}{\sqrt{2}}$

   (c) 0

4. Path integrals often involve Gaussians because

   (a) Path integrals are the square of an energy integral.

   (b) Path integrals do not involve Gaussians.

   (c) A path integral contains the Lagrangian, which has a quadratic dependence on momentum.

5. The relationship between the quantum mechanical path integral and classical mechanics can be stated as

   (a) There is no relation.

   (b) A path integral involves an exponential of the action S.

   (c) The action S can be recovered from a path integral by explicit calculation.

   (d) The square of the argument to the exponential gives the action.

6. The vacuum to vacuum amplitude with a source is

   (a) $Z[J] = e^{iW[J]} = \langle 0|0 \rangle_J = \int D\varphi \exp\left[\int d^4 x (L + \varphi^2(x) J^2(x)) \right]$

   (b) $Z[J] = e^{iW[J]} = \langle 0|0 \rangle_J = \int D\varphi \exp\left[\int d^4 x (L - \varphi(x) J(x)) \right]$

   (c) $Z[J] = e^{iW[J]} = \langle 0|0 \rangle_J = \int D\varphi \exp\left[\int d^4 x (L + J(x)) \right]$

   (d) $Z[J] = e^{iW[J]} = \langle 0|0 \rangle_J = \int D\varphi \exp\left[\int d^4 x (L + \varphi(x) J(x)) \right]$

CHAPTER 12

Supersymmetry

Latter progress in theoretical particle physics prior to string theory came about in the 1970s under the name of *supersymmetry* (SUSY). This is a symmetry that relates or mixes (unites) fermions and bosons. Fermions are particles with half-integral spin, while bosons are particles with integer spin. The idea of supersymmetry is that for every fermion, there is a corresponding boson. We know that the force-carrying or mediator bosons have spin-0 or spin-1, so what we would hope to find is that corresponding to spin-1/2 particles like quarks and electrons, there would be spin-0 or spin-1 particles (denoted the *selectron* and *squark*). Also, corresponding to each spin-0 or spin-1 particle there would be a half-integral spin particle. The proposed particles go by the fanciful names *photino*, *wino*, and *gluino*, which would correspond to the photon, *W*, and gluon, for example.

At the time of writing, there is no experimental evidence for supersymmetry despite diligent experiment. However, it is hoped that when the Large Hadron Collider (LHC) begins operation sometime in 2008, experimental evidence for supersymmetry will be found or alternatively that supersymmetry can be effectively ruled out, making it an exercise in mathematics and history.

In this chapter we will provide a very cursory introduction to the subject. Interested readers are urged to consult advanced physics books and published articles to learn more about the theory.

Basic Overview of Supersymmetry

As mentioned in the introduction, supersymmetry proposes that to each fermion there exists a boson and vice versa. So we can think of *supersymmetry as proposing that a symmetry exists between bosons and fermions, and that in nature, there are equal numbers of fermion and boson states.* Ideally, these would be the same mass. So as the positron is the antiparticle of the electron, essentially it's an electron with the same mass but opposite charge. There would be a bosonic selectron which would have the same mass as the electron but would have whole integer spin rather than being spin-1/2. Obviously, no such particle has been seen. This means that if supersymmetry is real and selectrons do exist, their mass must be much heavier than that of real electrons or they are cloaked by mechanisms unknown. This suggests why we have not detected them yet—we need larger particle accelerators to attain the higher energies necessary to create selectrons. Since the masses actually differ in nature, supersymmetry must be broken.

If supersymmetry is not observed, why does it garner so much attention? Because supersymmetry solves many outstanding "mysteries" in particle physics. One important problem that stands out is called the *hierarchy problem*. It is believed that the mass of the Higgs boson, which we will denote by m_h, is much smaller than a fundamental quantity that physicists call the *Planck mass*. The Planck mass is computed from the fundamental constants.

$$m_p = \sqrt{\frac{\hbar c}{G}} \approx 10^{19} \text{ GeV/c}^2 \tag{12.1}$$

This is an astonishingly large 22 μg[1]. At the present time it is believed that

$$m_h \ll m_p \tag{12.2}$$

In quantum field theory, there are quadratic corrections to the Higgs mass that cause its mass to diverge. Therefore the natural value for the mass of the Higgs boson would be extremely large, that is, we could expect it to be on the order of the Planck mass. This would make it much larger than what is expected to be the experimentally observed value of the Higgs mass, which we are denoting by m_h. This is the hierarchy problem. Although standard theory requires the mass of the

---

[1]A tiny grain of SiO_2 sand of dimension 0.1 mm weighs 2.6 μg.

Higgs boson to be as large as possible, this is not what is seen (or at least what we think will be seen). Why is the Higgs mass so much smaller than its natural value? It turns out that supersymmetry does away with the corrections to the Higgs mass, and allows a Higgs mass more like what is expected to be seen experimentally.

There are also other reasons to hope that supersymmetry turns out to be valid. One is the so-called *gauge unification problem*. It is believed that the coupling constants of the standard model would unify (become the same strength) at a certain energy. Using standard model quantum field theory, this does not quite work out. But supersymmetry solves this problem, giving an energy at which the coupling constants of the different interactions would converge to a single value.

Supercharge

We can start to understand supersymmetry by taking a look at the *supercharge operator Q*. The supercharge operator acts to transform fermions into bosons and bosons into fermions. For the moment let's denote fermionic states by $|F\rangle$ and bosonic states by $|B\rangle$. Then the action of the supercharge is as follows:

$$Q|F\rangle = |B\rangle \tag{12.3}$$

$$Q|B\rangle = |F\rangle \tag{12.4}$$

The number of supercharges in the theory characterizes the theory. If there is a single supercharge Q, then we say that we have an $N = 1$ supersymmetry. If there are two supercharges, then there is an $N = 2$ supersymmetry. If there are three supercharges, then we have an $N = 3$ supersymmetry and so on.

The operators Q and $Q^{\dagger}$ transform as spin-1/2 operators under Lorentz transformations. If we let P^{μ} be a conserved 4-momentum, then Q is a spinor that satisfies anticommutation relations of the form

$$\begin{aligned} \{Q, Q^{\dagger}\} &= P^{\mu} \\ \{Q, Q\} &= \{Q^{\dagger}, Q^{\dagger}\} = 0 \end{aligned} \tag{12.5}$$

In addition, the operators Q and $Q^{\dagger}$ commute with the 4-momentum P^{μ}.

$$[Q, P^{\mu}] = [Q^{\dagger}, P^{\mu}] = 0 \tag{12.6}$$

Let the fermion and boson states of a given type be defined by $|\psi_F\rangle$ and $|\psi_B\rangle$, respectively. For example, if the particle we are talking about is an electron, then it represents the electron state while $|\psi_B\rangle$ is the superpartner selectron state. They transform into one another as

$$Q|\psi_F\rangle = |\psi_B\rangle$$
$$Q|\psi_B\rangle = |\psi_F\rangle$$

The square of the 4-momentum operator gives the mass of the state. That is,

$$P^2|\psi_F\rangle = -m_F^2|\psi_F\rangle \qquad (12.7)$$

Since Q and $Q^\dagger$ commute with the 4-momentum P^μ, it follows using $[A, BC] = [A,B]C + B[A,C]$ that they also commute with the square of the 4-momentum operator. Hence

$$[P^2,Q] = P^2Q - QP^2 = 0$$
$$\Rightarrow P^2Q = QP^2$$

This result indicates that the superpartners will have the same mass. First we note that

$$P^2Q|\psi_F\rangle = P^2|\psi_B\rangle$$

However, since $P^2Q = QP^2$, we can write

$$P^2Q|\psi_F\rangle = QP^2|\psi_F\rangle = -Qm_F^2|\psi_F\rangle$$

but m_F^2 is just a scalar. Hence

$$P^2Q|\psi_F\rangle = -Qm_F^2|\psi_F\rangle = -m_F^2Q|\psi_F\rangle = -m_F^2|\psi_B\rangle$$

Putting this together with $P^2Q|\psi_F\rangle = -Qm_F^2|\psi_F\rangle$, we see that

$$P^2|\psi_B\rangle = -m_F^2|\psi_B\rangle$$

This proves that if the anticommutation relations are satisfied, the partner and superpartner states have the same mass. We've already noted that this is not what is seen in nature—otherwise low mass partners like the selectron would have been detected experimentally long ago. Supersymmetry, if it exists, is broken in nature and the superpartners have much larger mass.

The basic program of supersymmetry is to add one or more supercharges to the fields of the standard model, and determine what happens when we vary the action with respect to the supercharge. The result will be some leftover terms. We then add more terms to cancel the unwanted ones. In the end, the action remains invariant under the supersymmetry transformation.

The added terms are the new fields and associated particles we mentioned earlier. So, bosonic partners to each fermion need to be added to the standard model such as selectrons and squarks to keep it invariant under a supersymmetry transformation. There are also fermionic fields like the *Higgsino* that correspond to the bosonic particles (in this case the Higgs) added to the action to keep it invariant. The $N = 1$ supersymmetry case with a single supercharge is called the *minimally super-symmetric standard model* or MSSM.

We will explain supersymmetry in the standard model in more detail below. However, first let's take a look at a simpler way to introduce supersymmetry using what is called *supersymmetric quantum mechanics*.

Supersymmetric Quantum Mechanics

Supersymmetric quantum mechanics was developed to apply some of the ideas of supersymmetry to the simpler nonrelativistic quantum mechanics to gain insight into some of the problems of supersymmetry, such as some understanding into why supersymmetry is broken—meaning that superpartners of the same mass to known particles do not exist in nature. In supersymmetric quantum mechanics there exist N supercharges Q_i that commute with the Hamiltonian.

$$[Q_i, H] = 0 \tag{12.8}$$

These supercharges satisfy a set of anticommutation relations relating them to H

$$\{Q_i, Q_j\} = \delta_{ij} H \tag{12.9}$$

and there exists a *superpotential* $W(x)$ whose role will become clear in a moment. We consider a two-state system in one spatial dimension x with a wave function given by

$$\psi(x) = \begin{pmatrix} \alpha(x) \\ \beta(x) \end{pmatrix} \tag{12.10}$$

There exist two supercharges Q_1 and Q_2. They can be defined using the Pauli matrices together with the superpotential $W(x)$ as

$$Q_1 = \frac{1}{2}\left[\sigma_1 p + \sigma_2 W(x)\right]$$

$$Q_2 = \frac{1}{2}\left[\sigma_2 p - \sigma_1 W(x)\right]$$

(12.11)

EXAMPLE 12.1
Compute the Hamiltonian corresponding to the supercharges defined in Eq. (12.11).

SOLUTION
We find the Hamiltonian using $\{Q_i, Q_j\} = \delta_{ij} H$. Let us use Q_1. Now

$$\begin{aligned}\{Q_1, Q_1\} &= Q_1 Q_1 + Q_1 Q_1 \\ &= 2Q_1 Q_1 \\ &= 2Q_1^2\end{aligned}$$

Hence we see that the Hamiltonian can be written directly in terms of the supercharge as $H = 2Q_1^2$. Explicitly

$$\begin{aligned}Q_1 Q_1 &= \frac{1}{2}\left[\sigma_1 p + \sigma_2 W(x)\right]\frac{1}{2}\left[\sigma_1 p + \sigma_2 W(x)\right] \\ &= \frac{1}{4}\left[\sigma_1 p \sigma_1 p + \sigma_1 p \sigma_2 W(x) + \sigma_2 W(x)\sigma_1 p + \sigma_2^2 W^2(x)\right]\end{aligned}$$

However, we know that the Pauli matrices square to the identity. Therefore

$$\sigma_2^2 = I$$

Moreover, we can move the Pauli matrices and momentum operators past each other with impunity. Hence the first term is

$$\sigma_1 p \sigma_1 p = \sigma_1^2 p^2 = p^2$$

Now, recall the Lie algebra of these matrices, which is as follows:

$$\sigma_1\sigma_2 = \begin{pmatrix} 0 & 1 \\ 1 & 0 \end{pmatrix}\begin{pmatrix} 0 & -i \\ i & 0 \end{pmatrix} = \begin{pmatrix} i & 0 \\ 0 & -i \end{pmatrix} = i\sigma_3$$

$$\sigma_2\sigma_1 = \begin{pmatrix} 0 & -i \\ i & 0 \end{pmatrix}\begin{pmatrix} 0 & 1 \\ 1 & 0 \end{pmatrix} = \begin{pmatrix} -i & 0 \\ 0 & i \end{pmatrix} = -i\sigma_3$$

And, recall that the commutator of any function of position with the momentum operator is

$$[F(x), p] = F(x)p - pF(x) = i\frac{dF}{dx}$$

Therefore,

$$\sigma_1\sigma_2 PW + \sigma_2\sigma_1 WP = \sigma_1\sigma_2 PW + \sigma_2\sigma_1\left(PW + i\frac{dW}{dx}\right)$$

$$= (\sigma_1\sigma_2 + \sigma_2\sigma_1)PW + i\sigma_2\sigma_1\frac{dW}{dx}$$

$$= (i\sigma_3 - i\sigma_3)PW + \sigma_3\frac{dW}{dx}$$

$$= \sigma_3\frac{dW}{dx}$$

Putting all of these results together, we find

$$Q_1^2 = \frac{1}{4}\left(p^2 + \sigma_3\frac{dW}{dx} + W^2(x)\right)$$

Hence, the Hamiltonian is

$$H = 2Q_1^2 = \frac{1}{2}\left(p^2 + \sigma_3\frac{dW}{dx} + W^2(x)\right)$$

Take a look back at the action of a supercharge in a quantum field theory, as described schematically in Eqs. (12.3) and (12.4). When we look at the action of the supercharges in this example of nonrelativistic quantum mechanics, we will see how we are constructing a simple model of a supersymmetric theory. We define spin-up and spin-down states as

$$|+\rangle = \begin{pmatrix} 1 \\ 0 \end{pmatrix} \qquad |-\rangle = \begin{pmatrix} 0 \\ 1 \end{pmatrix}$$

Now we have

$$Q_1|+\rangle = \frac{1}{2}(\sigma_1 p + \sigma_2 W(x))|+\rangle$$

$$= \frac{p}{2}\begin{pmatrix} 0 & 1 \\ 1 & 0 \end{pmatrix}\begin{pmatrix} 1 \\ 0 \end{pmatrix} + \frac{W(x)}{2}\begin{pmatrix} 0 & -i \\ i & 0 \end{pmatrix}\begin{pmatrix} 1 \\ 0 \end{pmatrix}$$

$$= \frac{p}{2}\begin{pmatrix} 0 \\ 1 \end{pmatrix} + \frac{W(x)}{2}\begin{pmatrix} 0 \\ i \end{pmatrix} = \frac{1}{2}\begin{pmatrix} 0 \\ p + iW \end{pmatrix}$$

That is, the supercharge Q_1 turns a spin-up state into a spin-down state, analogous to a supercharge converting a fermion to a boson, say. Next consider the action of Q_1 on a spin-down state as follows:

$$Q_1|-\rangle = \frac{1}{2}(\sigma_1 p + \sigma_2 W(x))|-\rangle$$

$$= \frac{p}{2}\begin{pmatrix} 0 & 1 \\ 1 & 0 \end{pmatrix}\begin{pmatrix} 0 \\ 1 \end{pmatrix} + \frac{W(x)}{2}\begin{pmatrix} 0 & -i \\ i & 0 \end{pmatrix}\begin{pmatrix} 0 \\ 1 \end{pmatrix}$$

$$= \frac{p}{2}\begin{pmatrix} 1 \\ 0 \end{pmatrix} + \frac{W(x)}{2}\begin{pmatrix} -i \\ 0 \end{pmatrix} = \frac{1}{2}\begin{pmatrix} p - iW \\ 0 \end{pmatrix}$$

If supersymmetric quantum mechanics is unbroken, then the states $\psi(x)$ are left invariant by a unitary transformation of the form

$$U = e^{-i\alpha_i Q_i}$$

Moreover there exists a state $\psi_0(x)$ that is annihilated by the supercharges, that is,

$$Q_1\psi_0(x) = 0 \qquad (12.12)$$

This implies that the ground state of the system has zero energy since

$$H\psi_0(x) = 2Q_1^2\psi_0(x) = 0$$

EXAMPLE 12.2

Derive the form of the ground state $\psi_0(x)$.

SOLUTION

We use the fact that the supercharge annihilates the state Eq. (12.12) together with Eq. (12.11).

$$Q_1\psi_0(x) = 0$$

$$\Rightarrow 0 = \frac{1}{2}(\sigma_1 p + \sigma_2 W(x))\psi_0(x)$$

$$= \sigma_1 p\psi_0(x) + \sigma_2 W(x)\psi_0(x)$$

$$= -i\sigma_1 \frac{d\psi_0}{dx} + \sigma_2 W(x)\psi_0(x)$$

Multiplying by σ_1 and using $\sigma_1^2 = I$ together with $\sigma_1\sigma_2 = i\sigma_3$ gives

$$0 = -i\frac{d\psi_0}{dx} + i\sigma_3 W(x)\psi_0(x)$$

$$\Rightarrow \frac{d\psi_0}{dx} = \sigma_3 W(x)\psi_0(x)$$

Integrating and taking the initial condition to be $\psi_0(0)$ gives

$$\psi_0(x) = \exp\int_0^x dx' \sigma_3 W(x')\psi_0(0)$$

The Simplified Wess-Zumino Model

Now that we have gotten a taste of supersymmetry by taking a brief look at supersymmetric quantum mechanics, we are ready to dive into a simple supersymmetric quantum field theory. We will discuss the Wess-Zumino model, a supersymmetric theory proposed in 1974 that starts with a Lagrangian consisting of bosonic and fermionic fields and shows that it is possible to develop a transformation that mixes bosonic and fermionic fields.

THE CHIRAL REPRESENTATION

In supersymmetry, it is convenient to work in the representation where the Dirac matrices are given by

$$\gamma_\mu = \begin{pmatrix} 0 & \sigma_\mu \\ \bar{\sigma}_\mu & 0 \end{pmatrix} \tag{12.13}$$

where we have introduced the barred Pauli matrices which are obtained with a sign change as

$$\bar{\sigma}^0 = \sigma^0 = \begin{pmatrix} 1 & 0 \\ 0 & 1 \end{pmatrix}$$

$$\bar{\sigma}^1 = -\sigma^1 = \begin{pmatrix} 0 & -1 \\ -1 & 0 \end{pmatrix} \qquad \bar{\sigma}^2 = -\sigma^2 = \begin{pmatrix} 0 & i \\ -i & 0 \end{pmatrix} \tag{12.14}$$

$$\bar{\sigma}^3 = -\sigma^3 = \begin{pmatrix} -1 & 0 \\ 0 & 1 \end{pmatrix}$$

Note that lowering the index with the metric introduces a sign change when $i = 1, 2, 3$ so that

$$\bar{\sigma}_3 = -\sigma_3 = \begin{pmatrix} 1 & 0 \\ 0 & -1 \end{pmatrix}$$

for example. The following anticommutation relations are useful.

$$\{\sigma^\mu, \bar{\sigma}^\nu\} = \sigma^\mu \bar{\sigma}^\nu + \bar{\sigma}^\nu \sigma^\mu = -2g^{\mu\nu}$$
$$\{\bar{\sigma}^\mu, \sigma^\nu\} = \bar{\sigma}^\mu \sigma^\nu + \sigma^\nu \bar{\sigma}^\mu = -2g^{\mu\nu} \tag{12.15}$$

A Simple SUSY Lagrangian

The key to supersymmetry is writing down a Lagrangian and then relying on our old friend: computing a variation of the bosonic and fermionic states together and requiring that the variation of the action be zero. This will introduce a *supercurrent* that is conserved, meaning that if S is the supercurrent then the total divergence is 0.

$$\partial_\mu S^\mu = 0$$

The supercurrent will allow us to calculate the supercharges Q. A simple model that illustrates how supersymmetry works is the *Wess-Zumino model*. In the following, we will consider the simplest possible example, a model that consists of a single massless spin-0 boson field A and a single massless spin-1/2 fermion field ψ. If we let the scalar field for the spin-0 boson be a complex field, then the Lagrangian is

$$L_B = -\partial_\mu A^* \partial^\mu A$$

If we take the spinor field to be right handed, the Lagrangian for a massless spin-1/2 field can be written as

$$L_F = \frac{i}{2}\partial_\mu \psi^\dagger \bar{\sigma}^\mu \psi - \frac{i}{2}\psi^\dagger \bar{\sigma}^\mu \partial_\mu \psi$$

In order to close the SUSY algebra, it will be necessary to add an *auxiliary field F*. The auxiliary Lagrangian is the simple term

$$L_{\text{aux}} = F^* F$$

The supersymmetric Lagrangian is the sum of these individual Lagrangians, that is,

$$\begin{aligned} L &= L_B + L_F + L_{\text{aux}} \\ &= -\partial_\mu A^* \partial^\mu A + \frac{i}{2}\partial_\mu \psi^\dagger \bar{\sigma}^\mu \psi - \frac{i}{2}\psi^\dagger \bar{\sigma}^\mu \partial_\mu \psi + F^* F \end{aligned} \tag{12.16}$$

For the Lagrangian to be invariant under supersymmetry, we require that

$$\delta S = \delta \int L\, d^4 x = 0$$

where the supersymmetric variation converts bosons into fermions and vice versa. We need variations that will work with each field in the Lagrangian, so obtain

$$\begin{aligned} \delta A &= \sqrt{2}\varepsilon\psi \qquad \delta A^* = \sqrt{2}\varepsilon^\dagger \psi^\dagger \\ \delta\psi &= i\sqrt{2}\sigma^\mu \varepsilon^\dagger \partial_\mu A + \sqrt{2}\varepsilon F \qquad \delta\psi^\dagger = -i\sqrt{2}\varepsilon\sigma^\mu \partial_\mu A^* + \sqrt{2}\varepsilon^\dagger F^* \\ \delta F &= i\sqrt{2}\varepsilon^\dagger \bar{\sigma}^\mu \partial_\mu \psi \qquad \delta F^* = -i\sqrt{2}\partial_\mu \bar{\psi}\bar{\sigma}^\mu \varepsilon \end{aligned} \tag{12.17}$$

We have introduced a new quantity here, a two-component Weyl spinor ε. When a supersymmetry is global, ε does not depend on spacetime and so

$$\partial_\mu \varepsilon = 0 \tag{12.18}$$

This will not be the case when a supersymmetric transformation is local.

We proceed by considering the variation of the Lagrangian term by term. Starting with the first term in Eq. (12.16),

$$\delta L_B = \delta(\partial_\mu A^* \partial^\mu A)$$

$$= \delta(\partial_\mu A^*)\partial^\mu A + \partial_\mu A^* \delta(\partial^\mu A)$$

$$= \partial_\mu(\sqrt{2}\varepsilon^\dagger \psi^\dagger)\partial^\mu A + \partial_\mu A^* \partial^\mu(\sqrt{2}\varepsilon\psi)$$

$$= \sqrt{2}\varepsilon^\dagger \partial_\mu \psi^\dagger \partial^\mu A + \sqrt{2}\varepsilon \partial_\mu A^* \partial^\mu \psi \tag{12.19}$$

Now you remember that $\delta(\partial_\mu \varphi) = \partial_\mu(\delta\varphi)$, a trick we used in moving from the second line to the third line. Next, we compute the variation of the second term in Eq. (12.16), which is

$$\delta \frac{i}{2}\partial_\mu \psi^\dagger \bar{\sigma}^\mu \psi = \frac{i}{2}\partial_\mu(\delta\psi^\dagger)\bar{\sigma}^\mu \psi + \frac{i}{2}\partial_\mu \psi^\dagger \bar{\sigma}^\mu(\delta\psi)$$

$$= \frac{i}{2}\partial_\mu(-i\sqrt{2}\varepsilon\sigma^\nu \partial_\nu A^* + \sqrt{2}\varepsilon^\dagger F^*)\bar{\sigma}^\mu \psi$$

$$+ \frac{i}{2}\partial_\mu \psi^\dagger \bar{\sigma}^\mu(i\sqrt{2}\sigma^\nu \varepsilon^\dagger \partial_\nu A + \sqrt{2}\varepsilon F)$$

$$= \frac{1}{\sqrt{2}}\varepsilon\sigma^\nu \bar{\sigma}^\mu(\partial_\mu \partial_\nu A^*)\psi + \frac{i}{\sqrt{2}}\varepsilon^\dagger \partial_\mu F^* \bar{\sigma}^\mu \psi$$

$$- \frac{1}{\sqrt{2}}\partial_\mu \psi^\dagger \bar{\sigma}^\mu \sigma^\nu \varepsilon^\dagger \partial_\nu A + \frac{i}{\sqrt{2}}\partial_\mu \psi^\dagger \bar{\sigma}^\mu \varepsilon F \tag{12.20}$$

Continuing, the next term is similar.

$$\delta\left(-\frac{i}{2}\psi^\dagger \bar{\sigma}^\mu \partial_\mu \psi\right) = -\frac{i}{2}(\delta\psi^\dagger)\bar{\sigma}^\mu \partial_\mu \psi - \frac{i}{2}\psi^\dagger \bar{\sigma}^\mu \partial_\mu(\delta\psi)$$

$$= -\frac{i}{2}(-i\sqrt{2}\varepsilon\sigma^\nu \partial_\nu A^* + \sqrt{2}\varepsilon^\dagger F^*)\bar{\sigma}^\mu \partial_\mu \psi$$

$$- \frac{i}{2}\psi^\dagger \bar{\sigma}^\mu \partial_\mu(i\sqrt{2}\sigma^\nu \varepsilon^\dagger \partial_\nu A + \sqrt{2}\varepsilon F)$$

$$= -\frac{1}{\sqrt{2}}\varepsilon\sigma^\nu \bar{\sigma}^\mu \partial_\nu A^* \partial_\mu \psi - \frac{i}{\sqrt{2}}\varepsilon^\dagger F^* \bar{\sigma}^\mu \partial_\mu \psi$$

$$+ \frac{1}{\sqrt{2}}\psi^\dagger \bar{\sigma}^\mu \sigma^\nu \varepsilon^\dagger \partial_\mu \partial_\nu A - \frac{i}{\sqrt{2}}\psi^\dagger \bar{\sigma}^\mu \varepsilon \partial_\mu F \tag{12.21}$$

Remember, the Pauli matrices only operate on the spinors, so we can move them past the bosonic fields however we like. Finally, we compute δFF^*.

$$\delta FF^* = (\delta F)F^* + F(\delta F^*)$$

$$= (i\sqrt{2}\varepsilon^\dagger \bar{\sigma}^\mu \partial_\mu \psi)F^* + F(-i\sqrt{2}\partial_\mu \psi^\dagger \bar{\sigma}^\mu \varepsilon) \tag{12.22}$$

Now it is useful to consider terms dependent on ε only. Grouping them together using Eqs. (12.19) through (12.22), we have

$$\delta L_\varepsilon = -\sqrt{2}\varepsilon \partial_\mu A^* \partial^\mu \psi + \frac{1}{\sqrt{2}} \varepsilon \sigma^\nu \bar{\sigma}^\mu (\partial_\mu \partial_\nu A^*)\psi - \frac{1}{\sqrt{2}} \varepsilon \sigma^\nu \bar{\sigma}^\mu \partial_\nu A^* \partial_\mu \psi$$

$$- \frac{i}{\sqrt{2}} \partial_\mu \psi^\dagger \bar{\sigma}^\mu \varepsilon F - \frac{i}{\sqrt{2}} \psi^\dagger \bar{\sigma}^\mu \varepsilon \partial_\mu F \tag{12.23}$$

We wish to simplify this expression so that it can be written as a total derivative, which would allow us to satisfy $\delta S = \delta \int L\, d^4x = 0$. First, notice that Eq. (12.23) can be written as

$$\delta L_\varepsilon = -\sqrt{2}\varepsilon \partial_\mu A^* \partial^\mu \psi + \frac{1}{\sqrt{2}} \varepsilon \sigma^\nu \bar{\sigma}^\mu (\partial_\mu \partial_\nu A^*)\psi - \frac{1}{\sqrt{2}} \varepsilon \sigma^\nu \bar{\sigma}^\mu \partial_\nu A^* \partial_\mu \psi$$

$$- \frac{i}{\sqrt{2}} \partial_\mu \psi^\dagger \bar{\sigma}^\mu \varepsilon F - \frac{i}{\sqrt{2}} \psi^\dagger \bar{\sigma}^\mu \varepsilon \partial_\mu F$$

$$= \partial_\mu \left(-\frac{i}{\sqrt{2}} \psi^\dagger \bar{\sigma}^\mu \varepsilon F \right) - \sqrt{2}\varepsilon \partial_\mu A^* \partial^\mu \psi + \frac{1}{\sqrt{2}} \varepsilon \sigma^\nu \bar{\sigma}^\mu (\partial_\mu \partial_\nu A^*)\psi$$

$$- \frac{1}{\sqrt{2}} \varepsilon \sigma^\nu \bar{\sigma}^\mu \partial_\nu A^* \partial_\mu \psi$$

Now we can apply Eq. (12.15) to simplify this even further. Recall that

$$\{\bar{\sigma}^\mu, \sigma^\nu\} = \bar{\sigma}^\mu \sigma^\nu + \sigma^\nu \bar{\sigma}^\mu = -2g^{\mu\nu}$$

$$\Rightarrow \sigma^\nu \bar{\sigma}^\mu = -2g^{\mu\nu} - \bar{\sigma}^\mu \sigma^\nu$$

Also note that repeated indices are *dummy indices,* so we can change them. The next to last term can then be rewritten as

$$\frac{1}{\sqrt{2}}\varepsilon\sigma^\nu\bar{\sigma}^\mu(\partial_\mu\partial_\nu A^*)\psi = \frac{1}{\sqrt{2}}\varepsilon\sigma^\mu\bar{\sigma}^\nu(\partial_\nu\partial_\mu A^*)\psi$$

$$= \frac{1}{\sqrt{2}}\varepsilon\sigma^\mu\bar{\sigma}^\nu(\partial_\mu\partial_\nu A^*)\psi$$

Now, applying the anticommutation relation and raising an index with the metric,

$$\frac{1}{\sqrt{2}}\varepsilon\sigma^\mu\bar{\sigma}^\nu(\partial_\mu\partial_\nu A^*)\psi = \frac{1}{\sqrt{2}}\varepsilon(-2g^{\mu\nu}-\sigma^\nu\bar{\sigma}^\mu)(\partial_\mu\partial_\nu A^*)\psi$$

$$= -\sqrt{2}\,\varepsilon\,g^{\mu\nu}(\partial_\mu\partial_\nu A^*)\psi - \frac{1}{\sqrt{2}}\varepsilon\sigma^\nu\bar{\sigma}^\mu(\partial_\mu\partial_\nu A^*)\psi$$

$$= -\sqrt{2}\,\varepsilon(\partial_\mu\partial^\mu A^*)\psi - \frac{1}{\sqrt{2}}\varepsilon\sigma^\nu\bar{\sigma}^\mu(\partial_\mu\partial_\nu A^*)\psi$$

Therefore,

$$-\sqrt{2}\varepsilon\partial_\mu A^*\partial^\mu\psi + \frac{1}{\sqrt{2}}\varepsilon\sigma^\nu\bar{\sigma}^\mu(\partial_\mu\partial_\nu A^*)\psi - \frac{1}{\sqrt{2}}\varepsilon\sigma^\nu\bar{\sigma}^\mu\partial_\nu A^*\partial_\mu\psi$$

$$= -\sqrt{2}\varepsilon\partial_\mu A^*\partial^\mu\psi - \sqrt{2}\,\varepsilon(\partial_\mu\partial^\mu A^*)\psi - \frac{1}{\sqrt{2}}\varepsilon\sigma^\nu\bar{\sigma}^\mu(\partial_\mu\partial_\nu A^*)\psi$$

$$-\frac{1}{\sqrt{2}}\varepsilon\sigma^\nu\bar{\sigma}^\mu\partial_\nu A^*\partial_\mu\psi$$

$$= -\partial_\mu(\sqrt{2}\varepsilon\partial_\mu A^*\psi) - \partial_\mu\left(\frac{1}{\sqrt{2}}\varepsilon\sigma^\nu\bar{\sigma}^\mu\partial_\nu A^*\psi\right)$$

So here is the variation of the Lagrangian, considering terms dependent upon ε, written as a total derivative.

$$\delta L_\varepsilon = \partial_\mu\left(-\frac{i}{\sqrt{2}}\psi^\dagger\bar{\sigma}^\mu\varepsilon F\right) - \partial_\mu(\sqrt{2}\varepsilon\partial_\mu A^*\psi) - \partial_\mu\left(\frac{1}{\sqrt{2}}\varepsilon\sigma^\nu\bar{\sigma}^\mu\partial_\nu A^*\psi\right)$$

We use this to identify a current.

$$K^\mu_{\ \varepsilon} = -\frac{i}{\sqrt{2}} \psi^\dagger \bar{\sigma}^\mu \varepsilon F - \sqrt{2}\varepsilon \partial_\mu A^* \psi - \frac{1}{\sqrt{2}} \varepsilon \sigma^\nu \bar{\sigma}^\mu \partial_\nu A^* \psi \qquad (12.24)$$

A similar procedure can be used to define a current dependent upon $\varepsilon^\dagger$, giving us a total current.

$$K^\mu = K^\mu_{\ \varepsilon} + K^\mu_{\ \varepsilon^\dagger} \qquad (12.25)$$

By calculating the Noether current J^μ, we can arrive at a supercurrent that is conserved.

$$S^\mu = J^\mu - K^\mu$$
$$\partial_\mu S^\mu = 0 \qquad (12.26)$$

The Noether current is calculated as follows. We take the kinetic part of the Lagrangian, which is given by

$$J^\mu = \frac{i}{2}(\delta\psi^\dagger)\bar{\sigma}^\mu \psi - \frac{i}{2}\psi^\dagger \bar{\sigma}^\mu (\delta\psi) - (\delta A^*)\partial^\mu A - \partial^\mu A^* (\delta A)$$

The variation procedure is accomplished in the same manner as used to calculate δL, using Eq. (12.17). Once again, we can extract ε dependent and $\varepsilon^\dagger$ dependent Noether currents. It can be shown that

$$J^\mu_\varepsilon = \frac{1}{\sqrt{2}} \varepsilon \sigma^\nu \bar{\sigma}^\mu \psi \partial_\nu A^* - \frac{i}{\sqrt{2}} \psi^\dagger \bar{\sigma}^\mu \varepsilon F - \sqrt{2}\varepsilon \psi \partial^\mu A^*$$

Hence, the ε dependent supercurrent in Eq. (12.26) takes on a relatively simple form.

$$S^\mu_\varepsilon = \sqrt{2}\varepsilon \sigma^\nu \bar{\sigma}^\mu \psi \partial_\nu A^* \qquad (12.27)$$

The supercharges are computed by integrating the time-component of the supercurrent, in a manner analogous to electrodynamics. For example,

$$Q_\varepsilon = \int d^3x \, S^0_\varepsilon = \int d^3x \sqrt{2}\varepsilon \sigma^\nu \bar{\sigma}^0 \psi \partial_\nu A^* \qquad (12.28)$$

Summary

Supersymmetry is a proposition to introduce a symmetry between fermions and bosons. If such a symmetry exists, then there are supercharge operators that convert fermion states into boson states and vice versa. In its most basic form, the theory predicts that the partners of known particles, obtained by applying the supercharge operators and known as superpartners, have the same mass. This has not been experimentally observed. If supersymmetry is real, the masses of the superpartners are much larger than the masses of known particles, and this explains why they have not yet been detected experimentally. The difference in masses breaks the supersymmetry. Hence we know supersymmetry is broken. Theorists have great hopes for the theory because it solves many outstanding problems in theoretical physics, such as the mass of the Higgs particle (the hierarchy problem). Supersymmetry may also explain the existence of mysterious dark matter particles and it is of fundamental importance to string theory.

Quiz

1. Consider the supersymmetric quantum mechanics described in Examples 12.1 and 12.2. Compute $\{Q_1, Q_2\}$ and $\{Q_2, Q_2\}$.

   For problems 2 to 4, consider the Lagrangian

   $$L = \frac{i}{2}\partial_n\chi\sigma^n\bar{\chi} - \frac{i}{2}\sigma^n\partial_n\bar{\chi} - \partial_n\bar{A}\partial^n A + \bar{F}F$$

   which describes a left handed spinor χ that does not interact with a complex spin-0 field A.

2. Find the field equations.

3. If the supersymmetry transformation is

   $$\delta A = \sqrt{2}\chi\varepsilon$$
   $$\delta\chi = -i\sqrt{2}\bar{\varepsilon}\bar{\sigma}^m\partial_m A + \sqrt{2}\varepsilon F$$
   $$\delta F = -i\sqrt{2}\partial_m\chi\sigma^m\bar{\varepsilon}$$

   find the SUSY current.

4. Find an expression for the supercharge.

5. Let an operator A be given by $A = (-1)^{2S}$ where S is the spin operator. Given that

$$\{Q, A\} = \{Q^\dagger, A\} = 0$$

calculate $\sum_i \langle i | (-1)^{2S} P^\mu | i \rangle$ where $|i\rangle$ is a set of fermion or boson states belonging to the same multiplet. (*Hint:* Assume that the states satisfy a completeness relation

$$\sum_j |j\rangle \langle j| = I$$

The result of this calculation implies that each supermultiplet contains the same number of bosonic and fermionic degrees of freedom.)

Final Exam

1. Consider the following Einstein's equation relating energy, mass, and momentum.

$$E^2 = \vec{p}^2 + m^2$$

Make the usual operator substitutions from quantum mechanics, that is,

$$E \to i\frac{\partial}{\partial t} \qquad \vec{p} \to -i\vec{\nabla}$$

Determine the resulting field equation.

2. Let $\mathcal{L} = \frac{1}{2}\{(\partial_\mu \varphi)^2 - m^2 \varphi^2\}$ where φ is a real scalar field and determine the conserved quantity Q.

3. Consider an electromagnetic type field with mass, and let $S = \int(-\frac{1}{4}F_{\mu\nu}F^{\mu\nu} + \frac{m^2}{2}A_\mu A^\mu)$. Vary the action to determine and equation of motion.

4. Return to the action of the previous problem and the equation of motion you found. What condition on the vector potential does the equation of motion imply?

5. Suppose that $\mathcal{L} = \frac{1}{2}\partial_\mu\varphi\partial^\mu\varphi - \frac{m^2}{2}\varphi^2 + \frac{\lambda^3}{6}\varphi^3 + \frac{\rho^4}{24}\varphi^4$, but the field is invariant under a parity transformation $\varphi \to -\varphi$. What restrictions does this place on the Lagrangian?

6. Calculate $tr(\gamma^\mu\gamma^\nu)$.

7. Find $i\partial_\mu(\bar{\psi}\gamma^\mu\psi)$.

8. Calculate $\partial\!\!\!/\partial\!\!\!/$ where $a\!\!\!/ = \gamma^\mu a_\mu$.

9. The Dirac equation applies to

   (a) Spin-1 particles

   (b) Spin-3/2 particles

   (c) Spin-½ particles

   (d) Both (b) and (c)

10. If $\gamma^\mu\partial_\mu\psi = 0$ find $\gamma^\mu\partial_\mu(\gamma_5\psi)$.

11. Find $(I + \gamma_5)\psi$ if $\psi = \begin{pmatrix} \psi_L \\ \psi_R \end{pmatrix}$.

12. Supersymmetry is a symmetry that relates

    (a) Half-integral fields

    (b) Scalar fields to the higgs field

    (c) Quarks and antiquarks

    (d) Fermions and bosons

13. The Higgs field can be best described as responsible for

    (a) Dark energy

    (b) Dark matter

    (c) Giving mass to fundamental particles

    (d) Relating supersymmetry to $SU(5)$ transformations

14. If supersymmetry is unbroken then

    (a) The masses of a fermion and superpartner boson are the same.

    (b) The masses of a fermion and superpartner boson are inverse.

    (c) There is no relation between the mass of a fermion and its superpartner boson.

    (d) The mass of the Higgs particle is 100 GeV.

15. The commutation or anticommutation relation obeyed by a supercharge operator Q is

    (a) $[Q, Q^\dagger] = P^\mu$

    (b) $[Q, Q^\dagger] = 0$

(c) $\{Q,Q^\dagger\} = P^\mu$

(d) $\{Q,Q^\dagger\} = 0$

16. The commutation relation satisfied between the supercharge operator Q and 4-momentum is

    (a) $[P^2,Q] = 0$

    (b) $[P^2,Q] = \lambda$

    (c) $[P^2,Q] = -Q$

    (d) $[P^2,Q] = Q^\dagger$

17. In supersymmetric quantum mechanics, the supercharges satisfy

    (a) $\{Q_i,Q_j\} = 0$

    (b) $\{Q_i,Q_j\} = \delta_{ij}$

    (c) $[Q_i,Q_j] = \delta_{ij}H$

    (d) $\{Q_i,Q_j\} = \delta_{ij}H$

18. In the chiral representation, the Pauli matrices satisfy

    (a) $\{\sigma^\mu,\bar{\sigma}^\nu\} = \sigma^\mu\bar{\sigma}^\nu + \bar{\sigma}^\nu\sigma^\mu - 2g^{\mu\nu}$

    (b) $\{\sigma^\mu,\bar{\sigma}^\nu\} = 2g^{\mu\nu}$

    (c) $\{\sigma^\mu,\bar{\sigma}^\nu\} = \sigma^\mu\bar{\sigma}^\nu + \bar{\sigma}^\nu\sigma^\mu = -2g^{\mu\nu}$

19. The supercurrent satisfies

    (a) $\partial_\mu S^\mu = J^\mu$, where J^μ is the Noether current

    (b) $\partial_\mu S^\mu = 0$

    (c) The supercurrent cannot be conserved

    (d) An uncertainty relation with the supercharge

20. In order to close the supersymmetry algebra, it can be necessary to introduce

    (a) An auxiliary field

    (b) A supersymmetric Hamiltonian operator

    (c) The uncertainty relations

    (d) The Cauchy-Schwarz lemma

21. A group is abelian if group elements a and b satisfy

    (a) $\{a,b\} = ab + ba = 0$

    (b) $[ab - ba] = e$, where e is the identity element

    (c) $[ab - ba] = 0$

    (d) ab is closed

22. A Lie group

 (a) Depends on a finite set of continuous parameters θ_i

 (b) Depends on a finite set of discrete parameters that are periodic in 2π

 (c) Does not have derivatives with respect to group elements

 (d) Obeys an open algebra

23. In a Lie group, a generator X is related to a group element g through

 (a) $X = \dfrac{\partial g}{\partial \theta}\big|_{\theta=\pi}$

 (b) $X = \dfrac{\partial^2 g}{\partial \theta^2}\big|_{\theta=0}$

 (c) $X = \dfrac{\partial g}{\partial \theta}\big|_{\theta=0} + \displaystyle\int_o^{2\pi} g(\phi)d\phi$

 (d) $X = \dfrac{\partial g}{\partial \theta}\big|_{\theta=0}$

24. A representation D of a group can be related to the generators X using

 (a) $D(\varepsilon\theta) \approx 1 + i\varepsilon\theta X$

 (b) $D(\varepsilon\theta) = i\varepsilon\theta X$

 (c) $D(\varepsilon\theta) \approx \varepsilon \dfrac{\partial X}{\partial \theta}$

 (d) $D(\varepsilon\theta) \approx \lim_{\theta\to\infty} 1 - i\varepsilon\theta X$

25. If the generator of a group X is Hermitian then the representation D is

 (a) Anti-Hermitian

 (b) Unitary

 (c) Anti-unitary

26. The Lie algebra of a group is $[X_i, X_j] = if_{ijk}X_k$. We call the coefficients f_{ijk}

 (a) Representation constants

 (b) Group generators

 (c) The fine structure constants

 (d) The structure constants of the group

27. A group $SO(N)$ is

 (a) Orthogonal $N \times N$ matrices with determinant $+ 1$

 (b) Orthogonal $N \times N$ matrices with determinant $- 1$

(c) Unitary, orthogonal $N \times N$ matrices with determinant $+1$

(d) Unitary $N \times N$ matrices with determinant $+1$

28. The special unitary group $SU(2)$ has

(a) 1 generator

(b) 3 generators

(c) 2 generators

(d) 8 generators

29. The special unitary group $SU(3)$ has

(a) 1 generator

(b) 3 generators

(c) 4 generators

(d) 8 generators

30. The unitary group $U(1)$ can be represented by

(a) $U = e^{-i\theta}$

(b) $U = \int e^{-i\theta}$

(c) $U = \dfrac{dg}{d\theta}$

31. The Pauli matrices are a representation of

(a) $SU(3)$

(b) $U(1)$

(c) $SU(2)$

(d) $SU(1)$

32. The rank of the group is defined as

(a) The number of matrix representations of the generators that are diagonal

(b) The number of generators

(c) The number of generators minus one

(d) The number of matrix representations of the generators that are diagonal, minus one

33. A Casimir operator

(a) Forms a finite representation of the group

(b) Commutes with all the generators

(c) Does not commute with any generator

    (d) Commutes with the rank representation

34. Consider a quantum number. If the quantum number is multiplicative

    (a) For a composite system, the quantum number is the product of individual quantum numbers $\prod_i n_i$.

    (b) For a composite system, the quantum number is the sum of individual quantum numbers.

    (c) The quantum number is a product of fundamental constants.

    (d) The quantum number is a sum of fundamental constants.

35. The eigenvalues of parity are

    (a) $\alpha = 0, \pm 1$

    (b) $\alpha = 0, 1$

    (c) $\alpha = \pm 1$

36. If a wave function has even parity then

    (a) $\psi(-x) = \psi(x)$

    (b) $\psi(-x) = -\psi(x)$

    (c) $\psi(-x) = 0$

    (d) $\psi(-2x) = 2\psi(x)$

37. The parity operator P satisfies

    (a) $P^2 = iI$

    (b) $P^2 = 0$

    (c) $P^2 = -I$

    (d) $P^2 = I$

38. A parity operator acts on an angular momentum state as

    (a) $P|L,m_z\rangle = (-1)^{m_z L}|L,m_z\rangle$

    (b) $P|L,m_z\rangle = L|L,m_z\rangle$

    (c) $P|L,m_z\rangle = -|L,m_z\rangle$

    (d) $P|L,m_z\rangle = (-1)^{L}|L,m_z\rangle$

39. By convention

    (a) An electron has positive parity.

    (b) An electron has negative parity.

(c) An antielectron has positive parity.

(d) The parity of an electron is indeterminate.

40. The parity of a composite system *ab* each with individual parity operators P_a, P_b is

    (a) $-P_a P_b$

    (b) $P_a P_b$

    (c) $(-1)^{P_a P_b} P_a P_b$

    (d) $(-1)^{P_a + P_b} P_a P_b$

41. Parity is

    (a) Conserved in the electromagnetic and strong interactions

    (b) Not conserved in the weak interaction

    (c) Conserved in the weak interaction

    (d) Both a and b

    (e) Both a and c

42. A particle denoted by 0^-

    (a) Has spin-0 and negative parity

    (b) Has no parity and negative spin-½

    (c) Is a scalar particle with negative parity

    (d) Has zero charge and negative parity

43. Charge conjugation

    (a) Only applies to vector particles

    (b) Turns particles into antiparticles

    (c) Turns particles into antiparticles with a sign change

    (d) Flips charge for scalar bosons

44. The charge conjugation operator C acts on the electromagnetic field as

    (a) $CA^\mu C^{-1} = -A^\mu$

    (b) $CA^\mu C^{-1} = A^\mu$

    (c) $CA^\mu C^{-1} = -J^\mu$

    (d) $CA^\mu C^{-1} = J^\mu$

45. CP is

    (a) Never violated

    (b) Violated in the strong interaction

    (c) Violated in the weak interaction

    (d) Violated in the electroweak interaction

46. An operator A is antiunitary. It satisfies

    (a) $\langle A\phi | A\psi \rangle = \langle \phi | \psi \rangle$

    (b) $\langle A^\dagger \phi | A\psi \rangle = \langle \phi | \psi \rangle^*$

    (c) $\langle A\phi | A\psi \rangle = \langle \phi | \psi \rangle^*$

    (d) $\langle A\phi | A\psi \rangle = -\langle \phi | \psi \rangle^*$

47. An antilinear operator satisfies

    (a) $T(\alpha | \psi \rangle + \beta | \phi \rangle) = \alpha^* | \psi' \rangle + \beta^* | \phi' \rangle$

    (b) $T(\alpha | \psi \rangle + \beta | \phi \rangle) = -\alpha | \psi \rangle - \beta | \phi \rangle$

    (c) $T(\alpha | \psi \rangle + \beta | \phi \rangle) = \alpha^* \langle \psi | + \beta^* \langle \phi |$

48. The *CPT* theorem implies that

    (a) *CPT* is conserved in all interactions except the weak interaction.

    (b) *CPT* is conserved in all interactions.

    (c) *CPT* is conserved only in weak interactions.

    (d) *CPT* is conserved except in K meson decay.

49. The eigenvalues of charge conjugation are

    (a) $c = \pm 1$

    (b) $c = 0, \pm 1$

    (c) $c = \pm q$

    (d) $c = 0, \pm q$

50. If *CP* were conserved in weak interactions then

    (a) $| K_1 \rangle \rightarrow$ would only decay into 2π mesons

    (b) $| K_1 \rangle \rightarrow$ only decays into 3π mesons

    (c) Time invariance would not be satisfied

    (d) $| K_2 \rangle \rightarrow$ only decays into 2π mesons

51. When interpreted as a single particle wave equation, the Klein-Gordon equation

    (a) Is plagued by infinities

    (b) Leads to negative probability densities

    (c) Always gives zero

    (d) Gives the same results as the Schrödinger equation

52. Under a Lorentz transformation $\Lambda^{\mu}{}_{\nu}$, a scalar field transforms as

    (a) $\varphi'(x) = -\varphi(\Lambda^{-1}x)$

    (b) $\varphi'(x) = \varphi(\Lambda x)$

    (c) $\varphi'(x) = -\varphi(\Lambda x)$

    (d) $\varphi'(x) = \varphi(\Lambda^{-1}x)$

53. The Klein-Gordon equation was deemed incorrect because it leads to solutions for the energy as

    (a) $E = \pm\sqrt{p^2 + m^2}$

    (b) $E = -\sqrt{p^2 + m^2}$

    (c) $E = \sqrt{p^2 + m^2}$

    (d) $E = p^2 + m^2$

54. The probability density of the Klein-Gordon equation is given by

    (a) $\rho = \varphi^*\varphi$

    (b) $\rho = -i\left(\varphi^*\dfrac{\partial\varphi}{\partial t} - \varphi\dfrac{\partial\varphi^*}{\partial t}\right)$

    (c) $\rho = i\left(\varphi^*\dfrac{\partial\varphi}{\partial t} - \varphi\dfrac{\partial\varphi^*}{\partial t}\right)$

    (d) $\rho = -\varphi^*\varphi$

55. Second quantization

    (a) Imposes *equal time* commutation relations on the fields and their conjugate momenta

    (b) Imposes *no time* commutation relations on the fields and their conjugate momenta

    (c) Promotes time to an operator

    (d) Imposes *equal time* commutation relations on the fields at the same spacetime point

56. Particles can be created and destroyed at relativistic energies. In particular
    (a) High energy processes tend to create antiparticles where $E = mc^2$.
    (b) To create an antiparticle, we need the rest mass energy $E = mc^2$.
    (c) To create a particle, we need the rest mass energy $E = mc^2$.
    (d) To create a particle, we need twice the rest mass energy $E = mc^2$ and this creates a particle–antiparticle pair.

57. The creation and annihilation operators of the harmonic oscillator satisfy
    (a) $[\hat{a}, \hat{a}^\dagger] = i$
    (b) $[\hat{a}, \hat{a}^\dagger] = 1$
    (c) $[\hat{a}, \hat{a}^\dagger] = -1$
    (d) $[\hat{a}, \hat{a}^\dagger] = -i$

58. The number operator is defined by
    (a) $\hat{N} = \hat{a}\hat{a}^\dagger$
    (b) $\hat{N} = \hat{a}^\dagger \hat{a}$
    (c) $\hat{N} = -\hat{a}^\dagger \hat{a}$

59. The number operator satisfies
    (a) $[\hat{N}, \hat{a}^\dagger] = \hat{a}^\dagger$
    (b) $[\hat{N}, \hat{a}^\dagger] = -\hat{a}^\dagger$
    (c) $[\hat{N}, \hat{a}] = \hat{a}^\dagger$
    (d) $[\hat{N}, \hat{a}] = \hat{a}$

60. For a scalar field φ, the equal time commutation relation that is satisfied is
    (a) $[\hat{\varphi}(x), \hat{\pi}(y)] = 0$
    (b) $[\hat{\varphi}(x), \hat{\pi}(y)] = i\delta(\vec{x} - \vec{y})$
    (c) $[\hat{\varphi}(x), \hat{\pi}(y)] = i$
    (d) $[\hat{\varphi}(x), \hat{\pi}(y)] = -i\delta(\vec{x} - \vec{y})$

61. The Fourier expansion of a field is given by
$$\varphi(x) = \int \frac{d^3 p}{\sqrt{(2\pi)^3 2p^0}} \left[a(\vec{p})e^{ipx} + a^\dagger(\vec{p})e^{-ipx} \right]$$

Find the conjugate momentum.

62. In a field theory, the creation and annihilation operators satisfy

   (a) $\left[a(\vec{p}), a^{\dagger}(\vec{p}')\right] = \delta(\vec{p} - \vec{p}')$

   (b) $\left[a(\vec{p}), a^{\dagger}(\vec{p}')\right] = 0$

   (c) $\left[a(\vec{p}), a^{\dagger}(\vec{p}')\right] = i$

   (d) $\left[a(\vec{p}), a^{\dagger}(\vec{p}')\right] = \delta_{p,p'}$

63. To calculate the energy of the vacuum, we find

   (a) $\langle 0 | 0 \rangle$

   (b) $\langle 0 | \hat{a} + \hat{a}^{\dagger} | 0 \rangle$

   (c) It cannot be calculated

   (d) $\langle 0 | \hat{H} | 0 \rangle$

64. The normal product is

   (a) $: \hat{a}(\vec{k})\hat{a}^{\dagger}(\vec{k}) := \hat{a}^{\dagger}(\vec{k})\hat{a}(\vec{k})$

   (b) $: \hat{a}(\vec{k})\hat{a}^{\dagger}(\vec{k}) := \hat{a}(\vec{k})\hat{a}^{\dagger}(\vec{k})$

   (c) $: \hat{a}(\vec{k})\hat{a}^{\dagger}(\vec{k}) := -\hat{a}^{\dagger}(\vec{k})\hat{a}(\vec{k})$

   (d) $: \hat{a}(\vec{k})\hat{a}^{\dagger}(\vec{k}) := -\hat{a}(\vec{k})\hat{a}^{\dagger}(\vec{k})$

65. The time ordered product of two fields is

   (a) $T(\varphi(t_1)\psi(t_2)) = \psi(t_2)\varphi(t_1) - \varphi(t_1)\psi(t_2)$

   (b) $T(\varphi(t_1)\psi(t_2)) = \begin{cases} \varphi(t_1)\psi(t_2) \text{ if } t_1 > t_2 \\ \psi(t_2)\varphi(t_1) \text{ if } t_2 > t_1 \end{cases}$

   (c) $T(\varphi(t_1)\psi(t_2)) = \begin{cases} \psi(t_2)\varphi(t_1) \text{ if } t_1 > t_2 \\ \varphi(t_1)\psi(t_2) \text{ if } t_2 > t_1 \end{cases}$

66. When including particles and antiparticles

   (a) We only include annihilation operators for antiparticles.

   (b) We only include creation operators for particles.

   (c) To get the field operator, we sum up negative frequency parts for particles together with positive frequency parts for antiparticles.

   (d) To get the field operator, we sum up positive frequency parts for particles together with negative frequency parts for antiparticles.

67. The creation and annihilation operators for antiparticles satisfy

    (a) $\left[\hat{b}(\vec{k}),\hat{b}^\dagger(\vec{k}')\right]=\delta(\vec{k}-\vec{k}')$

    (b) $\left[\hat{b}(\vec{k}),\hat{b}^\dagger(\vec{k}')\right]=-\delta(\vec{k}-\vec{k}')$

    (c) $\left[\hat{b}(\vec{k}),\hat{a}^\dagger(\vec{k}')\right]=\delta(\vec{k}-\vec{k}')$, where $\hat{a}^\dagger$ creates particles

    (d) $\left[\hat{b}(\vec{k}),\hat{b}^\dagger(\vec{k}')\right]=0$

68. For a charged field, the charge operator can be written as

    (a) $\hat{Q}=\int d^3k\,\vec{k}\left[\hat{a}^\dagger(\vec{k})\hat{a}(\vec{k})+\hat{b}^\dagger(\vec{k})\hat{b}(\vec{k})\right]=\hat{N}_{\hat{a}}-\hat{N}_{\hat{b}}$

    (b) $\hat{Q}=\int d^3k\left[\hat{a}^\dagger(\vec{k})\hat{a}(\vec{k})+\hat{b}^\dagger(\vec{k})\hat{b}(\vec{k})\right]=\hat{N}_{\hat{a}}-\hat{N}_{\hat{b}}$

    (c) $\hat{Q}=\int d^3k\,k\left[\hat{a}^\dagger(\vec{k})\hat{a}(\vec{k})+\hat{b}^\dagger(\vec{k})\hat{b}(\vec{k})\right]=\hat{N}_{\hat{a}}-\hat{N}_{\hat{b}}$

    (d) $\hat{Q}=\int d^3k\left[\hat{a}^\dagger(\vec{k})\hat{a}(\vec{k})-\hat{b}^\dagger(\vec{k})\hat{b}(\vec{k})\right]=\hat{N}_{\hat{a}}-\hat{N}_{\hat{b}}$

69. Find the energy for the vacuum using

$$\hat{H}_R=\hat{H}-\int d^3k=\int d^3k\,\omega_k\hat{N}(\vec{k})=\int d^3k\,\omega_k\hat{a}^\dagger(\vec{k})\hat{a}(\vec{k})$$

70. State vectors in the interaction picture evolve in time according to

    (a) The interaction part of the Hamiltonian

    (b) The free part of the Hamiltonian

    (c) Are stationary in time

    (d) The full Hamiltonian

71. Interaction picture and Schrödinger picture operators are related by

    (a) $A_I=e^{iH_0t}A_Se^{-iH_It}$

    (b) $A_I=e^{iH_It}A_Se^{-iH_It}$

    (c) $A_I=e^{iH_0t}A_Se^{-iH_0t}$

    (d) $A_I=-e^{iH_0t}A_Se^{-iH_0t}$

72. In the interaction picture, the time evolution of operators is determined by

    (a) The full Hamiltonian

    (b) The interaction part of the Hamiltonian

    (c) The free part of the Hamiltonian

    (d) They are stationary in time

73. In quantum field theory, scattering

    (a) Results from the exchange of a force-carrying boson

    (b) Results from the exchange of a force-carrying fermion

74. In a Feynman diagram, if the arrow for a particle points against the direction of time flow

    (a) It is a force-carrying boson

    (b) It is an incoming or outgoing antiparticle

    (c) It is an incoming antiparticle or an outgoing particle

    (d) It is an incoming or outgoing particle

75. In a Feynman diagram, conservation of momentum at a vertex

    (a) Is enforced with a Dirac delta function $\delta(\Sigma p_i - q)$

    (b) Momentum is not conserved

    (c) Is enforced at the corresponding absorption vertex

    (d) 4-momentum is not conserved

76. At each vertex in a Feynman diagram

    (a) Add two factors of the coupling constant g

    (b) Take the product of two factors of the coupling constant g

    (c) Include one factor of the coupling constant g

    (d) Add one factor of the inverse coupling constant ig^{-1}

77. A propagator is associated with

    (a) An internal line in a Feynman diagram

    (b) Outgoing lines in a Feynman diagram

    (c) The coupling constant

78. The lifetime of a particle

    (a) Is proportional to the amplitude squared of a process

    (b) Is proportional to the magnitude of a process

    (c) Is proportional to the inverse of the amplitude squared of a process

    (d) Cannot be estimated by perturbation theory

79. The rate of decay of a process

    (a) Is proportional to the amplitude squared of the process

    (b) Cannot be calculated using perturbative expansions

    (c) Is proportional to the inverse of the amplitude squared of the process

80. Feynman diagrams can be best described as

    (a) A trick

    (b) Are a symbolic representation of a perturbative expansion

    (c) Are exact calculations

    (d) Can be used to exactly describe a process to second order

81. In quantum electrodynamics, the electromagnetic force results from

    (a) The exchange of photons

    (b) The exchange of photons and W particles

    (c) The exchange of photons, W particles, and Z particles

    (d) The field only

82. The 4-momentum and polarization vector of a photon state satisfy

    (a) $p_\mu \varepsilon^\mu = -1$

    (b) $p_\mu \varepsilon^\mu = 1$

    (c) $p_\mu \varepsilon^\nu = g_\mu{}^\nu$

    (d) $p_\mu \varepsilon^\mu = 0$

83. The gauge group of electromagnetism is

    (a) $SU(3)$

    (b) $SU(2) \otimes U(1)$

    (c) $SU(2)$

    (d) $U(1)$

84. The gauge group of electroweak theory is

    (a) $SU(3)$

    (b) $SU(2) \otimes U(1)$

    (c) $SU(2)$

    (d) $U(1)$

85. A Dirac particle is interacting with the electromagnetic field. The interaction Lagrangian is best written as

    (a) $L_{int} = -q\bar{\psi}\gamma^\mu \psi A_\mu$

    (b) $L_{int} = -\bar{\psi}\gamma^\mu \psi A_\mu$

    (c) $L_{int} = -q\bar{\psi}\psi A_\mu$

    (d) $L_{int} = m^2 \bar{\psi}\psi A$

86. A global $U(1)$ transformation can be written as

    (a) $\psi(x) \to e^{i\theta(x)}\psi(x)$

    (b) $\psi(x) \to e^{i\theta}\psi(x)$

    (c) $\psi(x) \to e^{-i\theta(x)}\psi(x)$

    (d) $\psi(x) \to e^{i\theta}\psi(x) + \partial_\mu\theta$

87. Consider a Feynman diagram for an electromagnetic process. An outgoing particle is represented by

    (a) $\bar{u}(p,s)$

    (b) $u(p,s)$

    (c) $-\bar{u}(p,s)$

    (d) $\bar{u}(-p,s)$

88. In a Feynman diagram for a QED process, at each vertex we add a factor of

    (a) $g_e = \sqrt{4\pi\alpha}$

    (b) $-ig_e\gamma^\mu$

    (c) $ig_e\gamma^\mu$

    (d) $ig_e\gamma^0$

89. For an internal line in a Feynman diagram in QED, an electron or positron is associated with a propagator of the form

    (a) $\dfrac{i\gamma^\mu q_\mu}{q^2 - m^2}$

    (b) $\dfrac{im}{q^2 - m^2}$

    (c) $\dfrac{i\gamma^\mu}{q^2 - m^2}$

    (d) $\dfrac{i(\gamma^\mu q_\mu + m)}{q^2 - m^2}$

90. Spontaneous symmetry breaking can be best described as

    (a) Setting the minimum potential energy to the coupling constant

    (b) Reducing the minimum potential energy by the ground state energy

    (c) Shifting the minimum of the potential energy such that the energy of the ground state is nonzero

91. Consider scalar fields. A mass term in the Lagrangian can be recognized

    (a) Because it is quadratic in the fields

    (b) Because it is quartic in the fields

    (c) It is linear

    (d) It is real

92. A covariant derivative

    (a) Ensures that the Euler-Lagrange equations are satisfied

    (b) Ensures the Lagrangian is invariant under a Lorentz transformation

    (c) Is not used in quantum electrodynamics

    (d) Can only be used in weak theory

93. Let $\psi = \begin{pmatrix} \psi_R \\ \psi_L \end{pmatrix}$ and compute $\frac{1}{2}(1-\gamma_5)\psi$.

94. The adjoint spinor is given by

    (a) $\psi^\dagger$

    (b) $\bar{\psi} = \psi^\dagger \gamma^0$

    (c) $\bar{\psi} = \gamma^0 \psi$

    (d) γ^0

95. For a spinor, the Lagrangian can be separated into left- and right-handed kinetic parts as

    (a) $L = i\bar{\psi}_L \gamma^\mu \partial_\mu \psi_L + i\bar{\psi}_R \gamma^\mu \partial_\mu \psi_R$

    (b) $L = i\bar{\psi}_L \gamma^\mu \partial_\mu \psi_L - i\bar{\psi}_R \gamma^\mu \partial_\mu \psi_R$

    (c) $L = i\bar{\psi}_L \gamma^\mu \partial_\mu \psi_L + i(1+\gamma_5)\bar{\psi}_R$

    (d) $L = i\bar{\psi}_L (1-\gamma_5)\gamma^\mu \partial_\mu \psi_L + i\bar{\psi}_R (1+\gamma_5)\gamma^\mu \partial_\mu \psi_R$

96. The isospin of the neutrino is

    (a) 0

    (b) $I_\nu^3 = +\frac{3}{2}$

    (c) $I_\nu^3 = +\frac{1}{2}$

    (d) $I_\nu^3 = -\frac{1}{2}$

97. A right-handed electron

    (a) Has isospin +1/2

    (b) Has isospin −1/2

(c) Has isospin +3/2

(d) Has 0 isospin

98. In electroweak theory, conservation of hypercharge corresponds with

(a) A symmetry describing three gauge fields, $SU(2): W_\mu^1, W_\mu^2, W_\mu^3$

(b) A single gauge field $U(1): B_\mu$

(c) A symmetry describing three gauge fields, $SU(2): W_\mu^+, W_\mu^-, Z$

99. The $SU(2)$ transformation of electroweak theory can be written as

(a) $U(\alpha) = \exp(i\alpha_j \tau_j/2)$, where the generators are the Pauli matrices.

(b) $U(\alpha) = \exp(i\alpha_j \lambda_j/2)$, where λ_j are the Gell-Mann matrices.

(c) There is no $SU(2)$ transformation that leaves electroweak theory invariant.

100. The Weinberg angle

(a) Mixes $SU(2)$ and $SU(3)$ symmetries in quantum chromodynamics

(b) Gives a scattering cross section

(c) Mixes the gauge fields in electroweak theory giving rise to the massless electromagnetic field and the massive Z vector boson

Solutions to Quizzes and Final Exam

Chapter 1

1. d	6. a
2. a	7. c
3. a	8. c
4. c	9. c
5. b	10. d

Chapter 2

1. $\dfrac{d^2x}{dt^2} + \omega^2 x = -\dfrac{\alpha}{m}$

2. (a) $\partial_\mu \partial^\mu \varphi - m^2 \varphi = \dfrac{\partial V}{\partial \varphi}$

   (b) $\pi = \dot{\varphi}$

   (c) $H = \displaystyle\int d^3x \left(\dfrac{1}{2}\pi^2 + \dfrac{1}{2}(\nabla\varphi)^2 + \dfrac{1}{2}m^2\varphi^2 + V(\varphi) \right)$

3. $J^\mu = \partial^\mu \varphi$

4. Each field separately satisfies a Klein-Gordon equation, that is, $\partial_\mu \partial^\mu \varphi + m^2\varphi = 0$, $\partial_\mu \partial^\mu \varphi^\dagger + m^2\varphi^\dagger = 0$. To get this result apply Eq. (2.14) to the Lagrangian Eq. (2.37).

5. $Q = i\displaystyle\int d^3x \left(\varphi^\dagger \dfrac{\partial\varphi}{\partial t} - \varphi \dfrac{\partial\varphi^\dagger}{\partial t} \right)$

6. The action is invariant under that transformation.

Chapter 3

1. $U = \cos\alpha + i\sigma_x \sin\alpha$

2. $2\delta_{ij}$

3. 1

4. Try $\vec{\sigma}^2 = \sigma_x^2 + \sigma_y^2 + \sigma_z^2$

5. $K_x = -i \begin{pmatrix} 0 & 1 & 0 & 0 \\ 1 & 0 & 0 & 0 \\ 0 & 0 & 0 & 0 \\ 0 & 0 & 0 & 0 \end{pmatrix}$

6. $K_y = -i \begin{pmatrix} 0 & 0 & 1 & 0 \\ 0 & 0 & 0 & 0 \\ 1 & 0 & 0 & 0 \\ 0 & 0 & 0 & 0 \end{pmatrix}$

7. No, because the algebra among the generators requires the introduction of the angular momentum operators. Therefore, Lorentz transformations together with rotations form a group.

Chapter 4

1. c   4. a

2. a   5. c

3. d

Chapter 5

1. $i\dfrac{\partial\bar{\psi}}{\partial x^{\mu}}\gamma^{\mu}+m\bar{\psi}$

2. 0

3. $v=\dfrac{\vec{k}\cdot\vec{\sigma}}{\omega_{k}+m}u$

4. $\psi(0)=\sqrt{2m}\begin{pmatrix}u\\v\end{pmatrix}$

5. $-\dfrac{i}{2}\begin{pmatrix}\sigma_{1}&0\\0&-\sigma_{1}\end{pmatrix}$

6. $\gamma^{\mu}(i\partial_{\mu}-qA_{\mu})\psi-m\psi=0$

Chapter 6

1. 0

2. $\left[n(\vec{k})+1\right]\hat{a}^{\dagger}(\vec{k})\left|n(\vec{k})\right\rangle$

3. 0

4. $[H,Q]=0$

Chapter 7

1. $-i\dfrac{g^2}{(p_2 - p_4)^2 - m_B^2}$

2. $\dfrac{1}{g^2}$

3. b

4. c

5. a

6. $\alpha = 1/137$

Chapter 8

1. $iqF_{\mu\nu}$

2. a

3. $-g_e^2 [v(k')\gamma^\mu \bar{v}(k)]\dfrac{g_{\mu\nu}}{(p-p')^2}\bar{u}(p')\gamma^\mu u(p)$

4. c

5. a

Chapter 9

1. It describes a massive particle. Use $\cosh ax = 1 + \dfrac{a^2 x^2}{2} + \dfrac{a^4 x^4}{4!} + O(x^6)$.

2. b

3. $L = -\partial_\mu \psi \, \partial^\mu \psi - \psi^2 \partial_\mu \theta \, \partial^\mu \theta - \dfrac{\lambda}{4}\left(\psi^4 - 2\dfrac{\mu^2}{\lambda}\psi + \dfrac{\mu^4}{\lambda^2}\right)$

4. $\dfrac{\mu^2}{2}\psi$

5. $-\dfrac{\lambda}{4}\psi^4$

Chapter 10

1. b   5. b

2. a   6. b

3. d   7. a

4. b   8. a

Chapter 11

1. b   4. c

2. a   5. b

3. a   6. d

Chapter 12

1. 0 $\dfrac{1}{2}p^2 + \dfrac{1}{2}W^2 + \dfrac{1}{2}\sigma_3 \dfrac{dW}{dx}$

2. $i\sigma^n \partial_n \overline{\chi} = 0 \qquad \partial_n \partial^n A = 0 \qquad F = 0$

3. $S_\varepsilon^n = \sqrt{2}\chi\sigma^n \overline{\sigma}^m \varepsilon \partial_n \overline{A}$

4. $Q^a = \sqrt{2}\int d^3x (\chi\sigma^0 \overline{\sigma}^m)^a \partial_m \overline{A}$

5. 0. The states are eigenstates of momentum, so that $P^\mu |i\rangle = p^\mu |i\rangle$. We have

$$\sum_i \langle i|(-1)^{2S} P^\mu |i\rangle = p^\mu Tr[(-1)^{2S}] = p^\mu (n_B - n_F) = 0$$

$$\Rightarrow n_B = n_F$$

Final Exam

1. This is the Klein-Gordon equation $\dfrac{\partial^2 \varphi}{\partial t^2} - \nabla^2 \varphi + m^2 \varphi = 0$.

2. There is no conserved quantity, $Q = 0$.

3. $\partial^\mu F_{\mu\nu} + m^2 A_\nu = 0$

4. $\partial^\mu A_\mu = 0$

5. The φ^3 term must be dropped, so $\mathcal{L} = \dfrac{1}{2} \partial_\mu \varphi \partial^\mu \varphi - \dfrac{m^2}{2} \varphi^2 + \dfrac{\rho^4}{24} \varphi^4$.

6. $4g^{\mu\nu}$

7. 0

8. $\partial_\mu \partial^\mu = \partial^2$

9. c

10. 0

11. $2\psi_R$

12. d

13. c

14. a

15. c

16. a

17. d

18. c

19. b

20. a

21. c

22. a

23. d

24. a

25. b

26. d

27. a

28. b

29. d

30. a

31. c

32. a

33. b

34. a

35. c

36. a

37. d

38. d

39. a

40. b

41. d

42. a

43. b

44. a

45. c

46. c

47. a

48. b

49. a

50. a

51. b

52. d

53. a

54. c

55. a

56. d

57. b

58. b

59. a

60. b

61. $i \int \dfrac{d^3 p}{\sqrt{(2\pi)^3}} \sqrt{\dfrac{p^0}{2}} \left[a(\vec{p}) e^{ipx} - a^\dagger(\vec{p}) e^{-ipx} \right]$

62. a	75. a	88. c
63. d	76. c	89. d
64. a	77. a	90. c
65. b	78. c	91. a
66. d	79. a	92. b
67. a	80. b	93. ψ_L
68. b	81. a	94. b
69. 0	82. d	95. a
70. b	83. d	96. c
71. c	84. b	97. d
72. c	85. a	98. b
73. a	86. b	99. a
74. b	87. a	100. c

References

Quantum Field Theory Demystified is a cursory, introductory treatment of this difficult and rich subject. Those seeking deeper knowledge of the theory should consult one of the many books and papers used in the production of this volume. These are listed below.

Burgess, C. P.: "A Goldstone Primer," http://arxiv.org/abs/hep-ph/9812468.

Cahill, K.: "Elements of Supersymmetry," http://xxx.lanl.gov/abs/hep-ph/9907295.

Cottingham, W. N., and D. A. Greenwood: *An Introduction to the Standard Model of Particle Physics,* Cambridge University Press, London (1998).

Griffiths, D.: *Introduction to Elementary Particles*, John Wiley & Sons, Inc., Hoboken, N.J. (1987).

Guidry, M.: *Gauge Field Theories, An introduction with Applications,* John Wiley & Sons, Hoboken, N.J. (1980).

Halzen, F. and A. Martin: *Quarks and Leptons: An Introductory Course in Modern Particle Physics,* John Wiley & Sons, Hoboken, N.J. (1984).

Itzykson, C. and J. B. Zuber: *Quantum Field Theory,* McGraw-Hill, Inc., New York, N.Y. (1980).

Martin, S.: "A Supersymmetry Primer," http://xxx.lanl.gov/abs/hep-ph/9709356.

Peskin, M. and D. Schroeder: *An Introduction to Quantum Field Theory,* Addison-Wesley, Reading, Mass. (1995).

Ryder, L. H.: *Quantum Field Theory,* Cambridge University Press, London (1996).

Seiden, A.: *Particle Physics: A Comprehensive Introduction,* Addison Wesley, San Francisco, Calif. (2005).

Weinberg, S.: *The Quantum Theory of Fields: Volume I Foundations,* Cambridge University Press, London (1995).

Zee, A.: *Quantum Field Theory in a Nutshell,* Princeton University Press, Princeton, N.J. (2003).

INDEX

A

abelian groups, 50
abstract generators, 53
action (S), 26–29
additive quantum numbers, 71–72
adjoint spinors, 94
amplitudes (M)
  calculating, 151–153
  constructing
    coupling constants, 153–154
    propagators, 154–159
angular momentum operators, 59, 81
annihilation operators, 119, 121,
    127, 134
anticommutation, 87, 103, 258
antilinear time-reversal
    operators, 81
antiparticles, 101, 114, 135
antisymmetry, 41
antiunitary time-reversal operators,
    81
auxiliary fields, 255

B

Baryon numbers, 77
baryons, 17
boosts, 9, 103–104
Bose-Einstein statistics, 131–133
bosons
  defined, 131

  gauge, 12, 63, 224–227
  Goldstone, 202
  Higgs, 18, 75
  supercharge operator, 247
  vector, 76
Brown, Lowell, 233

C

calculating amplitudes, 151–153
canonical equaltime commutation rule,
    168
canonical momentum, 29–30
canonical quantization, 117
Cartesian coordinates, 123
casimir operators, 67–68
charge conjugation (C)
  CP violation, 78–80
  CPT theorem, 81–82
  overview, 76–78
charge operators, 136
chiral representation, 91,
    105, 254
closure property, 50
color charge, 13, 16
complex scalar fields,
    135–137
confinement, 14
conjugate momentum, 122
conservation laws, 35–37, 39
conservation of energy, 37

conservation of probability, 115
conserved currents, 38–39
continuous symmetries, 71
contravariant vectors, 7
Coulomb gauge, 44, 168
coupling constants, 153–154, 164, 217, 219
covariant derivatives, 46
covariant vectors, 7
CP violation, 78–80
CPT theorem, 81–82
creation operators, 119, 121, 127, 134

D

D'Alembertian operators, 10, 112
derivative operators, 42
Dirac delta functions, 126, 151, 156, 176
Dirac equation
  adding quantum theory, 87–89
  adjoint spinors, 94
  boosts, 103–104
  fields, 4, 88
  free space solutions, 99–103
  helicity, 103–104
  matrices
    form of, 89–91
    properties of, 91–94
  overview, 85–87
  quiz
    questions, 108
    solutions, 283
  rotations, 103–104
  slash notation, 95
  solutions of, 95–99
  transformation properties, 94
  Weyl spinors, 104–107
Dirac Lagrangians, 170, 222
Dirac matrices
  classical field theory, 86–87
  form of, 89–91
  properties of, 91–94
  SUSY, 254
Dirac spinors, 210

Dirac-Pauli representation, 89, 99
discrete symmetries
  charge conjugation, 76–78
  CP violation, 78–80
  parity, 72–76
  quiz
    questions, 83
    solutions, 283
  time reversal, 80–81
dummy indices, 42, 258
Dyson series, 145

E

eigenstates, 79
eigenvalue, 72
Einstein summation convention, 7, 31
electrodynamics, 259
electromagnetic field tensor, 39, 165
electromagnetic fields, 39–43
electromagnetic force ($U(1)$), 12, 18–19, 78
electron neutrinos, 213, 218
electron state, 247–248
electron-electron scattering, 176
electron-lepton fields, 223
electron–muon scattering, 184–185
electroweak interactions
  charges of, 213–215
  leptonic fields of, 212–213
electroweak theory
  charges of electroweak interaction, 213–215
  defined, 19
  gauge masses, 224–231
  lepton fields, 212–213, 222–224
  massless Dirac Lagrangians, 211–212
  overview, 209–210
  quiz
    questions, 231–232
    solutions, 285
  right- and left-handed spinors, 210–211
  symmetry breaking, 220–222
  unitary transformations and gauge fields, 215–219
  weak mixing angle, 219–220

elementary particles
  generations of, 17
  leptons, 14–16
  quarks, 16–17
energy density, 136
energy differences, 133
energy-momentum 4-vectors, 11
energy-momentum tensor, 37
equal time commutation relations, 123, 137
equations
  Dirac
    adding quantum theory, 87–89
    adjoint spinors, 94
    boosts, 103–104
    fields, 4, 88
    free space solutions, 99–103
    helicity, 103–104
    matrices, 89–94
    overview, 85–87
    quiz, 108, 283
    rotations, 103–104
    slash notation, 95
    solutions of, 95–99
    transformation properties, 94
    Weyl spinors, 104–107
  Euler-Lagrange
    conserved currents, 38
    equations of motion, 24, 29
    Lagrangian field theory, 31
    symmetries, 36
  Klein-Gordon
    Dirac field, 85, 88
    mass terms in Lagrangians, 192
    overview, 3–4, 110–117
    reinterpreting field, 117
    scalar fields, 109
  Maxwell's, 39, 43, 165
  of motion, 26–29
  Schrödinger, 3, 72, 109
  sine-Gordon equation, 34–35
Euler-Lagrange equations
  conserved currents, 38
  equations of motion, 24, 29
  Lagrangian field theory, 31

$SO(N)$, 60
symmetries, 36

F

families, particle, 17
fermions, 247, 249
Feynman diagrams, 139, 164, 198
Feynman rules
  amplitudes
    calculating, 151–153
    constructing, 153–159
  basics of, 146–151
  interaction picture, 141–143
  lifetimes, 160
  overview, 139–141
  perturbation theory, 143–146
  for QED, 173–185
  quiz
    questions, 160–161
    solutions, 284
  rates of decay, 160
field quantization
  overview, 117, 121–126
  second quantization
    overview, 118–119
    simple harmonic oscillators, 119–121
final exam
  questions, 263–279
  solutions, 286–287
fine structure constant, 164
Fourier transforms, 122
free space solutions, 99–103
frequency decomposition, 128
functional action, 26

G

gamma matrices, 94
gauge bosons, 12, 63, 224–227
gauge fields, 202–204, 215–219
gauge invariance, 170–173
gauge masses, 224–231
gauge potential, 46
gauge transformations, 43–47

gauge unification problem, 247
Gaussian integrals, 233–238, 240
Gell-Mann matrices, 67
Gell-Mann-Nishijima relation, 214
generations, particle, 17
generators, 12, 53
Glashow, Sheldon, 209
global $U(1)$ transformations, 188
gluons, 13
Goldstone bosons, 202
grand unification energy, 19
grand unified theories (GUTs), 18–19
gravitons, 12
ground state, 121
group composition rule, 54
group theory
  casimir operators, 67–68
  Lie groups, 52–54
  overview, 49–50
  parameters, 52
  quiz
    questions, 68–69
    solutions, 282–283
  representations of, 50–52
  rotation groups
    overview, 54–55
    representing, 55–58
  $SO(N)$, 58–61
  unitary groups, 62–67
GUTs (grand unified theories),
  18–19

H

hadrons, 17
Hamiltonian density, 32, 37
Hamiltonian function, 2, 29–30, 141
Hamiltonian operators, 62, 85, 132
harmonic oscillators, 26,
  119–121, 133
Heaviside function, 175
helicity, 103–104
helicity operators, 104
Hermitian conjugate, 63
hierarchy problem, 246

Higgs bosons, 18, 75
Higgs fields, 18, 206, 220, 224
Higgs mass, 246
Higgs mechanism, 18, 202–207
Higgs, Peter, 202
Higgsino fermionic fields, 249
homomorphic, 50

I

indices, 7, 42, 258
infinitesimal rotation, 59
integrals, path
  basic, 238–242
  Gaussian, 233–238
integrands, 31
interaction picture, 141–143
internal symmetries, 35
intervals, 6
intrinsic parity, 75
invariance, 5, 61

K

kinetic energy
  gauge masses, 225
  Lagrangians, 190
  momentum, 29
  QED, 170
  spontaneous symmetry breaking, 196
Klein-Gordon equation
  Dirac field, 85, 88
  mass terms in Lagrangians, 192
  overview, 3–4, 110–117
  reinterpreting field, 117
  scalar fields, 109
Klein-Gordon Lagrangians, 44, 198,
  206–207
Kronecker delta function, 7, 123, 174

L

Λ (Lorentz transformations), 8–11, 82
Lagrangian field theory
  action, 26–29
  basic mechanics, 23–26
  canonical momentum, 29–30

conservation laws, 35–37
conserved currents, 38–39
electromagnetic fields, 39–43
equations of motion, 26–29
gauge transformations, 43–47
Hamiltonian function, 29–30
overview, 30–35
quiz
    questions, 47–48
    solutions, 282
symmetries, 35–37
Lagrangians
adjoint spinors, 94
density, 30
electromagnetic fields, 166
Higgs mechanism, 18
mass terms in, 192–195
massless Dirac, 211–212
with multiple particles, 199–202
overview, 4–5
SUSY, 254–259
symmetry breaking, 192–193
Large Hadron Collider (LHC), 76,
    231, 245
laws, conservation, 35–37, 39
left-handed spinors, 210–211, 217
lepton fields
of electroweak interactions, 212–213
giving mass to, 222–224
leptons, 14–16, 212
Levi-Civita tensor, 60
LHC (Large Hadron Collider), 76,
    231, 245
Lie algebra, 54, 64, 251
Lie groups, 52–54
lifetimes, 160
linear momentum, 81
local fields, 30
local gauge invariance, 202
local gauge transformation, 204
local symmetry, 188
local $U(1)$ transformations, 188
Lorentz condition, 166
Lorentz gauge, 44

Lorentz scalars, 94
Lorentz transformations (Λ),
    8–11, 82

M

mass
giving to lepton fields, 222–224
spontaneous symmetry breaking and,
    196–199
terms in Lagrangians, 192–195
units, 195–196
massless Dirac Lagrangians, 211–212
massless gauge fields, 202, 204
mathematical operators, 1
Maxwell's equation, 39, 43, 165
mesons, 17
metric, 6
minimal coupling prescription,
    172
Minkowski space, 112
Møller scattering, 176
momentum
canonical, 29–30
conjugate, 122
linear, 81
motion, equations of, 26–29
multiplicative quantum numbers,
    71–72, 75

N

natural units, 112
negative energy solution, 99
negative frequency decomposition,
    128
negative probability densities, 114
neutrino term, 223
neutrons, 16
Newtonian mechanics, 23
Noether current, 259
Noether's theorem, 35, 39
nonabelian groups, 50
non-zero charges, 214
nonzero structure constants, 67
normal ordering, 134

normalization of states, 130–131
number operators, 128–130
number states, 120

O

occupation numbers, 129
odd parity, 73
operator expansion, 132
orthogonal matrices, 58
oscillators, simple harmonic, 26,
    119–121, 133
outgoing positron states, 173

P

parameters, group, 52
parity (*P*)
    CP violation, 78–80
    CPT theorem, 81–82
    overview, 72–76
parity discrete symmetry, 71
parity operators, 73
parity violation, 76
particle physics
    electromagnetic force, 12
    elementary particles
        generations or families of, 17
        leptons, 14–16
        quarks, 16–17
    GUTs, 18–19
    Higgs mechanism, 18
    quiz
        questions, 20–22
        solutions, 281
    range of forces, 13–14
    string theory, 19–20
    strong force, 13
    supersymmetry, 19
    weak force, 12–13
particles and antiparticles,
    101, 135
particles, elementary
    generations or families of, 17
    leptons, 14–16
    quarks, 16–17

path integrals
    basic, 238–242
    Gaussian, 233–238
    quiz
        questions, 243
        solutions, 285
Pauli exclusion principle, 19
Pauli matrices
    Dirac equation, 98
    Lie groups, 54
    SUSY, 250, 257
    unitary groups, 64
    unitary transformations, 217
perturbation theory, 143–146, 164,
    189–190
photon polarization, 165
photon state, 78
Planck mass, 246
Planck's constant, 2
polarization, 169, 175
polarization vectors, 167
position operators, 117
positive frequency decomposition, 128
positive frequency solution, 99
probability current, 114
propagators, 137, 154–159
proper rotations, 57
pseudoscalars, 75

Q

QED. *See* quantum electrodynamics
quantization, field
    overview, 117, 121–126
    second quantization
        overview, 118–119
        simple harmonic oscillators, 119–121
quantum chromodynamics, theory of,
    13, 18, 66
quantum electrodynamics (QED)
    Feynman rules for, 141, 173–185
    gauge invariance and, 170–173
    overview, 163–164
    quantized electromagnetic field,
        168–169

quiz
  questions, 185–186
  solutions, 284
review of, 165–168
quantum field theory
  Dirac field and, 87–89
  overview, 1–5
  states in, 127
Quantum Field Theory text, 231
quantum mechanics, supersymmetric,
    249–253
quantum numbers, 71–72
quarks, 14, 16–17
quizzes
  Dirac equation, 108, 283
  discrete symmetries, 83, 283
  electroweak theory, 231–232, 285
  Feynman rules, 160–161, 284
  group theory, 68–69, 282–283
  Lagrangian field theory,
      47–48, 282
  particle physics, 20–22, 281
  path integrals, 243, 285
  QED, 185–186, 284
  scalar fields, 137–138, 283
  spontaneous symmetry breaking,
      207–208, 284
  SUSY, 260–261, 285

R

range of forces, 13–14
rapidity, 9
rates of decay, 160
references, 289–290
renormalized Hamiltonian operators, 133
representations, group, 50–52, 55–58
rest mass, 13
Riemann sum, 238
right-handed spinors, 210–211, 217
rotation groups
  overview, 49, 54–55
  representing, 55–58
rotation matrices, 54, 59
rotations, 61, 103–104

S

S matrices, 139
scalar fields
  Bose-Einstein statistics
    energy and momentum, 132–133
    overview, 131–132
  complex, 135–137
  field quantization of, 117–126
    overview, 121–126
    second quantization, 118–121
  frequency decomposition, 128
  Klein-Gordon equation
    overview, 110–117
    reinterpreting field, 117
  normal ordering, 134
  normalization of states, 130–131
  number operators, 128–130
  overview, 109–110
  quiz
    questions, 137–138
    solutions, 283
  states in quantum field theory, 127
  time-ordered products, 134
scalar product, 10
scalars, 10, 75, 94
scaling with mass terms, 195–196
scattering events, 149
Schrödinger equation, 3, 72, 109
Schrödinger picture, 141
second quantization
  overview, 118–119
  simple harmonic oscillators, 119–121
selectrons, 245, 248
simple harmonic oscillators, 26,
    119–121, 133
sine-Gordon equation, 34–35
slash notation, 95
SO(N), 58–61
spacetime metric, 87
special relativity
  Lorentz transformations, 8–11
  overview, 5–8
  string theory, 19–20

spinor vectors, 90
spinors
    adjoint, 94
    Feynman rules, 173
    left-handed, 210–211, 217
    right-handed, 210–211, 217
    Weyl, 104–107
spontaneous symmetry breaking
    in field theory, 189–192
    Higgs mechanism, 202–207
    Lagrangians with multiple particles, 199–202
    mass
        overview, 196–199
        scaling, 195–196
        terms in Lagrangians, 192–195
    overview, 187–189
    quiz
        questions, 207–208
        solutions, 284
squarks, 245
standard model, 12
string theory, 19–20
strong force ($SU(3)$), 13
structure constants, 54
supercharge (Q), 247–249
supercurrent, 254
supersymmetry (SUSY)
    Lagrangians, 254–259
    overview, 19, 245–247
    quiz
        questions, 260–261
        solutions, 285
    supercharge, 247–249
    supersymmetric quantum mechanics, 249–253
    Wess-Zumino model
        chiral representation, 254
        overview, 253–254
symmetries
    discrete
        charge conjugation, 76–78
        CP violation, 78–80
        parity, 72–76
        quiz, 83, 283
        time reversal, 80–81

overview, 35–37
spontaneous breaking
    electroweak theory, 220–222
    in field theory, 189–192
    Higgs mechanism, 202–207
    Lagrangians with multiple particles, 199–202
    mass, 192–199
    overview, 187–189
    quiz, 207–208, 284

T

Taylor expansion, 27, 53
tensors
    electromagnetic field, 39, 165
    energy-momentum, 37
    Levi-Civita, 60
theorems
    CPT, 81–82
    Noether's, 35, 39
    trace, 184
theories
    electroweak
        charges of electroweak interaction, 213–215
        defined, 19
        gauge masses, 224–231
        lepton fields, 212–213, 222–224
        massless Dirac Lagrangians, 211–212
        overview, 209–210
        quiz, 231–232, 285
        right- and left-handed spinors, 210–211
        symmetry breaking, 220–222
        unitary transformations and gauge fields, 215–219
        weak mixing angle, 219–220
    grand unified (GUTs), 18–19
    group
        casimir operators, 67–68
        Lie groups, 52–54
        overview, 49–50
        parameters, 52

quiz, 68–69, 282–283
representations of, 50–52
rotation groups, 54–58
$SO(N)$, 58–61
unitary groups, 62–67
Lagrangian field
action, 26–29
basic mechanics, 23–26
canonical momentum, 29–30
conservation laws, 35–37
conserved currents, 38–39
electromagnetic fields, 39–43
equations of motion, 26–29
gauge transformations, 43–47
Hamiltonian function, 29–30
overview, 30–35
quiz, 47–48, 282
symmetries, 35–37
perturbation, 143–146, 164,
    189–190
of quantum chromodynamics, 13,
    18, 66
quantum field
Dirac field and, 87–89
overview, 1–5
states in, 127
string, 19–20
time reversal (T)
CPT theorem, 81–82
overview, 80–81
time reversal discrete symmetry, 71
time-ordered products, 134
time-reversal operators, 81
trace theorems, 184
trajectory, 28–29
trigonometry, 56

U
unbroken symmetry, 204
uncertainty principle, 2
unit vectors, 61
unitary groups, 62–67
unitary operators, 62, 143
unitary time evolution operators, 140
unitary transformations, 215–219
units, mass, 195–196
unpolarized cross sections, 183

V
vacuum state, 121, 127
variance, 26
vector bosons, 76
vector calculus, 43
virtual photons, 163

W
wave function, 188, 249
weak force ($SU(2)$), 12–13, 18
weak hypercharge, 214
weak interactions, 209
weak isospin charge, 214, 217
weak mixing angle, 219–220, 229
Weinberg angle, 219–220
Weinberg-Salam model, 209, 222, 230
Wess-Zumino model, 253–254
Weyl spinors, 104–107

Y
Yukawa term, 222

Z
zero total divergence, 37

Key Quotations to Learn

Arthur Birling: 'It's one of the happiest nights of my life'

Gerald: '[*laughs*] You seem to be a nice well-behaved family'

Inspector: 'I'd like some information, if you don't mind'

Summary

- The Birlings are celebrating and they are very happy.
- The Inspector interrupts to discuss Eva Smith's suicide.
- Eva was sacked two years ago from Mr Birling's factory.
- The discussion of the suicide causes conflict on stage between Mr Birling, Eric and the Inspector.

Questions

QUICK TEST

1. Who are the four members of the Birling family?
2. What is the family celebrating?
3. Why does Inspector Goole arrive at the Birlings' house?
4. How is Eva Smith linked to Arthur Birling?
5. What different opinions do the characters have about Arthur's treatment of Eva?

EXAM PRACTICE

Using one or more of the 'Key Quotations to Learn', write a paragraph exploring the significance of the family's contentment at the start of the play.

Act 1 (part 2)

You must be able to: understand what happens in the second half of Act 1.

How does the focus shift to Sheila?

Sheila is upset by the news of the girl's suicide and the Inspector reveals that the case involves more than just Mr Birling.

The Inspector gives some background to Eva's life and her situation. Sheila agrees with him that the girl led a pitiful life and needed **compassion**.

The Inspector tells of how Eva took a job at Milwards in December 1910 but she was then sacked in January 1911 after a complaint by a customer. Sheila recognises the girl's picture and runs out of the room upset.

When she returns, Sheila admits that it is her fault that the girl was sacked.

She explains that she was in a bad mood and felt the girl was laughing at her. She also admits she was jealous of Eva's prettiness. She is clearly upset but partly for herself rather than just about Eva's death.

How does the focus shift to Gerald?

When Sheila leaves the room upset, Mr Birling departs to speak to her and his wife.

Eric and Gerald are left with the Inspector. Eric gets angry and tries to leave the room but the Inspector insists that he stays.

After Sheila returns and confesses, the Inspector tells the three young people that Eva Smith changed her name to Daisy Renton. From Gerald's reaction, he is clearly disturbed by this news.

The Inspector and Eric go to the drawing room to speak to Mr and Mrs Birling, leaving Sheila and Gerald alone.

Sheila realises that Gerald was having an affair with Daisy during the previous spring/summer of 1911.

Gerald thinks it can be hidden but Sheila realises that the Inspector knows everything already.

The Inspector returns to the dining room.

How to use your Snap Revision Text Guide

This 'An Inspector Calls' Snap Revision Text Guide will help you get a top mark in your Edexcel English Literature exam. It is divided into two-page topics so that you can easily find help for the bits you find tricky. This book covers everything you will need to know for the exam:

Plot: what happens in the play?

Setting and Context: what periods, places, events and attitudes are relevant to understanding the play?

Characters: who are the main characters, how are they presented, and how do they change?

Themes: what ideas does the author explore in the play, and how are they shown?

The Exam: what kinds of question will come up in your exam, and how can you get top marks?

To help you get ready for your exam, each two-page topic includes the following:

Key Quotations to Learn
Short quotations to memorise that will allow you to analyse in the exam and boost your grade.

Summary
A recap of the most important points covered in the topic.

Sample Analysis
An example of the kind of analysis that the examiner will be looking for.

Quick Test
A quick-fire test to check you can remember the main points from the topic.

Exam Practice
A short writing task so you can practise applying what you've covered in the topic.

Glossary
A handy list of words you will find useful when revising 'An Inspector Calls' with easy-to-understand definitions.

AUTHOR:
IAN KIRBY

ebook

To access the ebook version of this Snap Revision Text Guide, visit
collins.co.uk/ebooks
and follow the step-by-step instructions.

Published by Collins
An imprint of HarperCollins*Publishers*
1 London Bridge Street
London SE1 9GF

HarperCollins*Publishers*
1st Floor, Watermarque Building,
Ringsend Road, Dublin 4, Ireland

ISBN 978-0-00-835301-8

First published 2019

10 9 8 7 6 5 4

British Library Cataloguing in Publication Data.

A CIP record of this book is available from the
British Library.

Commissioning Editor: Fiona McGlade
Project Managers: Shelley Teasdale and
  Richard Toms
Author: Ian Kirby
Proofreader: Lauren Murray
Typesetting: Mark Steward and QBS Learning
Cover designers: Kneath Associates and
  Sarah Duxbury
Production: Karen Nulty
Printed and bound in the UK using 100%
Renewable Electricity at CPI Group (UK) Ltd

ACKNOWLEDGEMENTS
Quotations from An Inspector Calls taken
from An Inspector Calls and Other Plays by
J.B. Priestley, 2001, London. An Inspector Calls
copyright 1947 by J.B. Priestley. Reproduced by
permission of Penguin Books Ltd.

The author and publisher are grateful to the
copyright holders for permission to use quoted
materials and images.

Every effort has been made to trace copyright
holders and obtain their permission for the
use of copyright material. The author and
publisher will gladly receive information
enabling them to rectify any error or omission
in subsequent editions. All facts are correct at
time of going to press.

MIX
Paper from
responsible source
FSC
www.fsc.org   **FSC* C007454**

This book is produced from independently
certified FSC™ paper to ensure responsible
forest management.

For more information visit:
www.harpercollins.co.uk/green

Contents

Plot
Act 1 (part 1)   4
Act 1 (part 2)   6
Act 2 (part 1)   8
Act 2 (part 2)   10
Act 3 (part 1)   12
Act 3 (part 2)   14
Narrative Structure   16

Setting and Context
Class and Politics   18
J.B. Priestley and 1945   20
The Midlands, 1912   22
The Birlings' House   24

Characters
Inspector Goole   26
Arthur Birling   28
Sybil Birling   30
Sheila Birling   32
Eric Birling   34
Gerald Croft   36
Eva Smith and Edna   38
How the Characters Change   40

Themes
Morality   42
Survival of the Fittest   44
Social Responsibility   46
Personal Repsonsibility   48
Inequality   50
Young and Old   52
Time   54
Love   56

The Exam
Tips and Assessment Objectives   58
Practice Questions   60
Planning a Character Question
    Response   62
Grade 5 Annotated Response   64
Grade 7+ Annotated Response   66
Planning a Theme Question
    Response   68
Grade 5 Annotated Response   70
Grade 7+ Annotated Response   72

Glossary   74
Answers   76

Act 1 (part 1)

You must be able to: understand what happens at the start of the play.

The setting
The play is set in 1912, in the fictional industrial city of Brumley, North Midlands.

The opening stage directions reveal that the play takes place in the large suburban house of a wealthy businessman.

What is the situation?
The Birling family – Mr Arthur Birling, Mrs Sybil Birling and their two children, Eric and Sheila – have just finished dinner. There is a happy atmosphere.

The Birlings' guest is Gerald Croft, Sheila's wealthy fiancé, and they are all celebrating the couple's engagement.

Who are the characters?
The audience are briefly introduced to all the characters but the focus is on Arthur Birling.

After congratulating the young couple, he makes a speech about 1912's social and political climate.

He and Gerald are then left alone for a while and he talks of his social aspirations.

It is made clear that Gerald and the Birlings are pleased with their lives and see themselves as good people.

When Eric returns to the room, Mr Birling continues to talk to the young men about his experience of the world but is interrupted by the doorbell.

What changes the situation?
Inspector Goole arrives and the audience are told about the suicide of a young woman called Eva Smith.

Despite Mr Birling's status, the Inspector is confident, abrupt and mysterious.

Mr Birling sacked Eva from his factory two years ago, in September 1910. She led a group of girls to ask for a pay-rise of two-and-a-half shillings a week. This turned into an unsuccessful **strike**, after which the ringleaders, including Eva, were sacked.

Mr Birling finds himself defending his actions and Gerald supports him.

Hearing Eva's story, Eric feels sorry for her, which causes him to argue with his father.

The Inspector is clearly on Eva's side and this angers Mr Birling. He tries to intimidate the Inspector, which he ignores.

At this point, Sheila innocently enters the room.

Key Quotations to Learn

Sheila: 'What do you mean by saying that? You talk as if we were responsible – '

Sheila: '[*miserably*] So I'm really responsible?'

Gerald: 'So – for God's sake – don't say anything to the Inspector'

Summary

- Sheila feels sorry for Eva Smith.
- Sheila confesses that she complained about the girl and got her sacked from Milwards.
- Eva Smith changed her name to Daisy Renton.
- Gerald admits to Sheila that he had an affair with Daisy.

Questions

QUICK TEST
1. What does Sheila feel for Eva Smith when she first hears about her sacking and her suicide?
2. Why did Sheila get Eva sacked from Milwards?
3. What is significant about the reasons for Sheila being upset?
4. Who reacts strangely to the news that Eva changed her named to Daisy Renton?
5. What do we find out about Daisy's life during the spring/summer of 1911?

EXAM PRACTICE
Using one or more of the 'Key Quotations to Learn', write a paragraph exploring the importance of responsibility in the play.

Act 2 (part 1)

You must be able to: understand what happens in the first half of Act 2.

How is Mrs Birling presented?

Continuing from the end of Act 1, Gerald tries to get Sheila to leave so he can hide his involvement but she refuses.

As the Inspector talks about their joint responsibility, Mrs Birling enters very confidently. Sheila is instantly worried as she, Gerald and Arthur had all behaved the same way.

Mrs Birling behaves in a **superior** manner, referring to Eva's lower class status, talking down to Sheila, acting grandly in front of the Inspector and referring to her husband's high status in the community. Despite this, the Inspector remains calm and blunt when speaking to her.

When Sheila reveals that Eric drinks too much, a fact that Gerald confirms, Mrs Birling is shocked and annoyed.

Sheila repeats her warning that her parents are making the situation worse and they are shocked to discover that Gerald knew Eva/Daisy.

How did Gerald know Daisy?

Gerald says he met Daisy in March 1911 in a bar. Sheila's parents don't want her to hear his story but she refuses to leave.

The reality of Daisy's death suddenly hits Gerald and he's visibly distressed.

He describes taking Daisy for a drink. She wanted to talk, having been upset by an encounter with Joe Meggarty (a city councillor who Gerald describes as a **womanising** drunk, which shocks Mrs Birling).

Gerald explains to the Inspector how Daisy had no money and was hungry. He innocently moved her into a friend's apartment and they later became lovers. Mrs Birling is disgusted by his behaviour.

Sheila points out that Gerald should be explaining to her not to the Inspector.

Gerald says that he didn't love Daisy but enjoyed being loved. He broke off the relationship in September 1911. The Inspector reveals that Daisy then went to the seaside; she wrote in her diary that she wanted to be alone to pretend her time with Gerald was continuing.

Gerald asks to go for a walk. Before he leaves, Sheila returns the engagement ring. She says she respects his honesty but they are now both different people.

Key Quotations to Learn

Inspector: 'we'll have to share our guilt'

Mrs Birling: 'Girls of that class'

Gerald: 'All she wanted was to talk – a little friendliness'

Summary

- Gerald still hopes to hide his involvement with Eva/Daisy.
- The Inspector tells Gerald and Sheila that they are all responsible.
- Mrs Birling behaves in a superior way to the Inspector, Gerald and Sheila.
- Gerald reveals the details of his affair with Daisy, and Sheila breaks off the engagement.

Questions

QUICK TEST
1. Why does Gerald want Sheila to leave the room?
2. How does Mrs Birling criticise Eva Smith?
3. How does Mrs Birling behave in front of the Inspector?
4. Why is Sheila worried by her mother's behaviour?
5. What was Gerald's relationship with Daisy like?

EXAM PRACTICE
Using one or more of the 'Key Quotations to Learn', write a paragraph exploring the significance of Eva/Daisy in the play.

You must be able to: understand what happens in the second half of Act 2.

What was Mrs Birling's role in Eva Smith's death?

Mrs Birling says she doesn't recognise the photo of Eva. The Inspector and Sheila know she's lying.

The Inspector says that Mrs Birling is a prominent member of the Brumley Women's Charity Organisation and that Eva appealed to the charity two weeks ago.

Sybil reveals that Eva was using the name 'Mrs Birling', which she found insulting and so was instantly **prejudiced** against her.

Mrs Birling states that Eva only had herself to blame and says that it was her duty to use her influence to have the girl's claim refused.

The Inspector reveals that Eva was pregnant but clarifies that it wasn't Gerald's child.

Mrs Birling told the girl to look for the child's father as it was his responsibility. Sheila says her mother's behaviour was horrible, while Arthur suggests that it will look bad on them. In response, she points out that it was Arthur sacking Eva 'which probably began it all'.

How does the focus begin to turn on Eric?

At this point, it has not been made clear that Eric was the father of Eva Smith's child but the audience may be beginning to work it out. Just as Mrs Birling begins to be questioned, there is the sound of Eric leaving the house and the Inspector comments that he will be needed.

As the Inspector continues to question Mrs Birling, he pushes her into criticising the father of the **illegitimate** child. She also calls Eva a liar for claiming the man had offered her money but she thought it was stolen so didn't want to take it.

Mrs Birling refuses to accept any blame, saying it is the girl's fault first and the lover's second. She says the man should be made an example of and accepts that if the girl's story about stolen money is true then the lover is entirely to blame.

Sheila works out Eric's involvement and tries to stop her mother saying any more.

As Mr and Mrs Birling begin to realise the truth, Eric returns to the dining room.

Key Quotations to Learn

Mrs Birling: 'I'm very sorry. But I think she only had herself to blame'

Mrs Birling: 'As if a girl of that sort would ever refuse money!'

Sheila: '[*with sudden alarm*] Mother – stop! – stop!'

Summary

- Eva was pregnant and asked Mrs Birling's charity for help.
- Mrs Birling didn't like the girl and used her influence to have the claim rejected.
- The Inspector manipulates Mrs Birling into stating that the father of Eva's child is entirely to blame for her death.
- The audience and the characters on stage gradually realise that Eric is the father.

Questions

QUICK TEST
1. What immediately turned Mrs Birling against Eva?
2. Does Mrs Birling regret rejecting the girl's claim for charity?
3. How do Sheila and Mr Birling respond differently to the news that Mrs Birling didn't help a pregnant girl who was asking for charity?
4. What did Mrs Birling think the girl was lying about?
5. Why does Sheila try to stop Mrs Birling from criticising the father of Eva's child?

EXAM PRACTICE
Using one or more of the 'Key Quotations to Learn', write a paragraph exploring the importance of prejudice in the play.

Act 3 (part 1)

You must be able to: understand what happens in the first half of Act 3.

How did Eric affect Eva Smith's life?

The story picks up from the end of Act 2 and Eric realises that everyone knows the truth.

The family argue over Eric's drinking and he retells how he met Eva in a bar.

After he turned aggressive, she let him into her flat where they had sex. At this point, Mr Birling orders Sheila to take her mother out of the room.

Eric continues his **confession**. He met Eva again, by accident, and they slept together. Eva revealed that she was pregnant but knew Eric didn't love her. To help the girl, he stole money from Arthur's office but she refused to take it.

How is Mr Birling presented at this point?

Arthur loses his temper several times before Eric's confession: first with the Inspector, who refuses to be intimidated by him, and then with Sheila when she doesn't want to leave the room.

He also shows his aggression during Eric's confession. When Eric points out that some of Birling's supposedly **respectable** friends have affairs, the Inspector has to stop Arthur from interrupting. He has another angry reaction when Eric admits to stealing.

Arthur also comes across badly when Eric explains his poor relationship with his father, feeling that he couldn't have asked him for help.

In addition to this, his reaction is still to cover the events up and avoid public scandal. It is particularly damning when he says he would give thousands of pounds to make the problem go away, showing his **capitalist** values rather than genuine regret.

How does the Inspector make his exit?

When Sheila and Sybil return, the Inspector tells Eric how Eva was rejected by his mother's committee, causing Eric to accuse Sybil of killing Eva and her own grandchild.

The Inspector sums up, telling them they all killed Eva Smith. He goes through the family, one by one, finally reminding Arthur that he destroyed a girl over two-and-a-half shillings.

The Inspector then focuses on the state of the country, pointing out that there are 'millions of Eva Smiths and John Smiths'. Talking to the Birlings (and the audience) he says, 'We are members of one body. We are responsible for each other'. He leaves with a warning that change will have to come.

Key Quotations to Learn

Mr Birling (to Sheila): '[*very sharply*] You heard what I said'

Eric: 'You're not the kind of father a chap could go to when he's in trouble'

Mr Birling (to Eric): '[*furious, intervening*] Why you hysterical young fool – get back – or I'll – '

Inspector: 'We don't live alone. We are members of one body. We are responsible for each other'

Summary

- Eric met Eva and made her pregnant.
- Eric accuses his mother of killing Eva and the baby, and says his father is unapproachable.
- The Inspector reminds the Birlings that they are all responsible for Eva Smith's death.
- The Inspector states his belief that all the members of society need to look after each other.

Questions

QUICK TEST
1. What negative aspects of Eric's character are revealed?
2. Why wouldn't Eva marry Eric?
3. What does Eric accuse his mother of?
4. In what way does Arthur not seem to have changed during the play?
5. What is the Inspector's final message before he leaves?

EXAM PRACTICE
Using one or more of the 'Key Quotations to Learn', write a paragraph exploring the significance of power in the play.

You must be able to: understand what happens at the end of the play.

What happens after the Inspector leaves?

Arthur's reaction to the evening's events continues to be fear of the scandal. He also states his belief that he and Sybil can excuse their actions.

Sheila shows more guilt, pointing out that her parents haven't learnt anything and criticising them for not focussing on the actual victim.

Sheila and Sybil begin to suspect that the Inspector wasn't a real police officer. Sheila doesn't think it matters because they still killed Eva Smith. However, Mr and Mrs Birling focus on the possibility of the whole affair remaining private if the police don't actually know.

How does Gerald's return alter things?

When Gerald returns, Arthur tries to stop Sheila from telling him about Eric and Sybil's involvement in Eva's death.

Gerald reveals that the Inspector wasn't a real police officer.

Mr and Mrs Birling are relieved and believe the secret can be kept amongst them. Arthur telephones the chief constable and it is confirmed that Inspector Goole doesn't exist.

While Gerald agrees with Arthur and Sybil, Sheila and Eric still feel guilt for what has happened.

Gerald, Arthur, and Sybil begin to think the whole evening may have been a **hoax**. They ring the infirmary and find there is no dead girl.

How does the play end?

Arthur and Gerald relax and are pleased their bad experience is over. Arthur raises a toast to the family but Sheila and Eric refuse to take part.

Arthur feels that everything is back to normal. He laughs about the evening's events and suggests that Sheila asks Gerald for her engagement ring back. Gerald offers her the ring but she refuses.

Sheila and Eric realise that the others have learnt nothing. She and Eric have been affected by the Inspector's words of warning as he left.

Arthur laughs at his two children as the phone rings. It is the police: a girl has died after swallowing disinfectant and an inspector is on his way.

Key Quotations to Learn

Sheila: 'Everything we said had happened really had happened'

Arthur: '[*heartily*] Nonsense! You'll have a good laugh over it yet'

Sheila: 'You began to learn something. And now you've stopped'

Summary

- Gerald reveals that the Inspector wasn't a real police officer.
- They find out that there is no dead girl at the infirmary.
- Arthur, Sybil and Gerald relax, thinking everything can be covered up and forgotten.
- Sheila and Eric still feel guilty and cannot understand the others' behaviour.
- The play ends with a phone call from the police saying a girl has died and an inspector is on his way to the house.

Questions

QUICK TEST
1. What do Arthur and Sybil focus on after the Inspector leaves?
2. What do Sheila and Eric feel after the Inspector leaves?
3. How do Arthur, Sybil and Gerald feel when they realise that the Inspector wasn't real and that a girl hasn't died?
4. Why do Eric and Sheila feel differently to their parents?
5. How does the play end?

EXAM PRACTICE
Using one or more of the 'Key Quotations to Learn', write a paragraph exploring the significance of age in the play.

Narrative Structure

You must be able to: explain the significance of the different ways Priestley has structured the play.

How does Priestley maintain the dramatic focus?

The play takes place on one evening, with each act opening immediately where the other left off.

The play takes place in one setting – the dining room of the Birlings' house.

The play has one central plot – the events leading to Eva Smith's death.

By taking this very singular approach, Priestley can focus on the drama and **tension** of the situation in order to emphasise his message about society.

To achieve this, the characters' involvements in Eva Smith's death are also dealt with one at a time. Characters not needed leave the stage, allowing Priestley to focus the audience on each person's responsibility.

How does Priestley raise and lower the tension?

To make the play and his social message memorable, Priestley uses the narrative structure to keep the **atmosphere** on stage tense and engaging.

Priestley starts each act by lowering the tension. A relaxed mood is established at the start of the play through the happy family meal, while the different character confessions at the beginning of Acts 2 and 3 lower the tension.

Different family arguments create brief spikes in tension, as does the way the Inspector comes into **conflict** with other characters.

At the end of the first two acts, Priestley raises the tension to a climax and includes a **cliffhanger**.

Act 1 ends with the unravelling of Sheila and Gerald's engagement, while the **climax** of Act 2 is even greater with Sybil and Arthur realising the truth about Eric.

Why does Priestley include the twist at the end?

When the Inspector leaves, the play seems to be over.

This trick by Priestley allows the audience to see whether the individual characters have learned anything or changed in any way. It also gives them time to judge each character's attitude to the social questions that Priestley has raised during the play through the character of the Inspector.

The play finally ends on a memorable cliffhanger, ensuring that the audience will question what they have seen and think more about the play's message.

Key Quotations to Learn

Sheila: 'And I hate to think how much he knows that we don't know yet. You'll see. You'll see' (Act 1)

Arthur: '[*thunderstruck*] My God! But – look here – ' (Act 2)

'[*As they stare guiltily and dumbfounded, the curtain falls*]' (Act 3)

Summary

- Priestley uses a simple narrative structure (one plot, one set, one evening) in order to focus his social message.
- The drama is intensified by dealing with each character's guilt one by one, and removing them from the stage when they aren't needed.
- To make the play memorable, Priestley raises and lowers the tension throughout.
- Cliffhangers are used at the end of each act to keep the audience thinking about the events on stage.

Questions

QUICK TEST
1. In what way is the narrative structure of the play quite simple?
2. What narrative technique is used at the end of each act?
3. What things raise and lower the tension during the play?
4. What is Priestley trying to get the audience to focus on and think about?

EXAM PRACTICE
Using one or more of the 'Key Quotations to Learn', write a paragraph exploring the importance of mystery in the play.

Class and Politics

You must be able to: understand the concept of class and the political ideas that are explored in the play.

What is meant by class?

This is a way of grouping different people in society according to their social status and how wealthy they are.

The most familiar labels are upper, middle and working class.

The upper class are seen as having the most money and status in society. This group of people are usually born into an old, established family who own a lot of land and property, rather than gaining wealth through running a business. They often don't need to work. Upper class families would often be linked to a title such as Lord, Sir or Duke. In the play, Gerald is upper class; his parents are Sir George and Lady Croft, who Arthur refers to as 'an old county family – landed people'.

The middle class are people who have recently earned a lot of money. They are usually well-educated with a good profession (such as doctors and lawyers) but can be anyone who has become wealthy through hard work. They are home-owners and aspire to having more money and social status. The Birlings are presented as a middle-class family, having made a lot of money through Arthur's business.

Some of the middle classes are disliked by the upper classes because they have the same amount of money but don't have the same "breeding", meaning they aren't brought up to speak, behave and dress in the strict way that the upper classes see as respectable. This links to Arthur being aware that Lady Croft feels Gerald could have become engaged to someone "better" than Sheila.

The working class are the poorest in society. They work for others, rather than being the bosses, and their wages are fairly low. They are linked to manual labour. The working class also includes the unemployed as they need work to survive. Eva Smith is a good example of a working-class person; she has a job in a factory and, once she is unemployed, has no way to pay for food or rent.

What is capitalism?

Capitalism is a right-wing political belief in individual gain through hard work and a focus on profit. Capitalists accept that, for this to happen, there will always be people in society who are much better off than others.

Birling is a capitalist and talks happily about 'steadily increasing prosperity'. He disagrees with workers asking for more money and sacks Eva Smith and others for going on strike to try to achieve better wages.

What is socialism?

Socialism is a left-wing political belief in greater equality and fairness for all, especially the poorest and most needy in society. Socialists believe working class people should have more of a say in government and that wealth should be more evenly shared amongst the classes.

The Inspector represents socialist values, believing that people like Eva Smith should be better paid and, if they fall on difficult times, supported financially. He feels the poor are used or **exploited** in order to make the middle classes richer. When talking about the number of people in a similar position to Eva Smith, he adds, 'If there weren't, the factories and warehouses wouldn't know where to look for cheap labour. Ask your father'.

Summary

- Society can be divided into upper, middle and working class. In the play, these classes are represented by the Crofts, the Birlings and Eva Smith.
- The term capitalism refers to individual wealth through hard work. This links to Birling.
- Socialism refers to sharing wealth and looking after poorer people in society. This links to the Inspector.

Questions

QUICK TEST
1. In the play, which character is linked to the upper class?
2. Which characters are middle class?
3. What class is Eva Smith?
4. Is Arthur Birling presented as a capitalist or a socialist? How can you tell?

EXAM PRACTICE
In Act 1, Arthur says to Gerald:
'I have an idea that your mother – Lady Croft – while she doesn't object to my girl – feels you might have done better for yourself socially – [...] No Gerald, that's all right. Don't blame her. She comes from an old county family – landed people and so forth – and so it's only natural. But what I wanted to say is – there's a fair chance that I might find my way into the next Honours List. Just a knighthood, of course'.

Relating your ideas to the social context, write a paragraph explaining how Arthur is presented as having higher social aspirations and why he might want this.

J.B. Priestley and 1945

You must be able to: understand how the play's meaning has been shaped by the author's life and the time in which he was writing.

Who was J.B. Priestley and how did his life affect his play?

John Boynton Priestley was born in 1894.

Although not as wealthy as the Birlings in his play, his was also a middle-class **suburban** family.

He left school at 16, took a job as a junior clerk, and began writing in his spare time.

He fought in the First World War (1914–1918) and spent many months in hospital after being badly injured. During the war, he would have fought side by side with all kinds of men regardless of their class.

After the war, Britain and many other countries were plunged into economic **depression**, with widespread poverty and a rise of political extremism. Priestley was influenced by his father's socialist views and wanted the world to become a more equal place.

By the 1930s, he was a successful writer. During the Second World War (1939–1945), Priestley was a radio broadcaster and his programme was very popular. He shared his socialist views and his hopes for a better Britain with listeners until the show was taken off air (because the government thought it too left-wing). These views appear throughout the play via the voice of the Inspector.

Priestley wrote *An Inspector Calls* in 1945. He continued his career as a playwright, novelist, essayist and political commentator until his death in 1984.

How does the time the play was written affect the play?

The two world wars were major contributors to the gradual erosion of Britain's strict class and gender divisions.

At war, men of different classes fought side by side despite any preconceived prejudices. At home, women kept the country running by taking on jobs they had not previously been encouraged to do.

By the 1920s, the Labour Party had grown out of different trade union and socialist movements to become the main opposition party to the Conservatives. The Labour Party had formed minority and coalition governments before and during the war but swept to power with a landslide victory in 1945.

It was this government that established the **welfare state**, introduced the National Health Service and promoted new housing estates where people of different classes would live side by side. However, the country could not change overnight and socialists saw that there was still a lot of work to do.

First performed in 1946, *An Inspector Calls* would have taken an audience back to 1912 and a very different Britain with very different values. It reminded them of the past in order to praise the present and keep pushing for further changes to achieve complete equality.

Summary

- J.B. Priestley grew up in the early 1900s and developed strong socialist beliefs.
- The two world wars began to change society by breaking down the rigid class system.
- The writing of the play coincided with the popularity of left-wing views in Britain.
- *An Inspector Calls* reminds its audience of a very different Britain.

Questions

QUICK TEST
1. How do Priestley's ideas appear through the character of the Inspector?
2. How might Priestley's wartime experiences have helped to shape his socialist principles?
3. How did the post-Second World War Labour government try to make Britain a more equal place?

EXAM PRACTICE
In Act 1, Arthur Birling says, 'We employers at last are coming together to see that our interests – and the interests of Capital – are properly protected. And we're in for a time of steadily increasing prosperity'.

Relating your ideas to the historical context, write a paragraph explaining how Birling's 1912 views might have seemed outdated when the play was first performed in 1946.

The Midlands, 1912

You must be able to: link the events of the play to its setting.

What were the Midlands like in 1912?

Although written in 1945, the play is set in 1912 (two years before the First World War).

The Midlands was a key area during the **Industrial Revolution** of the early nineteenth century. Towns and cities like Birmingham, Wolverhampton, Coventry and Nottingham became centres for industries such as textiles, coal mining and car manufacture.

The growth of factories meant the need for more workers. Many people had moved from small countryside villages to take the low wages on offer and live in cramped, unhygienic accommodation. The increase in low-skilled workers meant that more goods could be produced which, in turn, made the factory owners very rich.

Although conditions were improving, if people fell ill or were made unemployed, there was still no benefits system to help so they would go hungry and become homeless.

Were men and women equal?

Attitudes towards gender were very different and women were considered to be 'the weaker sex'. Girls were brought up with the main aspirations of marriage and children.

Men dominated all aspects of life: home, workplace, church and government. This is often referred to as a patriarchy.

Men and women had specific roles and social expectations, and this actually became stricter the higher you moved up the class system.

Women in wealthy families were not expected to work. Instead, they looked after the home and took on charitable roles. Women in poorer families took unskilled jobs as domestic servants, shop assistants or factory workers. They were paid less than men and couldn't achieve the same level of promotion. Because of this, many women needed a man to support them so were easily seduced and used.

What political movements were important in 1912?

Trade unions had begun to form with the aim of securing better pay and conditions for workers. In 1912, a successful national strike secured a minimum wage for coal miners.

The Suffragettes were campaigning for gender equality, specifically trying to achieve the vote for women. They weren't successful until after the First World War.

The Labour Party was beginning to gain support.

What were the attitudes to morality in 1912?

Morality, having a clear sense of right and wrong, was central to society partly because the Christian church was still a major influence.

The middle classes wanted to appear especially respectable because they aspired to higher status. However, as the Inspector suggests, their morality didn't often lead to them helping other people, just making judgements against them. There were huge double standards, with people appearing righteous and proper while actually behaving badly in secret.

Big towns and cities were full of temptations – such as sex, gambling and alcohol – so men often enjoyed themselves and hid it from their families (especially if they were married!). Sex outside of marriage was disapproved of, especially for a woman. Eric points out to his father that plenty of Arthur's apparently respectable friends get drunk and womanise.

Many women were unaware of this side of life because a respectable woman didn't hang around in bars or walk the streets after dark. Sybil is a good example of this: she doesn't like Sheila knowing slang like 'squiffy' and has no idea about Eric's real behaviour.

Summary

- Industrial areas had lots of poor workers and a few rich businessmen.
- Different groups were campaigning for class and gender equality.
- Despite the strong influence of the Christian Church on society, people (especially men) weren't always as moral as they seemed.

Questions

QUICK TEST
1. Why were people vulnerable if they fell ill or were unemployed?
2. What different areas of life and society did men take charge of?
3. In what ways were women unequal to men?

EXAM PRACTICE
Look at this conversation from Act 3 between Eric and Arthur:

Eric   I wasn't in love with her or anything – but I liked her – she was pretty and a good sport –

Birling [*harshly*] So you had to go to bed with her?

Eric   Well, I'm old enough to be married, aren't I, and I'm not married, and I hate these fat old tarts round the town – the ones I see some of your respectable friends with –

Birling [*angrily*] I don't want any of that talk from you –

Relating your ideas to the historical and social context, write a paragraph explaining how Priestley presents different attitudes to morality in this extract.

The Birlings' House

You must be able to: comment on how the staging of the play reveals things about the characters and themes.

How should stage directions be written about?

Remember that this is a play and don't analyse the specific language of the stage directions. This is because stage directions are there to be performed not spoken aloud.

Instead, think about the effect of the stage directions. How do they affect the mood on stage? What do different things represent or **symbolise**? How do they show what characters are thinking and feeling?

How does J.B. Priestley describe the dining-room set?

The dining-room of a fairly large suburban house, belonging to a prosperous manufacturer. It has good solid furniture of the period. The general effect is substantial and heavily comfortable, but not cosy and homelike. [...]

The lighting should be pink and intimate until the Inspector arrives, and then it should be brighter and harder.

At rise of curtain, the four BIRLINGS and GERALD are seated at the table, with ARTHUR BIRLING at one end, his wife at the other. ERIC downstage, and SHEILA and GERALD seated upstage. EDNA, the parlour-maid, is just clearing the table, which has no cloth, of dessert plates and champagne glasses, etc., and then replacing them with decanter of port, cigar box and cigarettes. Port glasses are already on the table.

How does the setting reflect the family?

The set should show the audience that the Birlings have a high social status. This comes across through the size of the room. This is additionally shown in the expensive furniture, with its sturdiness also **symbolising** how confident the family are in their lives.

It is significant that all five characters are seated, which shows they are currently relaxed. In addition, it underlines their social superiority by contrasting with their maid, Edna, who is on her feet and clearing up after them.

However, the 'suburban' house (which means town or city rather than the countryside) shows the audience that this status is limited. They aren't upper class like the Crofts (which is why Arthur is so desperate for a status-changing knighthood).

Similarly, the general look of the stage shouldn't be warm and comfortable; this implies that there is something missing in this family: love.

How do the props reflect the family?

Different props reflect the family's wealth, in particular the champagne glasses. This is useful to represent the difference between Arthur Birling and Eva Smith: they show the Birlings' wealth but champagne is also a stark contrast to the bleach that Eva drank to kill herself.

How is lighting used on stage?

Priestley is very specific about the lighting, which creates a rosy glow. This symbolises how the Birlings view themselves by making the audience think of the saying, "rose-tinted glasses", about only seeing the best things in life. The Birlings are self-satisfied and ignore the immoral things that they do, partly because they don't see those things as wrong.

The entrance of the Inspector requires the lighting to change, with its increased brightness reflecting how the Birlings' lives are being examined and exposed. This altered lighting should help to create a mood of discomfort on stage.

Key Quotations to Learn

'substantial and heavily comfortable, but not cosy and homelike'

'pink and intimate ... brighter and harder'

'champagne glasses'

Summary

- The dining-room set and its props reflect the Birlings' wealth and self-satisfaction. However, the set also reveals underlying problems in their lives.
- The set creates a visual contrast between the Birlings and the life of Eva Smith.
- The lighting emphasises how the Inspector shines a spotlight on their behaviour.

Questions

QUICK TEST
1. What features of the setting show the Birlings are wealthy?
2. What fault is pointed out about the appearance of the dining-room?
3. How does the set establish that the characters are relaxed and happy?
4. When, how, and why should the lighting alter?

EXAM PRACTICE
Using one or more of the 'Key Quotations to Learn', write a paragraph explaining how the set shows the Birlings' social superiority. Remember to focus on the effect that is created on stage and what this shows the audience.

Inspector Goole

You must be able to: analyse how Inspector Goole is presented in the play.

What does the Inspector represent?

The Inspector is described in his first stage directions as not needing physical size but projecting '[*an impression of massiveness, solidity and purposefulness*]'. This is to highlight that he brings with him symbolic powers of knowledge, morality and judgement.

He embodies the socialist viewpoint that members of society should care about each other. He is also a figure of morality and judgement.

Priestley uses the Inspector to expose one family's **hypocrisy** and their lack of morals and compassion. The change in lighting on stage ('[*brighter and harder*]') symbolises how he has come to reveal their secrets.

How is the Inspector presented as mysterious?

At the end of the play, we find out that there is no such Inspector Goole at the local police station.

However, he is mysterious throughout the play. Like an **omniscient** narrator in a novel, he appears to know everything about the Birlings already. Sheila refers to this at the end of Act 1.

His behaviour also establishes mystery as he doesn't behave quite as the audience expect from a police inspector. He stares at characters before he speaks to them for the first time, refuses to let them see the photograph of Eva Smith until it is their turn and is clearly judgemental rather than **objective**.

Priestley's use of wordplay in Goole/ghoul suggests the character might not be a man but a higher power come to stand in judgement on the Birlings.

How does the Inspector create conflict and tension?

Priestley creates tension through the Inspector by having him ignore social **conventions**. Despite being of a lower class, he has all the power on stage. For example, he is not intimidated by Arthur or Gerald: '[*cutting through massively*]', '[*coolly, looking hard at him*]'.

He doesn't mind upsetting characters or challenging them, such as his effect on Sheila or his questioning of Sybil.

Priestley also uses the Inspector to turn the characters on each other. His presence creates arguments between the parents and their children, as well as ending Sheila and Gerald's engagement.

Key Quotations to Learn

'Yes, but you can't. It's too late. She's dead' (Act 1)

'Apologize for what – doing my duty?' (Act 2)

'their lives, their hopes and fears, their suffering, and chance of happiness' (Act 3)

Summary

- Inspector Goole is a mysterious and powerful figure.
- He is not intimidated by the Birlings' superior class.
- He displays socialism and morality.
- He creates conflict and exposes the other characters on stage.

Sample Analysis

Part of the Inspector's importance is the way in which he challenges Mr and Mrs Birling. Despite the Birlings being of a higher class, the Inspector seems more powerful. This is particularly shown through his use of orders and questioning, 'Don't stammer and yammer at me again, man. *What did she say?*', with Priestley using the Inspector to assert that no-one is above the law. The way the Inspector ignores the social conventions of the time could also suggest his contempt for the family's sense of superiority and that he won't let the Birlings evade their responsibilities.

Questions

QUICK TEST
1. What socialist view does the Inspector represent?
2. How does he seem to go against social conventions?
3. What is unusual about his surname?
4. In what way does he expose the other characters?

EXAM PRACTICE
Using one or more of the 'Key Quotations to Learn', write a paragraph exploring the significance of Inspector Goole in the play.

You must be able to: analyse how Arthur Birling is presented in the play.

How is Birling presented as self-satisfied?

At the start of the play, Birling shows off his wealth and experience.

He boasts about the quality of his port and cigars, and praises his own qualities (twice repeating the phrase 'hard-headed business man').

His values of selfishness and survival of the fittest are shown in his words to Eric and Gerald: 'a man has to make his own way – has to look after himself'.

How does Priestley show Birling is the head of the house?

Arthur Birling is a **patriarch**. He is placed symbolically at *'one end'* of the table to show his importance.

He gives orders, such as telling Sybil to praise the cook. Significantly, the cook isn't named (he may not know or care) and he presumably considers it below himself to speak to her personally. This shows his attitude to the working class, as well as to women and their roles.

Birling interrupts other characters and makes long speeches. Priestley gives him the most lines in the opening of the play to demonstrate his dominance.

Because he expects to be respected, he is regularly angered by the way the Inspector interrupts or ignores him.

How are Birling's aspirations revealed?

Birling has gained lots of money but not the social status he wants. His *'provincial'* accent shows he isn't upper class but his hope for a knighthood reveals he wants to be.

His lack of natural class is a personal grievance, acknowledging that Gerald's mother thinks the Birlings are socially **inferior**.

He is happy for Sheila's engagement, partly due to the status it gives him through the Croft name: 'Crofts Limited are both older and bigger than Birling and Company – and now you've brought us together'.

How can the audience tell that Birling isn't as wise as he thinks?

Although written in 1945, the play is set in 1912, which allows Birling to be wrong in his confident speeches about the future.

This is **dramatic irony** and is most obvious in his lines about war ('And I say there isn't a chance of war') and the Titanic ('unsinkable, absolutely unsinkable'). Techniques such as short sentences and repetition are used to emphasise his errors. Priestley also uses the Titanic to **foreshadow** the sinking of the Birlings: the ship's unsinkable luxury mirrors the family's self-assured, wealthy appearance.

Key Quotations to Learn

'working together – for lower costs and higher prices' (Act 1)

'You've a lot to learn yet. And I'm talking as a hard-headed, practical man of business' (Act 1)

'And we don't guess – we've had experience – and we *know*' (Act 1)

Summary

- Arthur Birling is a dominant, patriarchal figure.
- He is wealthy and pleased with himself but wishes for greater social status.
- He believes he has wisdom and experience but Priestley uses dramatic irony to undermine this.

Sample Analysis

Birling is used by Priestley to criticise the role of the **traditional**, early 20th century patriarch. Birling is made to seem dominant in the way he controls conversations, 'Now you three young people, just listen to this – ', and demands the attention of the younger generation. However, as the play progresses we begin to realise that his sense of experience and wisdom is deeply flawed. Priestley is suggesting that young people need new ideas and values, rather than simply maintaining the old order.

Questions

QUICK TEST

1. Why does Priestley give Birling the greatest number of lines and have him interrupt and give orders?
2. What phrase does he repeat to show his self-confidence?
3. As well as his love for his daughter, why is he pleased about her engagement to Gerald?
4. Why is Birling's confident belief that there will be no war important?

EXAM PRACTICE

Using one or more of the 'Key Quotations to Learn', write a paragraph exploring the significance of Arthur Birling in the play.

Sybil Birling

You must be able to: analyse how Mrs Birling is presented in the play.

What kind of woman is Sybil?

Sybil is presented as a traditional early-twentieth century middle class woman, focussed on the household and her children. Although she is his social superior, and sometimes tells him off for a lack of correct **etiquette**, she usually defers to her husband and follows his lead. She is told to pass on a message to the cook, accepts that men 'spend nearly all their time and energy on their business', leaves the room with Sheila so the men can talk together and looks to Arthur to take charge.

How is Sybil presented as feeling superior?

When she meets the Inspector, she thinks she is above him and speaks '[*haughtily*]' and '[*rather grandly*]'. She is surprised by his lack of respect, 'That – I consider – is a trifle impertinent, Inspector'.

To assert herself, she refers to her husband's social position.

She suggests she is better than other characters, taking a moralistic view of Gerald's affair when calling it 'disgusting' and, in Act 3, boasting that she was the 'only one of you who didn't give in' to the Inspector.

How does Priestley show her hypocrisy?

Sybil presents herself as moralistic, with a prominent role in the Brumley Women's Charity Organisation. She talks about helping others and doing her duty.

However, she turned down Eva's application for support, looking down on her for being single and pregnant. She refers to 'that sort' of girl, showing her dislike of the lower classes and what she sees as Eva's lack of sexual decency.

Throughout the play, she refuses to accept any blame and excuses herself by repeating the word '**justified**'.

She criticises Eva's lover's lack of morals but these turn out to be the qualities of her own son.

How is Sybil presented as a bad mother?

Sybil does not really know her children, particularly Eric. She doesn't realise he has a drinking problem and cannot believe he could ever have an affair. He subsequently accuses her of having killed her own grandchild and tells her she never tried to understand him.

Mr Birling calls Eric 'spoiled' and this reflects on Sybil's parenting skills.

She also belittles her children, not wanting them to grow up or develop their own views: 'He's only a boy', 'It would be much better if Sheila didn't listen to this story at all'.

Key Quotations to Learn

'Girls of that class – ' (Act 2)

'I think she only had herself to blame' (Act 2)

'Besides, you're not that type – you don't get drunk – ' (Act 3)

Summary

- Mrs Birling displays a traditional female role in the house.
- She sees herself as superior to people of the lower classes.
- She presents herself as moral and charitable.
- She is revealed to be a hypocrite and a bad mother.

Sample Analysis

Priestley uses the character of Sybil to explore what he saw as a lack of social responsibility amongst the middle classes at the time. This is shown throughout the play by her refusal to accept any blame for what happened to Eva Smith, 'So I was perfectly justified in advising my committee not to allow her claim', displaying her stubborn and selfish attitude as she excuses her actions. Priestley encourages the audience to judge Sybil's lack of compassion through the way he has her show off her social power and the fact that she could have helped Eva if she had wanted to.

Questions

QUICK TEST
1. In what ways does Sybil conform to the gender expectations of the time?
2. Why does she feel superior to others?
3. What does her charity role show about her?
4. What is her relationship with Eric like?

EXAM PRACTICE
Using one or more of the 'Key Quotations to Learn', write a paragraph exploring the significance of Sybil Birling in the play.

You must be able to: analyse how Sheila is presented in the play.

What is Sheila's life like?

The opening stage direction says that Sheila should be presented as '[*very pleased with life and rather excited*]'.

She is in her 20s and has an easy life. Following the traditional social expectations for a daughter of a wealthy family in the early twentieth century, Sheila doesn't work. She is obedient to her parents and engaged to marry Gerald.

She still refers to her parents as 'Mummy' and 'Daddy', suggesting her class and also her childishness.

What does her treatment of Eva Smith show about her?

She has Eva sacked because she felt she was mocking her. Sheila admits she over-reacted because she was in a bad mood and was jealous that the girl was prettier than her.

She uses her social power, as a good customer who is about to marry a wealthy man, to have a lower-class girl punished.

How does Sheila respond to her actions?

When she first hears of Eva Smith's death, she is sympathetic but more focussed on how it's made her sad, 'I've been so happy tonight. Oh I wish you hadn't told me'.

This self-centred response continues when she realises she had Eva sacked from Milwards, 'I feel now I can never go there again'.

These responses reflect the selfish values that she has been brought up with.

However, in Act 2, she begins to reflect more on her responsibility: 'And I know I'm to blame – and I'm desperately sorry – but I can't believe – I won't believe – it's simply my fault that in the end she – committed suicide'. Her guilt is still **juxtaposed** with a refusal to take ultimate responsibility but the use of dashes indicates that her speech should sound uncertain, as if struggling with her **conscience**.

How is Sheila presented as more insightful than her family?

Whereas Arthur, Sybil, and Gerald deny their responsibility and try to cover up their involvement, Sheila quickly realises that they cannot hide the truth.

At the end of Act 2, she contributes to the increasing tension by trying to stop her mother from incriminating herself and Eric (having already worked out that he fathered Eva's child).

Key Quotations to Learn

'I'll never, never do it again to anybody' (Act 1)

'No, he's giving us rope – so that we'll hang ourselves' (Act 2)

'Mother – I begged you and begged you to stop' (Act 2)

Summary

- Sheila has an easy happy life and a high social status.
- She used her status to have Eva sacked.
- She shows some guilt for what happened but is also busy feeling sorry for herself.
- Unlike the rest of the family, she realises that the Inspector already knows the full story.

Sample Analysis

Sheila is used many times by Priestley to create tension on stage. She realises the family's position long before the others, '[urgently, cutting in] Mother, don't – please don't. For your own sake, as well as ours', and tries to warn her mother. As well as the sense of danger, Priestley is creating conflict because, at the time, the younger generation weren't expected to interrupt their parents or claim greater wisdom. In addition, Priestley maintains the audience's disapproval of the Birlings as, despite feeling some guilt about Eva, Sheila is still clearly focussed on her own well-being.

Questions

QUICK TEST

1. Why does Sheila have social status?
2. What feelings cause her to have Eva sacked?
3. How does Sheila begin to change?
4. What does she realise about Eric, long before the rest of the family?

EXAM PRACTICE

Using one or more of the 'Key Quotations to Learn', write a paragraph exploring the significance of Sheila Birling in the play.

Eric Birling

You must be able to: analyse how Eric is presented in the play.

What first impressions are the audience given of Eric?

At the start of the play, Eric seems an outsider. He is the last character to speak and is absent from the second act.

The opening stage direction describes him as '[*not quite at ease*]' and this appears throughout Act 1, such as when he stops himself mid-sentence during a discussion about women and his reaction to Gerald's joke about misbehaviour.

He first gains the audience's attention when he bursts out laughing for no reason. This, and his squabble with Sheila, makes him seem immature. Mr and Mrs Birling regularly talk down to him, for example, 'Silly boy!', despite him being in his 20s.

Priestley draws attention to his drinking through the stage directions and this becomes more important later on.

How does he feel about what he has done?

Unlike the others, Eric makes no attempt to excuse his behaviour.

The stage directions repeatedly point out that he should speak '[*miserably*]'. He is clearly ashamed of his actions and this helps to explain his strange and distant behaviour at the start of the play.

How does Eric clash with his parents?

Early on, he displays different values to his father. He suggests that Eva was right to go on strike and receives an angry response from Arthur.

When Eric returns in Act 3, the family have been discussing his behaviour and, especially Sybil, making the situation worse for him. Instead of supporting him, his parents are angry about the impending scandal. When he asks for a drink, his father refuses '[*explosively*]' but the Inspector agrees, saying 'look at him. He needs a drink now just to see him through'.

He later tells his father that he's never been able to talk to him and he says a similar thing to his mother: 'You don't understand anything. You never did. You never even tried – you – '.

When he finds out that his mother turned Eva away from the charity, he accuses her of killing her own grandchild.

After the Inspector has left, Mr Birling singles Eric out as the person to blame.

Key Quotations to Learn

'Why shouldn't they try for higher wages?' (Act 1)

'I hate these fat old tarts round the town – the ones I see some of your respectable friends with – ' (Act 3)

'you killed them both – damn you, damn you – ' (Act 3)

Sample Analysis

Priestley uses Eric to explore the theme of shame. When he describes his first night with Eva, 'And I didn't even remember – that's the hellish thing. Oh – my God! – how stupid it all is!', Eric seems genuinely distressed and repentant. The use of religious language could link to the significance of Christianity in society at the time, suggesting that he sees himself as sinful and is confessing to the Inspector. Although Priestley is criticising male attitudes towards lower class women, he is also using Eric's shame to contrast with the lack of regret shown by Mr and Mrs Birling.

Summary

- Eric acts strangely at the start of the play.
- His behaviour is explained when it is revealed that he made Eva pregnant.
- The play gradually builds up the fact that he has a drink problem.
- He clashes with his parents and this reaches a climax just before the Inspector leaves.

Questions

QUICK TEST
1. Why does Eric seem different to the other characters at the start of the play?
2. How does he disagree with his father about the factory workers?
3. How does he feel about his behaviour?
4. What does he feel about his relationship with his parents?

EXAM PRACTICE
Using one or more of the 'Key Quotations to Learn', write a paragraph exploring the significance of Eric Birling in the play.

You must be able to: analyse how Gerald is presented in the play.

What is Gerald's relationship with Arthur like?

Gerald has a fairly equal relationship with Arthur. This is mainly because he is of a higher class, being the son of Sir George and Lady Croft, so Arthur respects him and hopes to gain status through him.

Like Arthur, Gerald is a businessman. When Eric argues with his father about the factory strike, Gerald sides with Arthur's viewpoint: 'I should say so!'

Although Birling lectures Gerald, he also confides in him about the possibility of his knighthood. It is clear that Mr and Mrs Birling like and trust their future son-in-law.

How does Priestley present Gerald's relationship with Eva Smith?

Gerald knew Eva as Daisy Renton. He initially denies it but Sheila sees through his lies. He plans to keep it a secret from the Inspector but Sheila says he won't be able to.

When he met Daisy, he found her attractive and felt sorry for her so he gave her money and a place to stay. He just wanted to help her but she became his mistress as well: 'I suppose it was inevitable. She was young and pretty and warm-hearted – and intensely grateful'. The dash creates a pause to suggest Gerald realises that Eva may have felt obliged because he was her only support.

He admits that she loved him more than he loved her and he eventually broke off the relationship. He is distressed when he thinks about her death.

How do the other characters' opinions of him change?

In contrast to their happiness at the start, once Sheila works out about the affair she loses respect for Gerald: 'Oh don't be stupid. We haven't much time'. As he tells his story, she mocks him, 'You were the wonderful Fairy Prince. You must have adored it, Gerald'.

After his confession, Sheila says she believes his intentions were initially good but she can no longer be his fiancée: 'You and I aren't the same people who sat down to dinner here'.

Sybil calls his story a, 'disgusting affair', which contrasts with the start of the play when she stops Sheila from teasing Gerald for having worked so much over the summer (when he was actually with Daisy).

Perhaps for selfish reasons, Arthur doesn't particularly change his opinion of Gerald. When Sheila breaks off the engagement, he tries to reason with her, 'Now, Sheila, I'm not defending him. But you must understand that a lot of young men – '.

Key Quotations to Learn

'we're respectable citizens and not criminals' (Act 1)

'We can keep it from him' (Act 1)

'She didn't blame me at all. I wish to God she had now' (Act 2)

Summary

- The Birlings are happy to have Gerald as Sheila's fiancé.
- He tries to hide his relationship with Eva/Daisy from Sheila and the Inspector.
- Gerald wanted to help Eva/Daisy but it turned into an affair.
- He is upset by the news of Eva's death and understands when Sheila breaks off the engagement.

Sample Analysis

Gerald's temporary regret is an important feature of the play as Priestley uses it to explore middle class selfishness. Gerald is clearly upset by Eva Smith's death, '[*distressed*] Sorry – I – well, I've suddenly realised – taken it in properly – that she's dead –', and appears to recognise the consequences of his actions. However, by the end of the play, when he believes the evening has been a hoax, it is clear that he was concerned about the consequences to his own life rather than that of Eva. This suggests that the class issues Priestley is highlighting are deep-rooted and difficult to change.

Questions

QUICK TEST
1. In what ways does Gerald pretend to be something that he's not?
2. Why does Arthur like Gerald?
3. What were the honourable aspects of his part in Eva Smith's life?
4. What were the dishonourable aspects of his part in Eva Smith's life?

EXAM PRACTICE
Using one or more of the 'Key Quotations to Learn', write a paragraph exploring the significance of Gerald Croft in the play.

Eva Smith and Edna

You must be able to: analyse how Eva Smith and Edna are presented in the play.

How does Priestley present Eva Smith?

Eva Smith never appears on stage; instead, the audience are given details about her by the other characters. This is important because it symbolises how the working classes play a large role in the Birlings' life and wealth but they are almost invisible.

The Inspector explains that she has committed suicide by drinking disinfectant. He later reveals that she was pregnant.

Arthur describes her as a lively, hard-working girl. Like Gerald and Sheila, he also comments on her good looks.

She is presented as moral (she won't take Eric's stolen money) and having a strong belief in fairness and equality (organising the strike at the factory). It is significant that she comes across as having better values than the people who ruined her life.

Eva also used the name Daisy Renton.

What is Eva Smith being used to represent?

Eva is used by Priestley as an 'every-woman' and he draws attention to this by giving her the common surname 'Smith'.

She represents the working classes: both their financial struggles and the prejudice against them from those of a higher status.

She also shows the audience some of the obstacles that women faced in society:

- The **dominance** of men (Arthur is her employer, Eric forces his way into her lodgings).
- Being treated like a sexual object (Joe Meggarty's advances, Eric only wanting her for sex).
- The difficulty in being independent or self-sufficient (she is unemployed and hungry when she meets Gerald and she is dependent on his support).
- Prejudice about sexual activity (Sybil is prejudiced against her partly because she is pregnant but unmarried).

What is the significance of Edna?

Although she is not on stage a great deal, Edna is another representation of the working classes. She is a servant, cleaning away the table and fetching drinks.

The Birlings mostly ignore her but Arthur and Sybil acknowledge her presence in relation to the jobs they want doing.

She is addressed informally as 'Edna' but she responds to her employers as 'ma'am' or 'sir', showing the differences in status between the classes.

Inspector: 'There are a lot of young women living that sort of existence in every city and big town in this country' (Act 1)

Inspector: 'She was here alone, friendless, almost penniless, desperate'
(Act 2)

Inspector: 'Just used her [...] as if she was an animal, a thing, not a person' (Act 3)

Summary

- Eva is only described by others, never seen.
- She has committed suicide.
- She represents the plight of working class women.
- She is shown to have better values than the seemingly more respectable Birlings.

Eva Smith's Timeline

Sep 1910	After helping to organise a strike over wages at Birling and Company, Eva is sacked.
Dec 1910	She gets a job at Milwards.
Jan 1911	Sheila complains about Eva and she is sacked.
March 1911	Eva, who has now changed her name to Daisy Renton, meets Gerald. He finds her a place to live and she becomes his lover.
Sep 1911	Gerald breaks off his relationship with Daisy. She goes to live by the seaside for two months.
Nov 1911	Eva meets Eric. They become lovers and she gets pregnant. After she realises that Eric is supporting her with stolen money, she leaves him.
Spring 1912	Eva, using the name Mrs Birling, asks for help from the Brumley Women's Charity Organisation but it is refused.
Spring 1912	Two weeks later, Eva Smith commits suicide.

Questions

QUICK TEST

1. Which two sections of society does Eva Smith represent?
2. In what different ways is she treated badly by characters in the play?
3. In what ways does she appear to have better values than the Birlings?
4. What does Edna show about the Birlings' attitude to the working class?

EXAM PRACTICE

Using one or more of the 'Key Quotations to Learn', write a paragraph exploring the importance of the working class in the play.

How the Characters Change

You must be able to: analyse how the characters change during the play.

Arthur and Sybil Birling

Mr and Mrs Birling change the least in the play. Their treatment of Eva Smith is exposed and their relationship with their children is damaged. The audience see their confidence and status weakened. However, once the Inspector leaves and they decide it was all a hoax, they go back to their old, self-centred ways. Arthur even makes a joke about the evening: '[Imitating Inspector in his final speech] *You all helped to kill her*. [Pointing at Sheila and Eric, and laughing]'.

Gerald

Like Arthur and Sybil, once Gerald decides it is a hoax, he cheers up and tries to get Sheila to wear the engagement ring again. The audience are shown his disregard for the working classes when he says, 'What girl? There were probably four or five different girls'. He does not mind how many girls' lives have been ruined so long as it isn't going to be a public scandal.

Sheila and Eric

The Birling children are most affected by the events of the play. Priestley does this to reflect society's changing values between the young and the old, suggesting that the young represent hope for a more equal, compassionate future.

Sheila's distress and regret can be seen as early as Act 1 but it is often focussed on her feeling sorry for herself.

By the end of the play, Sheila feels genuine remorse and she realises that it doesn't matter if the Inspector was real or not, 'I tell you – whoever that Inspector was, it was anything but a joke'.

She wants her family to change their ways and Priestley has her repeat the Inspector's words, 'Fire and blood and anguish', to show that she understands his revolutionary warning.

Eric is similarly altered and describes himself as frightened by the Inspector's final words.

Priestley focuses on Eric's broken relationship with his parents: '[*quietly, bitterly*] I don't give a damn now whether I stay here or not'.

Key Quotations to Learn

Arthur: 'there'll be a public scandal – unless we're lucky – and who here will suffer from that more than I will?'

Sheila: 'you don't seem to have learnt anything' (Act 3)

Eric: 'You're beginning to pretend now that nothing's really happened'

Mrs Birling: 'In the morning they'll be as amused as we are'

Gerald: 'Everything's alright now, Sheila'

Summary

- Arthur, Sybil and Gerald do not really change. Although disturbed by the night's events, once they decide it's a hoax, they return to normal.
- Sheila and Eric are upset and regret their actions. They can't understand their parents' attitudes and begin to see they must all change their behaviour.

Sample Analysis

Priestley uses the lack of change in Mr Birling's character to assert the need for social change, criticising class prejudice and selfish social attitudes. In Act 1, Mr Birling believes that the working classes need to be kept in their place, 'If you don't come down sharply on some of these people', showing that he sees the workers as all the same rather than individuals to be cared for. This outlook is still held at the end of the play, 'Probably a Socialist or some sort of crank', with ideas of equality and compassion being ridiculed. Priestley promotes the virtues of socialism by having it disparaged by a character whose lack of values has been exposed to the audience.

Questions

QUICK TEST
1. What is Arthur most worried about once the Inspector has left?
2. In what way do Eric and Sheila's attitudes at the end of the play differ to those of their parents?
3. What does Gerald ask Sheila that shows he thinks nothing has changed?
4. What has changed between Eric and his parents?

EXAM PRACTICE
Using one of the 'Key Quotations to Learn', write a paragraph exploring the importance of change in the play.

You must be able to: analyse how the theme of morality is presented in the play.

What is morality?

Morality is a sense of right and wrong. It is often linked to religion but Priestley explores it in terms of society, considering personal behaviour and the treatment of others.

Which characters see themselves as 'moral'?

Gerald and the Birlings base their sense of morality on their class. Because they have status they assume they are superior. Priestley challenges this view in Act 1:

Gerald: We're respectful citizens and not criminals.

Inspector: Sometimes there isn't as much difference as you think.

Sybil presumes moral superiority when she feels her charity work gives her the right to judge others. When she is called a 'prominent' member of the charity, Priestley is suggesting the role is more about making herself look good in society.

How does Priestley use the seven deadly sins?

Priestley draws on Christian ideas to shows his characters' lack of social morality.

When the play begins, the audience see the sin of *gluttony* (eating and drinking wastefully or to excess). The table is being cleared of '[*dessert plates and champagne glasses, etc.*]' and everyone is drinking port. Priestley uses this scene to contrast, in Act 2, with Gerald describing Eva as being hungry. Eric, through his drinking, becomes the particular focus of this sin.

Arthur is guilty of *greed* through his desire for more money and status. When he describes his 'duty to keep labour costs down', Priestley uses the **noun** 'duty' ironically to show Birling's lack of moral understanding.

He is also full of *wrath* or anger throughout the play, sacking Eva partly out of anger and losing his temper with the Inspector and his children.

Sheila represents *envy*, for having Eva sacked partly out of jealousy over her prettiness. This links to wrath, as she admits to being in a bad mood, while her lifestyle suggests *sloth* or laziness. Sloth also relates to the other characters' social laziness as they are unwilling to support people in need.

Sybil's *pride* is obvious in the way she speaks, regularly emphasising her status, such as the stress on the personal pronoun in the line, '*I'm* talking to the Inspector now'.

Gerald and Eric display *lust* when each makes Eva their mistress and loves her less than she loves them. This sin is extended to Joe Meggarty and Arthur's friends to criticise their attitudes to women.

Key Quotations to Learn

Inspector: 'you might be said to have been jealous of her' (Act 1)

Sybil: 'she called herself Mrs Birling [...] a piece of gross impertinence' (Act 2)

Eric: 'I wasn't in love with her or anything' (Act 3)

Summary

- The characters consider themselves to be moral because of their social status.
- They lack social morality and compassion for the people around them.
- Priestley links the Birlings and Gerald to the seven deadly sins.

Sample Analysis

Priestley presents a lack of morality amongst Brumley's middle classes in order to expose the hypocrisy he saw in society's class prejudices. When Gerald reveals that a town councillor, Joe Meggarty, is 'a notorious womanizer as well as being one of the worst sots and rogues', Priestley links him to the Christian sins of lust and gluttony. By making it clear that people know about Meggarty's behaviour and yet he still holds a position of responsibility, Priestley criticises the higher classes' abuse of power and how they ignore in themselves what they condemn in people of lower status.

Questions

QUICK TEST
1. How is Arthur linked to the sin of greed?
2. How is Eric linked to lust and gluttony?
3. What was Sheila envious of?
4. Where is Sybil shown to consider herself as morally superior to others?

EXAM PRACTICE
Using one or more of the 'Key Quotations to Learn', write a paragraph exploring the importance of morality in the play.

Survival of the Fittest

You must be able to: analyse how the theme of survival is presented in the play.

What is survival of the fittest?

This term was first used in the late 1800s to describe how, in the animal kingdom, weak species die out and the strongest survive. In a capitalist society, people with power and money will usually do better than those without. The term is sometimes used by capitalists to justify selfish economic behaviour (such as very low wages or trying to put others out of business).

What is Arthur's view?

Arthur promotes the idea of survival of the fittest to Eric and Gerald, telling them that a man 'has to look after himself [...] so long as he does that he won't come to much harm'.

His words are selfish and, near the end of the play, Eric mocks how these words returned to haunt his father: '[*laughs bitterly*] I didn't notice you told him that it's every man for himself'.

What does the Inspector think?

It is symbolic that, when he gives his selfish speech to Eric and Gerald, Arthur is interrupted by the Inspector's arrival: '[*We hear the sharp ring of a front door bell. Birling stops to listen*]'.

Part of the Inspector's role is to challenge Arthur's view of the survival of the fittest. He can be seen as representing Priestley's socialist views.

When Arthur explains his refusal to pay his workers more, the Inspector says 'Why?' This simple question implies how unfair Arthur has been. Their clash of viewpoints is seen in Arthur's response: '[*surprised*] Did you say "why?"?' His tone of voice and the repetition of the Inspector's question show that, to Arthur, placing one's own interests above those of others is the normal way to behave.

The Inspector suggests a different, more compassionate life based on empathy, 'it would do us all a bit of good if sometimes we tried to put ourselves in the place of these young women counting their pennies in their dingy little back bedrooms'.

How does this link to Eva Smith?

Eva represents the consequences of the survival of the fittest attitude.

Blaming the Birlings for putting their own lives above that of Eva, the Inspector describes her dead body as, 'A nice little promising life there, I thought, and a nasty mess someone's made of it'.

Key Quotations to Learn

Arthur: 'a man has to make his own way' (Act 1)

Inspector (about Eva): 'But she died in misery and agony – hating life' (Act 2)

Inspector: 'all intertwined with our lives, with what we think and say and do' (Act 3)

Summary

- Arthur believes in survival of the fittest: looking after your own interests at the expense of others.
- The Inspector provides the opposite view that members of society should look after each other.
- Eva Smith is used to represent the consequence of Arthur's viewpoint.

Sample Analysis

Priestley uses Mr Birling's belief in the survival of the fittest to offer a criticism of the social values of the time. When Eric and Gerald are advised that, 'a man has to mind his own business', the audience sees Mr Birling's belief in the importance of making money and ignoring the plight of others. He emphasises the need to, 'look after himself and his own', promoting selfishness as well as the idea that poorer people are somehow different or below him. This refusal to see himself as part of a wider society contrasts with Priestley's own socialist values and is eventually Mr Birling's downfall.

Questions

QUICK TEST

1. Who does Arthur encourage to follow self-interest?
2. What is symbolic about the arrival of the Inspector?
3. How does Eva represent the consequences of self-interest?
4. How are Arthur's words about self-interest reused by Eric in Act 3?

EXAM PRACTICE

Using one or more of the 'Key Quotations to Learn', write a paragraph exploring the importance of compassion in the play.

Social Responsibility

You must be able to: analyse how social responsibility is explored in the play.

What is social responsibility?

Priestley uses the play to express his belief that we should help those people in society who are less fortunate than us. This is a key principle of socialism. He believes that capitalism focuses too much on individual gain.

How does Arthur Birling present social responsibility?

Arthur doesn't believe in social responsibility. His views are based on his capitalist ideals. He feels a responsibility to his family and his business, not to other people: 'it's my duty to keep labour costs down'.

He believes the poor are greedy and that the rich need to be careful, not caring, about them: 'they'd soon be asking for the earth'.

For him to be wealthy and happy, he thinks others must naturally be poor and unhappy.

How does Sybil Birling present social responsibility?

Despite being a prominent member of the Brumley Women's Charity Organisation, Sybil also lacks social responsibility. They help girls who are 'deserving cases', suggesting the organisation is judgemental rather than compassionate.

She doesn't like Eva Smith and makes sure her case is refused. It is implied that she doesn't want to help because Eva is an unmarried mother, linking to the moral values of the time.

She refers to Eva as 'a girl of that sort' who was giving herself 'ridiculous airs'. This shows that she thinks working class girls are below her and should know their place.

How does the Inspector present social responsibility?

The Inspector is used by Priestley to promote social responsibility, 'their lives, their hopes and fears, their suffering, and chance of happiness, all intertwined with our lives, with what we think and say and do'.

The Inspector focuses on the vulnerable in society and how they need help: 'She needed not only money, but advice, sympathy, friendliness'.

When talking to Gerald and Sheila, he highlights his belief in social responsibility by including himself in the blame for Eva Smith's death, 'we'll have to share our guilt'.

He wants the Birlings to change, 'Remember what you did, Mrs Birling'.

Although the older members of the family don't seem to change, Sheila and Eric's outlook is altered by the Inspector.

Key Quotations to Learn

Arthur: 'like bees in a hive – community and all that nonsense' (Act 1)

Inspector: 'it would do us all a bit of good if sometimes we tried to put ourselves in the place of these young women' (Act 1)

Inspector: 'You must have known what she was feeling. And you slammed the door in her face' (Act 2)

Summary

- Priestley uses the Inspector to express the importance of social responsibility.
- He uses Arthur and Sybil Birling to represent how the higher classes focus on improving their own lives and status, rather than helping others.
- The Inspector changes Sybil and Eric's social viewpoint but Arthur, Sybil and Gerald quickly return to their old selfishness.

Sample Analysis

The importance of social responsibility is presented through the character of the Inspector. He becomes a mouthpiece for Priestley's socialist values, especially towards the end of the play. His comments about society, 'We don't live alone. We are members of one body. We are responsible for each other', emphasise socialist values of joint responsibility. The message that we should all respect and help each other seems shaped by the events of the first and second world wars when the country had to pull together and overcome individual differences.

Questions

QUICK TEST
1. What does the Inspector believe we should do for the less fortunate people in society?
2. What is Arthur's opinion of social responsibility?
3. In what way does Sybil only pretend to have social responsibility?

EXAM PRACTICE
Using one or more of the 'Key Quotations to Learn', write a paragraph exploring the importance of social responsibility in the play.

Personal Responsibility

You must be able to: analyse how Priestley shows the different characters' attitudes to personal responsibility.

What is personal responsibility?

This means accepting the consequences of your own actions. As well as wanting the Birlings to understand why they should behave differently towards other people, the Inspector challenges each character to admit, and show guilt for, their individual role in Eva's suicide. In Act 3, Sheila refers to this as being made to 'confess'.

This is highlighted in one of the Inspector's final speeches: 'This girl killed herself – and died a horrible death. But each of you helped to kill her. Remember that. Never forget it'. The **emotive** language emphasises the effects of the characters' behaviour, while the two short imperative sentences focus on the necessity of accepting guilt.

Who accepts responsibility?

Sheila and Eric accept that they wronged Eva Smith.

In Act 3, Sheila says, 'I behaved badly too. I know I did. I'm ashamed of it'. Priestley emphasises her guilt through the repetition of the personal pronoun 'I' and uses short sentences that add a tone of certainty about her guilt.

The two children also point out the responsibility of the rest of the family, with Eric stating: 'the fact remains that I did what I did. And Mother did what she did. And the rest of you did what you did to her'. The use of **parallelism** at the end of each sentence emphasises the personal responsibility of everyone in the room, with the repetition of 'and' building up a sense of their negative impact on Eva's life.

Who doesn't accept responsibility?

While Gerald feels guilty in Act 2 for cheating on Sheila and for eventually deserting Eva, by the end of the play – because he thinks no one else knows – he seems to have forgotten his shame. He asks Sheila to remain engaged to him and doubts the story of Eva's death.

Throughout the play, Arthur claims that he was right to sack Eva Smith and justifies it using the word 'duty'. Sybil takes a similar point of view about refusing to help Eva, saying 'I had done no more than my duty'.

Priestley highlights Arthur and Sybil's lack of personal responsibility by also having them point out the failings of others. This is most obvious in the way Sybil, not realising she is criticising Eric, focuses on the responsibility of the man who made Eva pregnant, saying he should be made an example of and be forced to make a public confession.

Key Quotations to Learn

Inspector: 'Remember' (Act 3)

Sybil: 'He certainly didn't make me *confess* – as you call it' (Act 3)

Eric: 'It's what happened to the girl and what we all did to her that matters' (Act 3)

Summary

- The Inspector encourages each character to confess and take personal responsibility for the death of Eva Smith.
- Sheila and Eric accept responsibility.
- Gerald seems to show remorse but this fades when he thinks the story can be kept secret.
- Arthur and Sybil refuse to accept that they did anything wrong.

Sample Analysis

Priestley asserts the importance of personal responsibility by encouraging the audience to deplore Mr Birling's refusal to take any blame for Eva Smith's death. When he says, 'There's every excuse for what both your mother and I did – it turned out unfortunately, that's all', his apparent dismissal of her death shows he has no genuine sympathy. Despite the audience having heard the terrible consequences of his actions, the character attempts to justify what he has done and the additional way that he includes Sybil in the acknowledgment of his involvement shows his reluctance to accept any individual guilt.

Questions

QUICK TEST
1. Which characters take responsibility for their actions?
2. What word does Sheila use to describe the process of admitting what they did?
3. What do Arthur and Sybil do instead of taking responsibility?

EXAM PRACTICE
Using one or more of the 'Key Quotations to Learn', write a paragraph exploring the importance of personal responsibility in the play.

Inequality

You must be able to: analyse how Priestley shows the idea of inequality.

What is inequality?

Priestley's key socialist message is a response to the lack of equality in society. In the play, he explores how a lack of equality makes Eva Smith's life difficult and contributes to her death. Because the Birlings and Gerald have more money, status and power, their lives are much easier.

How does Eva represent inequality?

In terms of social **hierarchies**, Eva is at the bottom because she is working class and a woman. She always has people above her, controlling her life.

This can be seen in her working life, where she was easily fired from the factory and Milwards. It can also be seen in her personal life, forming relationships with Gerald and Eric where she is dependent on them financially and emotionally.

Because of her low status, Eva has a lack of opportunities. When Arthur asks what she did after being sacked, 'Get into trouble? Go on the streets?', crime and prostitution are presented as her most obvious options.

Her life is a struggle and references are made to her being penniless, hungry, lonely and preyed upon – sexually – by more powerful men.

How do the Birlings represent inequality?

In contrast to Eva, the Birlings represent power and fortune. They benefit from inequality.

The dining-room set and the meal they have just finished suggest wealth and an easy lifestyle.

Birling refers to their better opportunities when he mentions Eric's 'public-school-and Varsity life'.

Their power comes from money but also their roles in society: Arthur is a magistrate, has been Lord Mayor, and is expecting a knighthood. In comparison, Eva would not even have had the vote (women didn't have the same voting rights as men until 1928).

As a wealthy client, Sheila has the power to have people sacked from Milwards. Similarly, Sybil has the power to influence who is helped by her charity.

How does the Inspector criticise inequality?

The Inspector criticises how working class girls are seen as 'cheap labour'.

As he investigates the family, he reveals the different ways they made Eva a victim of their social superiority.

He blames inequality for her suicide, saying her reason was 'Because she had been turned out and turned down too many times'. Priestley's use of the **passive voice** and repetition emphasise how Eva wasn't in control of her own life.

Key Quotations to Learn

Arthur: 'we're in for a time of steadily increasing prosperity' (Act 1)

Sheila: 'these girls aren't cheap labour – they're *people*' (Act 1)

Gerald: 'she was desperately hard up' (Act 2)

Summary

- Eva Smith suffers from social inequality due to her class and her gender. There are always more powerful people controlling her life.
- Arthur and his family benefit from inequality as cheap labour has made them rich. Their abuse of power has led to Eva's death.

Sample Analysis

Priestley highlights the effects of inequality through the depiction of Eva's life after she is sacked by Mr Birling. The Inspector's description of her having, 'no work, no money coming in, [...] no relatives to help her, few friends, lonely, half-starved, she was feeling desperate', builds up a sense of Eva's physical, emotional, and economic struggle. Importantly, Priestley uses Eva to represent all the victims of social inequality, making the play as relevant today as it was in the 1940s. Because Priestley places this speech soon after the Birlings' family celebration, the references to hunger and a lack of money and relatives also creates a stark contrast between Eva and the people who have ruined her life.

Questions

QUICK TEST
1. Why is Eva socially unequal?
2. What things suggest the Birlings have greater power and opportunities?
3. How do the Birlings benefit from social inequality?

EXAM PRACTICE
Using one or more of the 'Key Quotations to Learn', write a paragraph exploring the importance of inequality in the play.

You must be able to: analyse how Priestley explores the differences between youth and age.

How are traditional ideas about age shown in the play?

Traditionally, young people are expected to respect and obey their elders. These expectations were even more rigid at the start of the twentieth century.

When the play begins, Priestley gives Arthur the greatest number of lines to show he is in charge. He toasts the engagement, gives his opinions about the state of the world and goes on to lecture Eric and Gerald about life.

When he talks, he expects people to pay attention ('Are you listening, Sheila?'), doesn't like to be interrupted ('Just let me finish Eric') and gives opinions as facts ('The world's developing so fast it'll make war impossible').

Mrs Birling, although she lets her husband take the lead, also shows this dominance over the young, 'Please don't contradict me like that'.

Before the Inspector arrives, Eric and Sheila behave as tradition demands, for example, stopping squabbling when Sybil tells them off. Even when Eric tells his father not to make a speech, it is said '[not too rudely]' so the atmosphere on stage follows normal expectations.

Where are traditional age roles challenged?

When the Inspector arrives, Eric and Sheila begin to challenge the values of their parents. Both criticise Arthur's decision to sack Eva and, in Act 2, Sybil disapprovingly notes, 'You seem to have made a great impression on this child, Inspector'.

Sheila continues to challenge their behaviour, such as telling her father not to 'interfere' when he suggests she forgives Gerald or calling her mother's actions 'cruel and vile'. It is clear that she understands what the Inspector is doing whereas her parents do not.

Eric's response is more extreme. He curses his mother ('damn you'), is aggressive towards her ('[almost threatening her]') and mocks the **irony** of Arthur's after-dinner speech about self-interest.

How is Priestley using ideas about age?

By the end of the play, Eric and Sheila offer future hope. They represent the idea that society can change because they challenge their parents' refusal to alter their behaviour, with Sheila saying, 'You're ready to go on in the same old way. [...] And it frightens me the way you talk'.

The Inspector's words have had a positive effect on them and they are beginning to develop a social conscience.

Key Quotations to Learn

Arthur: 'But you youngsters just remember what I said' (Act 1)

Sheila (to Arthur): 'Oh – sorry. I didn't know. Mummy sent me in to ask you' (Act 1)

Eric: '[*shouting*] And I say the girl's dead and we all helped to kill her – and that's what matters – ' (Act 3)

Summary

- The young are expected to respect and obey their elders.
- At first, the Birlings display traditional expectations of age.
- After the Inspector's arrival, Sheila and Eric begin to challenge their parents.
- The children represent hope. Unlike their parents, they can learn and change.

Sample Analysis

Priestley explores the conflict between youth and age through Mr Birling and his son, Eric. In response to having his authority challenged, Mr Birling says, 'So hold your tongue if you want to stay here' to show his anger that the traditional expectations of respect for one's elders are being ignored. The play looks back to a time before the social changes of the 1940s and Priestley presents this questioning of tradition as a good thing while also showing how Mr Birling is alarmed at being challenged by the younger generation.

Questions

QUICK TEST
1. What does Arthur show about the older generation at the start of the play?
2. What does Sheila show about the younger generation at the start of the play?
3. Why does Priestley make Sheila and Eric come into conflict with their parents?

EXAM PRACTICE
Using one or more of the 'Key Quotations to Learn', write a paragraph exploring the importance of generational conflict in the play.

Time

You must be able to: analyse how Priestley uses ideas about time in the play.

How is the play linked to its time?

By setting the play several decades before he was writing, Priestley can use time to explore key ideas about its characters and themes.

In particular, Arthur's confidence in the Titanic and the impossibility of war creates dramatic irony that signals to the audience he isn't as wise as he thinks.

The Inspector's final words about 'fire and blood and anguish' refer to the two world wars that took place between the play's setting and when it was written. This allows the audience to reflect on the past and how and why society has changed.

How does Priestley use time to structure the play?

The play takes place over one evening with each act leading directly into the next, intensifying the destruction of the Birlings' cosy, self-centred way of life.

The series of **flashbacks** over a two-year period, as told by the characters, present Eva's life as a continual struggle against inequality. This emphasises the idea of a chain of events, consequences and the fact they cannot change the past.

By the end, the play has almost come full circle. Arthur, Sybil and Gerald are feeling safe and secure when they receive news that a police inspector is on his way. Priestley does this to convey the idea that, unless people change, society is doomed to keep making the same mistakes. He creates an intriguing cliffhanger ending that explores the idea of second chances; the audience are left wondering how the characters will respond to the consequences of their actions a second time around.

How does time affect the mood on stage?

The Inspector talks about needing to hurry in Act 3 and the quickened pace raises the tension on stage. It also adds to the story's mystery as the audience later realise the Inspector needed to finish before the real police arrived.

The play shows the past returning to haunt the characters. This allows Priestley to create a range of atmospheres on stage, such as Sheila's distress at how her previous actions have affected a girl's future, Gerald's desperation to keep his past secret, Sybil's shock at her son's past and Arthur's fear of future scandal.

While Eric and Sheila regret the past, Arthur and Sybil want to deny it. The Inspector says the characters will never forget what they did.

Key Quotations to Learn

Inspector: 'A chain of events' (Act 1)

Inspector (to Sybil): 'you're going to spend the rest of your life regretting it' (Act 2)

Inspector: 'I haven't much time' (Act 3)

Summary

- The short time-scale intensifies the events on stage.
- The flashbacks create an idea of past events and their consequences.
- Different responses to the past allow Priestley to create different moods on stage.
- Eric and Sheila want to mend the past while Arthur and Sybil deny it.

Sample Analysis

Priestley explores time through the idea that the past affects the present. For example,

Sybil: Fine feelings and scruples that were simply absurd in a girl in her position.

Inspector: Her position now is that she lies with a burnt-out inside on a slab.

This dialogue expresses how Mrs Birling's past actions contributed to Eva Smith's suicide. Whereas Mrs Birling is dismissive of Eva, the Inspector's brutal description asserts that she doesn't give enough weight to the consequences of her actions. Priestley's message works on two levels: the consequences of individual actions and the need for society as a whole to build a better, more compassionate world.

Questions

QUICK TEST
1. How is the past used to show the audience something about Arthur?
2. What is the effect of setting the play over one evening?
3. Why does Priestley use flashbacks in the play?
4. How do Sheila and Eric respond to the past differently to Arthur and Sybil?

EXAM PRACTICE
Using one or more of the 'Key Quotations to Learn', write a paragraph exploring the importance of time in the play.

You must be able to: analyse how Priestley presents the theme of love.

How is love presented at the start of the play?

The play opens with the family celebrating Sheila and Gerald's engagement. The stage directions show that Sheila should talk '[*gaily*]' and in a '[*playful*]' manner and this presents love in a happy way.

They seem deeply in love when they look at each other during the toast and Gerald says, 'And I drink to you – and hope I can make you as happy as you deserve to be'.

However, the audience might notice possible uncertainty in the **verbs** 'hope' and 'deserve', while Sheila's reaction to the ring ('Now I feel really engaged') could imply that money plays a role in their love. This returns when we realise Arthur's own reasons for approving of their engagement.

How are men and love presented?

Love is presented as meaning different things to men and women.

For Sheila and Eva, it represents faithfulness. When describing Eva's feelings for him, Gerald says, 'I became at once the most important person in her life', and the superlative suggests constancy and attachment.

Love is also linked to romance for women, with Sheila using fairy tale **imagery** to describe how Eva must have felt towards Gerald, 'You were the wonderful Fairy Prince'.

However, the men see love differently with women being viewed as a commodity. Most obviously, Gerald has cheated on Sheila but expects her to resume their engagement at the end.

Gerald and Eric's relationships are both linked to sex more than love. Gerald admits about Eva, 'I didn't feel about her as she felt about me', and the relationship is ended within six months.

In particular, Eric says of Eva, 'she was pretty and a good sport'. The use of the word 'sport' suggests that Eric sees love as a game rather than a serious commitment.

How are marriage and love presented?

Linking to the gender inequality of the time, it is suggested that the values of the young men are passed down by their parents.

Arthur tries to defend Gerald's behaviour to Sheila and it is suggested that his married friends have affairs.

Sybil points out that men focus more on business than marriage but says that Sheila will 'have to get used to that, just as I had'.

Key Quotations to Learn

Gerald (about Eva): 'She told me she'd been happier than she'd ever been before – but that she knew it couldn't last' (Act 2)

Eric (about Eva): 'I wasn't in love with her or anything' (Act 3)

Gerald: 'Everything's alright now, Sheila. [*Holds up the ring.*] What about this ring?' (Act 3)

Summary

- Priestley presents Gerald and Sheila as being in love.
- However, Gerald has had an affair.
- Men and women are presented as having different views of love.
- The men are criticised for wanting sex more than love.

Sample Analysis

Priestley uses Gerald and Sheila to explore the idea of honesty in a relationship. In his reply to her question about Daisy Renton, 'All right. I knew her. Let's leave it at that', Gerald is hiding the truth from Sheila. Rather than any apology, he tries to close the discussion. This is also an assertion of male dominance, when he makes a decision for 'us', linking to attitudes of gender at the time. Gerald sees himself as beyond reproach; he thinks Sheila should not question his love but instead just accept whatever he says as the truth.

Questions

QUICK TEST
1. Where is marriage presented positively?
2. How does Sybil present a negative aspect of marriage?
3. What is different about men and women's attitudes to love in the play?
4. How is Eric's relationship with Eva presented as being about sex?

EXAM PRACTICE
Using one or more of the 'Key Quotations to Learn', write a paragraph exploring the importance of love in the play.

Tips and Assessment Objectives

You must be able to: understand how to approach the exam question and meet the requirements of the mark scheme.

Quick tips

- You will get a choice of two questions. Do the one that best matches your knowledge, the quotations you have learned and the things you have revised.
- Make sure you know what the question is asking you and underline key words to help keep you focused.
- You should spend about 50 minutes on your *An Inspector Calls* response. Allow yourself five minutes to plan your answer so there is a clear structure to your essay.
- All your paragraphs should contain a clearly explained idea about the play, a relevant reference to the play (ideally a quotation) and links to the play's context.
- It can sometimes help, after each paragraph, to quickly re-read the question to keep yourself focussed on the exam task.
- Keep your writing concise and aim for sophisticated expression.
- It is a good idea to remember what the mark scheme is asking of you…

AO1: Understand and respond to the play (16 marks)

This is all about coming up with a range of points that match the question, making clear interpretations of the play, supporting your ideas with references from the text and writing your essay in a mature, academic style.

Lower	Middle	Upper
The essay has some good ideas that are mostly relevant. Some quotations and references are used to support the ideas.	A clear essay that always focuses on the exam question. Quotations and references support ideas effectively. The response refers to different points in the play.	A convincing, well-structured essay that answers the question fully. Quotations and references are well-chosen and integrated into sentences. The response covers the whole play (not everything, but ideas from all three acts rather than just focussing on one or two sections).

AO3: Understand the relationship between the play and its contexts (16 marks)

For this part of the mark scheme, you need to show your understanding of how the characters or Priestley's ideas relate to when he was writing (1945) or when the play was set (1912).

Lower	Middle	Upper
Some awareness of how ideas in the play link to its context.	References to relevant aspects of context show a clear understanding.	Exploration is linked to specific aspects of the play's contexts to show a detailed understanding. Context is fully integrated.

AO4: Written accuracy (8 marks)

You need to use a range of accurate vocabulary, punctuation and spelling in order to convey your ideas clearly and effectively.

Lower	Middle	Upper
Reasonable level of accuracy. Errors do not get in the way of the essay making sense.	Good level of accuracy. Vocabulary and sentences help to keep ideas clear.	Consistent high level of accuracy. Vocabulary and sentences are used to make ideas clear and precise.

Practice Questions

1. **Gerald:** *We are respectful citizens and not criminals.*

   Inspector: *Sometimes there isn't as much difference as you think.*

   Explore the importance of morality in the play.

   You must refer to the context of the play in your answer.

2. **Sybil:** *That – I consider – is a trifle impertinent, Inspector.*

   Explore the significance of Sybil Birling in *An Inspector Calls.*

   You must refer to the context of the play in your answer.

3. **Sybil:** *As if a girl of that sort would ever refuse money.*

   Explore the importance of class differences in *An Inspector Calls.*

   You must refer to the context of the play in your answer.

4. **Arthur:** *There's a fair chance that I might find my way into the next Honours List. Just a knighthood, of course.*

   Explore the significance of Arthur Birling in the play.

   You must refer to the context of the play in your answer.

5. **Inspector:** *But each of you helped to kill her. Remember that. Never forget it.*

   Explore the importance of responsibility in *An Inspector Calls.*

   You must refer to the context of the play in your answer.

6. **Eric:** *You're beginning to pretend now that nothing's really happened.*

   Explore the significance of Eric in the play.

   You must refer to the context of the play in your answer.

7. **Eric:** *I don't give a damn now whether I stay here or not.*

   Explore the relationship between Eric and Arthur Birling in *An Inspector Calls.*

   You must refer to the context of the play in your answer.

8. **Sheila:** *I behaved badly too. I know I did. I'm ashamed of it.*

   In what ways is regret important in the play?

   You must refer to the context of the play in your answer.

9. **Inspector:** *We don't live alone. We are members of one body. We are responsible for each other.*

   Explore the significance of the Inspector in the play.

   You must refer to the context of the play in your answer.

10. **Sybil:** *I consider I did my duty.*

    In what ways are attitudes to duty important in *An Inspector Calls*?

    You must refer to the context of the play in your answer.

11. **Arthur:** *Why you hysterical young fool – get back – or I'll –*

    Explore how characters come into conflict in *An Inspector Calls.*

    You must refer to the context of the play in your answer.

12. **Inspector:** *It would do us all a bit of good if sometimes we tried to put ourselves in the place of these young women counting their pennies in their dingy little back bedrooms.*

    Explore the importance of inequality in the play.

    You must refer to the context of the play in your answer.

13. **Arthur:** *[heartily] Nonsense! You'll have a good laugh over it yet. Look, you'd better ask Gerald for that ring you gave back to him, hadn't you? Then you'll feel better.*

    Sheila: *[passionately] You're pretending everything's just as it was before.*

    In what ways are differences between children and parents important in *An Inspector Calls*?

    You must refer to the context of the play in your answer.

14. **Arthur:** *'a man has to mind his own business and look after himself'.*

    Explore the importance of self-interest in the play.

    You must refer to the context of the play in your answer.

15. **Sheila:** *You and I aren't the same people who sat down to dinner here.*

    Explore the relationship between Sheila and Gerald in *An Inspector Calls*.

    You must refer to the context of the play in your answer.

16. **Arthur:** *Haven't I already said there'll be a public scandal – unless we're lucky – and who here will suffer from that more than I will?*

    Explore the importance of respectability in *An Inspector Calls*.

    You must refer to the context of the play in your answer.

17. **Arthur:** *[angrily] And you'd better keep quiet anyhow. If that had been a police inspector and he'd heard you confess –*

    Sybil: *[warningly] Arthur – careful!*

    Arthur: *[hastily] Yes, yes.*

    Explore the relationship between Arthur and Sybil Birling in the play.

    You must refer to the context of the play in your answer.

18. **Sheila:** *You began to learn something. And now you've stopped.*

    Explore the significance of Sheila in *An Inspector Calls*.

    You must refer to the context of the play in your answer.

Planning a Character Question Response

You must be able to: understand what an exam question is asking you and prepare your response.

How might an exam question on character be phrased?

A typical character question will read like this:

Sheila: *You began to learn something. And now you've stopped*.

Explore the significance of Sheila in *An Inspector Calls*.

You must refer to the context of the play in your answer.

[40 marks, including 8 AO4 marks]

How do I work out what to do?

The focus of this question is clear: Sheila and her role in the play.

'Explore', 'significance' and 'context' are important elements of this question.

For AO1, these words show that you need to display a clear understanding of what Sheila is like and how this relates to the themes of the play and Priestley's intentions.

For AO3, you need to be linking your interpretations to the play's social, historical or literary context.

You also need to remember to write in an accurate and sophisticated way to achieve your 8 AO4 marks for spelling, punctuation, grammar and expression.

How can I plan my essay?

You have approximately 50 minutes to write your essay.

This isn't long but you should spend the first five minutes writing a quick plan. This will help you to focus your thoughts and produce a well-structured essay.

Try to come up with five or six ideas. Each of these ideas can then be written up as a paragraph.

You can plan in whatever way you find most useful. Some students like to just make a quick list of points and then re-number them into a logical order. Spider diagrams are particularly popular; look at the example on the opposite page.

In Act 1 seems self-satisfied, obedient. Follows traditional expectations of class and gender. Forms contrast with Eva to create a comment on the middle classes.
'[very pleased with life and rather excited]'

Eventual guilt and understanding. Contrasts with parents to indicate that the younger generation can change. She understands the need for equality – link to world wars and possible revolution.
'began to learn something. And now you've stopped'
'Fire and blood and anguish'

Significance of Sheila

Has Eva sacked out of jealousy and anger. Middle class lack understanding of working class. Theme of morality and literary tradition of morality play. Lack of Christian values.
'She was a very pretty girl too – and that didn't make it any better'

Growing guilt but resists personal responsibility, reflection on middle class and attitude to socialism.
'I know I'm to blame – and I'm desperately sorry – but I can't believe – I won't believe – it's simply my fault that in the end – she committed suicide'

Initial lack of guilt, based on selfishness and capitalist values. Link to Priestley's socialist values and looking back to a different time (1912).
'I feel I can never go there again'

Summary

- Make sure you know what the focus of the essay is.
- Remember to interpret the character: what does he/she represent and what ideas are being conveyed?
- Try to relate your ideas to the play's context and Priestley's intentions.

Questions

QUICK TEST
1. What key skills do you need to show in your answer?
2. What are the benefits of quickly planning your essay?
3. Why do you need to take care with your writing?

EXAM PRACTICE
Plan a response to the following exam question:
Eric: *You're beginning to pretend now that nothing's really happened.*
Explore the significance of Eric in the play.
You must refer to the context of the play in your answer. [40 marks, including 8 AO4 marks]

Sheila: *You began to learn something. And now you've stopped.*

Explore the significance of Sheila in *An Inspector Calls*.

You must refer to the context of the play in your answer. [40 marks, including 8 AO4 marks]

Sheila is a significant character in An Inspector Calls as Priestley uses her to explore different themes and ideas about society [1].

The stage directions at the start of the play tell us that Sheila is self-satisfied. 'very pleased'. This is because she has just got engaged to Gerald. She does not work and she is obedient to her parents [2]. Priestley establishes Sheila as a typical daughter of a wealthy middle class family, following the social expectations at the time [3]. Sheila is like the opposite of Eva Smith who is poor and lonely and represents the working class at the time [4].

Sheila is important in the fate of Eva Smith because it is her fault that she is sacked from Milwards. Sheila has Eva sacked because she is in a bad mood and is jealous of her prettiness. This links to the play's theme of morality because Sheila's behaviour is cruel and unfair [5]. This shows how powerful the middle class were because shops needed their money so if they weren't happy the shop would do what they wanted, like sacking someone. It shows that some people really didn't care about what happened to the poor as long as they were happy themselves. The play is set in 1912 which was before the growth of socialism so Priestley is showing how different and bad things were in the past [6].

Sheila's guilt is an important part of the play. In comparison, her parents don't feel guilty. At first, she is selfish. This is because of the way she has been brought up and she moans that she can't go to Milwards anymore. 'I feel I can never go there again'. This shows she doesn't really feel guilty because she is thinking about herself and her life being worse rather than thinking about Eva Smith and how her life is over because she committed suicide. This is all about the values of the middle class being bad and how they just wanted more and more money [7].

Sheila is different by the end of the play because she has accepted full responsibility so Priestley is suggesting that people can change. Sheila is a contrast to Mr and Mrs Birling who still have the same attitudes so Priestley is saying that it is the younger generation who can change. 'You began to learn something. And now you've stopped'. She actually challenges her parents unlike at the start of the play. She has changed her values and wants her parents to do the same. She understands that it is important to care about the working class which shows that she has listened to what the Inspector has been saying and has learnt from him. She even repeats his words at the end as if she is becoming a bit socialist and will grow up better than her parents [8].

Sheila is a significant character. She shows what middle class people were like and how they could change [9].

1. A basic introduction that would benefit from more specific details about theme and intention. AO1

2. Clear ideas about Sheila's character that would benefit from stronger textual evidence. The quotation should be embedded. AO1

3. Relevant links to social context, although the actual time could be specified more clearly. AO3

4. Some interpretation of Sheila's character; Priestley's intentions could be explored more clearly. AO1/AO3

5. Some interpretation of Sheila's character with a fairly clear textual reference, although a fully embedded quotation would be better. AO1

6. Some good social context and acknowledgement of Priestley's intentions. It could be written in a more concise way with stronger links to the interpretation of Sheila's character. AO1/AO3/AO4

7. There is some interpretation of Sheila's character. This is supported by a clear, but not embedded, quotation. Context is used to develop the point. Writing is fairly accurate but sentences could be constructed more carefully and vocabulary could be more precise and sophisticated. AO1/AO3/AO4

8. Although it could be expressed in a more sophisticated way, there is clear interpretation of Sheila's character. It is supported by textual evidence, making good use of the exam question, and there is an attempt to integrate context. AO1/AO3/AO4

9. A basic conclusion that could be developed further. AO1/AO3

Questions

EXAM PRACTICE
Choose a paragraph from this essay. Read it through a few times then try to rewrite and improve it. You might:
- Improve the sophistication of the language or the clarity of expression.
- Replace a reference with a quotation.
- Ensure quotations are embedded in the sentence.
- Provide a clearer interpretation of Sheila's character.
- Link context to the interpretation more effectively.

Grade 7+ Annotated Response

A proportion of the best top-band answers will be awarded Grade 8 or Grade 9. To achieve this, you should aim for a sophisticated, fluent and nuanced response that displays flair and originality.

Sheila: *You began to learn something. And now you've stopped.*

Explore the significance of Sheila in *An Inspector Calls*.

You must refer to the context of the play in your answer. [40 marks, including 8 AO4 marks]

Sheila's significance in the play comes from the way Priestley uses the character to expose what he saw as the social irresponsibility of the middle classes as well as his hope that the younger generation can bring about change [1].

Sheila is initially presented as self-satisfied. Stage directions suggest she should be performed as '[very pleased with life and rather excited]', she does not work, is obedient to her parents, and her reference to them as 'Mummy and Daddy' indicates some childishness [2]. Priestley establishes Sheila as a typical daughter of a wealthy middle class family, following the social expectations of class and gender in the early twentieth century [3]. He does this to create a contrast with the life of working class girls like Eva Smith, later described as 'counting their pennies in their dingy little back bedrooms', to highlight the theme of inequality [4].

In terms of narrative, Sheila is important in the fate of Eva Smith as she has her sacked from Milwards. Drawing on the literary genre of the morality play, Sheila represents envy and wrath, 'She was a very pretty girl too – and that didn't make it any better', angrily complaining about Eva because she is jealous of her prettiness [5]. Priestley is pointing out the hypocrisy of a Christian society that does not follow Christian values. He is also showing the power of the middle classes and their lack of awareness of the precarious nature of working class lives. Writing in 1945, Priestley is looking back to a time before class structures began to erode, before the rise of socialism, and before the suffragette movement started to achieve female equality and sisterhood, allowing him to indirectly celebrate modern social changes [6].

Sheila's guilt, in comparison with her parents, is a key feature of the play. At first, she displays the selfishness that she has been brought up with, commenting that she cannot now return to Milwards. Her line, 'I feel I can never go there again', could also link to how her family values have been established through the capitalist ideals which were dominant at the start of the twentieth century. she is more concerned about material goods than the life of a young woman [7].

Although her guilt increases, her feelings are still mixed. When she says, 'I know I'm to blame – and I'm desperately sorry – but I can't believe – I won't believe – it's simply my fault', Priestley shows someone who is unwilling to take ultimate personal responsibility. She rejects the socialist values that the Inspector later summarises as 'We are members of one body. We are responsible for each *other', but it is also clear that she is struggling with her conscience in a way that her parents do not, introducing Priestley's assertion that the young can change [8].*

This change comes towards the end of the play where Sheila accepts full responsibility, in contrast to her parents. She points out that they, 'began to learn something. And now you've stopped', showing that she has changed her values and wants them to do the same. She will not return to her traditional and materialistic way of life, symbolised by the way she doesn't take back Gerald's ring. When she repeats the Inspector's words, 'Fire and blood and anguish', it underlines the idea that she has understood his warning about the need for social change. Priestley is making a reference to the two world wars, to highlight some of the reasons why class and inequality had begun to change, but it can also be an image of revolution with Sheila understanding the possible consequences of the wealthy not caring about the working classes [9].

Sheila is a significant character as Priestley uses her to illustrate middle class selfishness and prejudice, while also celebrating how people can change. Placed in 1912, Sheila can be seen as the start of a new generation, breaking away from traditional beliefs and ways of life in order to pave the way for the socialist politics that Priestley believed could improve Britain [10].

1. Clear overview of Sheila's character. AO1
2. Clear ideas about Sheila's character, supported by embedded textual evidence. AO1
3. Relevant links to social context. AO3
4. Interpretation of Sheila's character and what Priestley is trying to achieve. AO1/AO3
5. Interpretation of character integrated evidence and literary context. AO1/AO3
6. Interpretation of Sheila's importance integrated with social context and Priestley's intentions. AO1/AO3
7. This paragraph is a good example of accurate and varied writing to make points effectively. This includes fully punctuated sentence structures, integrated quotations, discourse markers, and precise complex vocabulary. AO4
8. The essay explores the importance of more complex aspects of Sheila's character. AO1
9. Clear interpretation of Sheila's character, supported by integrated textual evidence and historical context. Good use of quotation from the exam question. AO1/AO3
10. Clear conclusion linked to historical context and Priestley's intentions. AO1/AO3

> ## Questions

EXAM PRACTICE
Spend 50 minutes writing an answer to the following question:
Eric: *You're beginning to pretend now that nothing's really happened.*
Explore the significance of Eric in the play.
You must refer to the context of the play in your answer.
[40 marks, including 8 AO4 marks]
Remember to use the plan you have already prepared.

Planning a Theme Question Response

You must be able to: understand what an exam question is asking you and prepare your response.

How might an exam question on a theme be phrased?

A typical theme question will read like this:

Sybil: *As if a girl of that sort would ever refuse money.*

Explore the importance of attitudes towards the working class in *An Inspector Calls*.

You must refer to the context of the play in your answer.

How do I work out what to do?

The focus of this question is clear: different attitudes towards the working class in the play.

'Explore', 'importance' and 'context' are important elements of this question.

For AO1, these words show that you need to display a clear understanding of what attitudes are displayed towards the working class and how these relate to the themes of the play and Priestley's intentions.

For AO3, you need to be linking your interpretations to the play's social, historical or literary context.

You also need to remember to write in an accurate and sophisticated way to achieve your eight AO4 marks for spelling, punctuation, grammar and expression.

How can I plan my essay?

You have approximately 50 minutes to write your essay.

This isn't long but you should spend the first five minutes writing a quick plan. This will help you to focus your thoughts and produce a well-structured essay.

Try to come up with five or six ideas. Each of these ideas can then be written up as a paragraph.

You can plan in whatever way you find most useful. Some students like to just make a quick list of points and then re-number them in a logical order. Spider diagrams are particularly popular; look at the example on the opposite page.

Arthur and Gerald see working class as less important. They are just a resource for making money. Link to Priestley's beliefs in socialism and equality.
'it's my duty to keep labour costs down'

Arthur's distrust of the working class. Believes that equality will reduce the quality of life of the middle classes. 1912 labour strikes.
'If you don't come down sharply on these people, they'd soon be asking for the earth'

Inspector's view to contrast with Mr and Mrs Birling. Link to socialism and the creation of the welfare state. Sheila and Eric changing to agree with the Inspector shows hope for the younger generation. Reference to war shows working class have earned the right to equality.
'We don't live alone. We are members of one body. We are responsible for each other'
'fire and blood and anguish'

Importance of attitude towards the working class

Sybil's snobbery and prejudice. Belief that the working class are less important and lack morality. Irony of Christian charities without Christian values.
'As if a girl of that sort would ever refuse money'

Summary

- Make sure you know what the focus of the essay is.
- Remember to interpret the theme: how is it shown and what ideas are being conveyed?
- Try to relate your ideas to the play's context and Priestley's intentions.

Questions

QUICK TEST
1. What key skills do you need to show in your answer?
2. What are the benefits of quickly planning your essay?
3. Why do you need to take care with your writing?

EXAM PRACTICE
Plan a response to the following exam question:
Eric: *It's what happened to the girl and what we all did to her that matters.*
Explore the importance of guilt in *An Inspector Calls*.
You must refer to the context of the play in your answer.
40 marks, including 8 AO4 marks]

Grade 5 Annotated Response

Sybil: *As if a girl of that sort would ever refuse money.*

Explore the importance of attitudes towards the working class in *An Inspector Calls*.

You must refer to the context of the play in your answer.

[40 marks, including 8 AO4 marks]

Priestley shows different attitudes towards the working class through the characters in the play. Because he is writing about the past, Priestley is showing that things have changed [1].

In Act 1, Priestley uses Arthur and Gerald to represent traditional middle class attitudes towards the working class. Arthur believes that working class people are just there to work in factories. He thinks they shouldn't be paid much and that they need to be kept in their place [2]. He doesn't see them as the same as him. They are different to middle class businessmen who he thinks are better. This shows the need for equality which was starting when Priestley wrote the play [3].

Arthur also distrusts the working class. 'If you don't come down sharply on these people, they'd soon be asking for the earth'. He thinks they must be treated harshly or they will get above themselves and this would take money away from the middle class and cause their lives to be less luxurious. This links to the fact there were lots of labour strikes in 1912. Priestley was all for this because it could bring about equality. However, Birling is worried that the working class will start to be expensive and so his factory won't make as much profit. Gerald agrees with him which shows that this view was widespread among the middle class, it's not just Arthur thinking these things [4].

Priestley also uses Sybil to show that the middle class were really snobby about the working class and looked down on them like they were dirt and when she talks about Eva Smith she shows no sympathy or respect. It's like she thinks they are less important, almost like animals rather than people [5]. Sybil's view is wrong because the working class are good people and Eva Smith seemed really nice. Priestley shows Sybil up because she runs a charity that is linked to the church but she doesn't actually show any values of a religious person, she just judges people even though it's her who needs to be judged for looking down on innocent people like Eva Smith [6].

Priestley also includes the viewpoint of the Inspector as it's different to the viewpoint of the Birlings and suggests that the working class need to be treated better. This is the same as Priestley's socialist values which were around when the play was written. 'We don't live alone. We are members of one body. We are responsible for each other'. This is the attitude that society needs to unite and look after its less fortunate groups. In 1945 they set up the welfare state [7].

Overall, the play starts by showing traditional attitudes towards the working classes. Then it shows the Inspector's view which is better. Instead of being selfish and seeing the working class as slaves, we should be compassionate and try to make everyone equal [8].

1. Basic introduction. Ideas about attitudes could be more specific. There is a useful but vague link to historical context. AO1/AO3

2. A fairly clear interpretation of attitudes towards the working class. It would benefit from stronger textual evidence. AO1

3. Relevant but rather vague links to social context. AO3

4. Some interpretation of attitudes but ideas could be more clearly expressed. Strong evidence is included but could be integrated into the sentence. There is a generally successful attempt to link interpretation to historical context and consideration of what Priestley is trying to achieve. AO1/AO3

5. Some development of exploration through a new viewpoint, although the quotation from the exam question could be used to provide evidence. AO1

6. Writing is fairly accurate but sentences could be constructed more carefully. Vocabulary could be more precise and sophisticated, and there is too much repetition. AO4

7. Interpretation of a different viewpoint to add depth to the essay. Evidence is included but could be integrated more effectively. There are links to Priestley's aims and social/historical context but some of this is tagged on rather than being linked to the interpretation. AO1/AO3

8. Clear but basic conclusion that identifies the message of the play. AO1

 Questions

EXAM PRACTICE

Choose a paragraph from this essay. Read it through a few times then try to rewrite and improve it. You might:

- Improve the sophistication of the language or the clarity of expression.
- Replace a reference with a quotation.
- Ensure quotations are embedded in the sentence.
- Provide a clearer interpretation of attitudes towards the working class.
- Link context to the interpretation more effectively.

Grade 7+ Annotated Response

A proportion of the best top-band answers will be awarded Grade 8 or Grade 9. To achieve this, you should aim for a sophisticated, fluent and nuanced response that displays flair and originality.

Sybil: *As if a girl of that sort would ever refuse money.*

Explore the importance of attitudes towards the working class in *An Inspector Calls*.

You must refer to the context of the play in your answer.

[40 marks, including 8 AO4 marks]

Priestley criticises a lack of empathy for the working class through the attitudes of the Birlings, contrasting their viewpoint with those of the Inspector. Writing in 1945, at the time of a new socialist government, Priestley looks back to 1912 to show the importance of social change.

In Act 1, Priestley uses Arthur and Gerald to represent traditional middle class attitudes towards the working class. When Arthur says, 'it's my duty to keep labour costs down', he represents the viewpoint that the working class are there to be exploited and need to be kept in their place [2]. He sees them as a different group of people, in comparison to middle class businessmen, and Priestley uses this to assert the importance of new socialist ideas about equality [3].

Arthur is also used to show the idea that the working class are distrusted. He argues that, 'If you don't come down sharply on these people, they'd soon be asking for the earth', implying that they should be treated harshly in order to safeguard the privileges of the middle class. Priestley is drawing on the labour strikes of 1912 and using the audience's dislike of Birling to suggest the value of industrial action in order to achieve greater rights and fair pay. During Birling's speeches, Gerald's agreement is used to make it clear that the audience are listening to a majority view rather than an isolated opinion [4].

As well as showing how the working class were viewed as an exploitable resource, Priestley uses Sybil to convey the prejudice and snobbery that many of the middle class displayed towards the working class. In Act 2, she says of Eva Smith, 'as if a girl of that sort would ever refuse money!', suggesting she has no sympathy or respect for the working class because they are of a different breed [5]. This implies the working class were seen as lacking proper morals and Priestley uses the irony of Sybil's lack of Christian values to point out the hypocrisy of church-based charities (like the one Sybil chairs) that claimed to help the poor but actually sat in judgement on them [6].

Through the Inspector, Priestley challenges these attitudes and offers an alternative view that the working class need to be treated better. Reflecting the socialist views that were gathering popularity in the early 20th century, the Inspector says, 'We don't live alone. We are members of one body. We are responsible for each other', emphasising the need for society to unite and look after its less fortunate groups. This notion could be particularly seen, at the time the play was written, in the establishing of the welfare state [7]. The way in which Sheila and Eric change their attitudes towards the working class to show compassion is used by Priestley to represent hope for the future and the importance of educating and mobilising the younger generation [8].

Overall, the play establishes traditional attitudes towards the working classes, uses the story of Eva Smith to show their damaging consequences, and then offers a better alternative. First performed after World War Two, the Inspector's emotive reference to 'fire and blood and anguish' reminds the audience of the role the working classes played in maintaining the wealth and freedom of the rest of society, thereby highlighting the rightfulness of social equality to the audience, not just to the characters on stage [9].

1. Clear overview of attitudes towards the working class, incorporating historical context. AO1/AO3

2. Clear interpretation of attitudes towards the working class, supported by embedded textual evidence. AO1

3. Relevant links to social context. AO3

4. Interpretation of attitudes, supported by an embedded quotation, integrated with historical context and consideration of what Priestley is trying to achieve. AO1/AO3

5. Good development of exploration through a new viewpoint, making effective use of the quotation from the exam question. AO1

6. This paragraph is a good example of accurate and varied writing to make points effectively. This includes fully punctuated sentence structures, integrated quotations, discourse markers, and precise complex vocabulary. AO4

7. Interpretation of a different viewpoint to add depth to the essay. Textual evidence, Priestley's aims and social/historical context are well integrated. AO1/AO3

8. Reference to other characters is used to develop ideas further. AO1

9. Clear conclusion, supported by another embedded quotation, is linked to historical context and Priestley's intentions. AO1/AO3

> **Questions**

EXAM PRACTICE
Spend 50 minutes writing an answer to the following question:
Eric: *It's what happened to the girl and what we all did to her that matters.*
Explore the importance of guilt in *An Inspector Calls*.
You must refer to the context of the play in your answer.
[40 marks, including 8 AO4 marks]
Remember to use the plan you have already prepared.

Glossary

Atmosphere – the mood or emotion in a play.

Capitalist – a right-wing political belief in individual gain, through hard work and a focus on profit.

Cliffhanger – ending an act with a shock or problem.

Climax – the most intense part of an act or the whole play.

Compassion – sympathetic concern.

Confession – a statement admitting guilt; (in religion) acknowledging sin.

Conflict – a disagreement or argument.

Conscience – a person's sense of right and wrong.

Convention – an expected way in which something is usually done.

Depression (economic) – when less spending causes businesses to close and unemployment to rise, leading to even less spending.

Dominance – having power and influence over others.

Dramatic irony – when the audience of a play is aware of something that a character on stage isn't.

Emotive – creating or describing strong emotions.

Etiquette – a code of polite behaviour in society.

Exploited – used in an unfair way (for the user's benefit).

Flashback – a dramatic device taking a play back in time.

Foreshadow – hint at future events in the play.

Hierarchy – a ranking according to status and power.

Hoax – a humorous or nasty trick.

Hypocrisy – claiming to have better standards or behaviour than is true.

Illegitimate – a person whose parents aren't married.

Imagery – words used to create a picture in the imagination.

Industrial Revolution – a period when machinery started to be used much more in production, leading to big factories rather than small businesses.

Inferior – being less important than others.

Irony – something that seems the opposite of what was expected; deliberately using words that mean the opposite of what is intended.

Justified – something done for a good, fair reason.

Juxtapose – place two contrasting things side by side.

Noun – an object or thing.

Objective – not influenced by personal feelings or opinions.

Omniscient – all-knowing.

Parallelism – repeating similar word orders or sentence constructions to emphasise an idea.

Passive voice – where the subject of a sentence has something done to it (in contrast to active voice, where the subject of the sentence does something).

Patriarch – the male head of a family.

Prejudiced – being against someone because of an already-formed idea, rather than fact or experience.

Provincial – living beyond a major city, suggesting a lack of sophistication.

Respectable – seen as good and proper by society.

Sin – an immoral act.

Strike – when workers refuse to work as a form of protest.

Suburb – a (usually quite well-off) residential area on the outskirts of a town or city.

Superior – being better or more important than others.

Symbolise – when an object or colour represents a specific idea or meaning.

Tension – a feeling of anticipation, discomfort, or excitement in a play.

Traditional – long-established or old-fashioned.

Verb – a doing or action word.

Welfare state – a system in which the government looks after the poor and vulnerable in society.

Womanising – having various casual affairs with women.

Answers

Answers

Pages 4–5
Quick Test
1. Arthur and Sybil, and their children, Eric and Sheila.
2. Sheila's engagement to Gerald Croft.
3. To investigate the suicide of a young girl, Eva Smith.
4. Eva worked at Arthur's factory and was sacked after helping to organise a strike in demand of higher wages.
5. Arthur thinks the sacking was justified and Gerald agrees. Eric and the Inspector show sympathy for Eva's situation.

Exam Practice
Answers might explore how the Birlings' happiness contrasts with Eva's death, suggesting the differences between middle class and working class life; the Inspector's interruption could represent the need for the middle classes to be more aware of the people around them.

Pages 6–7
Quick Test
1. Sheila feels sorry for Eva.
2. She thought Eva was laughing at her. She was in a bad mood and jealous of her looks.
3. She's upset about Eva, which shows some guilt, but she's also upset for herself and how her evening has been spoiled, which highlights the selfish side of her character.
4. Gerald.
5. She was having an affair with Gerald.

Exam Practice
Answers might explore Sheila's unwillingness to accept responsibility and Gerald's attempt to avoid responsibility, linking to selfish attitudes of the middle class and the lack of interest in equality.

Pages 8–9
Quick Test
1. To hide his involvement from the Inspector and possibly hide further details from Sheila.
2. She refers to her lower-class status.
3. She acts in a grand and superior manner.
4. Sybil is behaving how the others behaved before the Inspector exposed them.
5. It started as him innocently helping her but then they became lovers.

Exam Practice
Answers might explore Sybil's prejudice against Eva, representing traditional attitudes held by the middle classes; Gerald and the Inspector convey the inequality that the working class suffered, linking to the need for sympathy and welfare.

Pages 10–11
Quick Test
1. She was using the false name Mrs Birling.
2. No, she thinks it was justified.
3. Sheila thinks it was horrible; Arthur worries it might make them look bad.
4. She thought Eva was lying about the lover offering money but her refusing because she thought it was stolen.
5. Sheila works out that Eric is the father of Eva's child before her parents do.

Exam Practice
Answers might explore Sybil's prejudice against the working class and the irony of her lack of Christian values despite running a Christian charity; Sheila could represent the need to change.

Pages 12–13
Quick Test
1. He gets drunk and aggressive.
2. She knew he didn't love her.
3. Murdering Eva along with her own grandchild.
4. He wants to cover up the evening's revelations.
5. That we are all responsible for each other in society and change must come.

Exam Practice
Answers might explore the traditional patriarchal power shown by Arthur, as well as his abuse of economic power; the Inspector presents the idea of using such economic power to solve inequality.

Pages 14–15
Quick Test
1. They want to avoid any scandal.
2. Their sense of responsibility for Eva's death.
3. Relieved with a sense that things can go back to normal. They make a joke of it all.
4. They still know that, individually, they ruined a girl's life.
5. A phone call reveals that a girl has been found dead and an inspector is coming to interview them.

Exam Practice
Answers might explore how Sheila's viewpoint contrasts with her father's, suggesting that change will come through the younger generations.

Pages 16–17
Quick Test
1. There's only one set, the play takes place (without changes in time between the Acts) over one evening and characters are generally focussed on one at a time.
2. Use of cliffhangers to raise tension.
3. Character confessions lower the tension whilst family conflict raises it.
4. His social message about having personal and social responsibility.

Exam Practice
Answers might explore how mystery is used to gradually deconstruct and criticise the values of this middle class family.

Pages 18–19

Quick Test

1. Gerald Croft.
2. The Birlings.
3. Working class.
4. Capitalist. He is a businessman and focuses on making money.

Exam Practice

Answers might explore how the pause before Lady Croft's name, the use of her title and the defence of her snobbery shows Arthur's respect for the upper classes and his wish to be like them. As a capitalist, he believes in the importance of money and status. We also know, from Act 1, that he hopes a marriage between the Crofts and the Birlings will help to improve his business and his social position. He points out that the Crofts are from an old, land-owning family, reminding us that Birling represents the new wealth of the middle classes benefitting from the Industrial Revolution. He boasts about the possibility of the Honours List and a knighthood, showing that he sees a title as a way to become equal in status if not breeding.

Pages 20–21

Quick Test

1. The Inspector represents Priestley's socialist views about looking after everyone in society.
2. Fighting alongside men of all different classes may have influenced Priestley's belief in equality.
3. They established the welfare state, the National Health Service and new housing plans where people of different classes would live side by side.

Exam Practice

Arthur's noun phrase 'steadily increasing prosperity' would seem ironic given the economic depression that followed the First World War. The capitalist comments would also have conflicted with the growth of trade unions and the popularity of socialism after the Second World War, signalled by the Labour Party's landslide election win in 1945.

Pages 22–23

Quick Test

1. There was no social security or free health care.
2. Home, workplace, church and government.
3. Women were seen as unequal to men and were expected to stay at home and bring up a family, they had less access to jobs and promotion and had fewer legal rights.

Exam Practice

Answers might explore how Arthur is critical of sex outside of marriage, which links to Christian values; Eric's ironic adjective 'respectable' when discussing Arthur's friends who have affairs suggest that the double standards of the middle classes, often ignoring their own immoral behaviour to focus on criticising others; how this is emphasised when Arthur interrupts Eric to keep him quiet; the implication that women are less moral (and less important) than men.

Pages 24–25

Quick Test

1. The furniture, the glasses, the cigar box.
2. It isn't very cosy or homely.
3. The lighting is warm, the props suggest quality or luxury and the characters are sitting down.
4. The lighting becomes brighter and harder, to emphasise that they are being interrogated and exposed by the Inspector.

Exam Practice

Answers might focus on the look of the furniture and how it suggests both wealth and self-confidence, the size of the room and the expensive props such as champagne glasses or how the wealthy characters are seated whilst the maid clears up after them.

Pages 26–27

Quick Test

1. People in society should care for each other.
2. He doesn't show respect for the Birlings' higher status.
3. Goole/ghoul suggests mystery and perhaps the idea that the Inspector isn't 'real'.
4. He reveals their secrets and their bad behaviour, challenging their view of themselves.

Exam Practice

Answers might explore how the Inspector represents Priestley's belief in socialism, standing up to the traditional power of the middle classes and encouraging equality.

Pages 28–29

Quick Test

1. To establish his dominant character and his status as the head of the family.
2. 'hard-headed business man'.
3. He hopes it will link his factory with Croft's factory and improve his social status.
4. Because we know there was a war two years later, it shows he isn't as clever as he thinks and doesn't really understand society/the world.

Exam Practice

Answers might explore how Arthur represents capitalism and the traditional views of the middle classes, seeing the working class as a resource to be exploited; Priestley suggests that his wish to pass on these views to the younger generation should be resisted.

Pages 30–31

Quick Test

1. She usually takes Arthur's lead & follows his instructions.
2. Because she believes she lives a moral life and she thinks her wealth and social status make her more important. She draws on Arthur's status to support this.
3. She's a hypocrite. She pretends to care but just likes to judge.
4. She treats him like a child, doesn't understand his feelings and is unaware of his problems.

Exam Practice

Answers might explore how Sybil represents prejudice and snobbery against the working classes; she feels that their problems are their own and sees no social responsibility to help; she sees herself as better and believes, despite the evidence, that her children are better simply because they are middle class.

Pages 32–33

Quick Test

1. She is the daughter of a wealthy businessman and is dating a man from an upper-class family.
2. She thinks she is being laughed at, is in a bad mood and is jealous of Eva's prettiness.
3. She begins to accept her responsibility and sympathise with Eva's life.
4. Sheila realises that Eric was the father of Eva's child.

Answers

Quick Test

1. Public scandal.
2. Eric and Sheila accept their guilt and know they need to change; their parents refuse to accept any responsibility.
3. He asks her to take the engagement ring again.
4. Their lack of closeness and understanding has now been exposed and their relationship has deteriorated.

Exam Practice

Answers might explore how the older generation are unwilling to change, wanting to cling on to tradition and power; their selfishness is contrasted with the willingness of the young to see the importance of different, more compassionate values.

Page 42–43

Quick Test

1. His wish for more money and status.
2. He used Eva for sexual gratification and he drinks too much.
3. Eva's prettiness.
4. When she refers to her status or her belief that she has the right to judge others.

Exam Practice

Answers might explore how Priestley exposes the hypocrisy and immorality of the middle class; despite seeing themselves as better, they display jealousy, pride, and desire; they don't care for the life of an individual, they care for themselves.

Pages 44–45

Quick Test

1. Eric and Gerald.
2. It interrupts Arthur's promotion of self-interest.
3. Her life is ruined because people followed their own interests not hers.
4. He points out the irony of them, given what the Inspector exposed.

Exam Practice

Answers might explore how Arthur's traditional capitalist values lack compassion, whereas the Inspector's socialist values suggest a responsibility for society and the importance of equality.

Pages 46–47

Quick Test

1. Offer financial, emotional and educational support.
2. It's nonsense.
3. She does charitable work but it is to make her look good; she looks down on the girls who ask for help.

Exam Practice

Answers might explore how the Inspector represents ideas of empathy and helping others; he embodies ideas of the new welfare state at the time Priestley was writing; in contrast, Arthur thinks such views are ridiculous and self-interest is more important.

Pages 48–49

Quick Test

1. Sheila, Eric and – to some extent – Gerald.
2. 'Confess'.
3. Deny involvement and blame others.

Exam Practice

Answers might explore how the Inspector encourages the characters to take personal responsibility; the younger generation accept this while the older generation avoid it, considering themselves beyond judgement.

Pages 32–33 continued

Exam Practice

Answers might explore how Sheila represents a change of values; despite an initial unwillingness to accept responsibility, she recognises her guilt and encourages the older generation to change.

Pages 34–35

Quick Test

1. He isn't paired with another character (like Arthur/Sybil and Sheila/Gerald) so seems isolated, he is the last to speak and he behaves strangely.
2. He thinks they had every right to go on strike and his father treated them badly.
3. He is ashamed.
4. He dislikes them and feels they've never tried to understand him.

Exam Practice

Answers might explore how Eric represents a change in values, turning against the traditions and hypocrisy of middle class society to see the need for equality and welfare.

Pages 36–37

Quick Test

1. He refers to himself as 'respectable' and has hidden his affair from his fiancée.
2. Arthur appreciates Gerald as a businessman but is more interested in his social status as a member of the upper classes.
3. He gave her food and found her somewhere to live.
4. He started an affair with her and then broke her heart.

Exam Practice

Answers might explore how Gerald represents middle class arrogance and the wish to avoid social and personal responsibility; his change of values is only temporary.

Pages 38–39

Quick Test

1. The working class and women.
2. Arthur sacks her for wanting more money, Sheila has her sacked unfairly, Gerald helps her but then makes her his lover and breaks her heart, Eric gets her pregnant and Sybil refuses to help her.
3. She is hardworking (as opposed to Sheila and Eric) and she refuses to take Eric's money once she thinks it is stolen.
4. Edna's role as a servant shows that the Birlings are happy to employ people to fulfil their needs; the idea of having some below them, in a position of inequality, does not trouble them.

Exam Practice

Answers might explore how the life of working class people is presented by the Inspector in order to emphasise the need for equality and the importance of socialist values.

Pages 50–51

Quick Test

1. Because she is working class and a woman.
2. Their home, their money, Eric's education and Arthur's status.
3. Being able to keep labour costs down means that the family business makes more money.

Exam Practice

Answers might explore how the characters display contrasting views of inequality; Arthur sees it as a good thing, believing in capitalist values of profit and self-interest, whereas Sheila recognises that the working class are more than just a resource - they are individuals; it can be argued that Gerald takes advantage of his economic superiority to fulfill his sexual desires.

Pages 52–53

Quick Test

1. They expect to be in charge.
2. They follow social expectations of respect for their elders.
3. To show that the older generation have old-fashioned views that are harmful to society as a whole, whereas the young people can change and contribute to a better future.

Exam Practice

Answers might explore how the play begins with the younger generation showing traditional obedience to the older generation, but later shows the values of the older generation being challenged.

Pages 54–55

Quick Test

1. By setting the play in the past, we can see that Arthur isn't as wise as he thinks.
2. It intensifies the situation and the emotions or tension being created.
3. To emphasise particular events and their consequences.
4. Sheila and Eric admit and want to remedy the past, whereas Arthur and Sybil want to deny and conceal the past.

Exam Practice

Answers might explore how the play focuses on the consequences of actions and the urgent need for social change; written in 1945 and looking back to 1912, there is the constant sense of changes to come and the effects of the two world wars.

Pages 56–57

Quick Test

1. At the start of the play with the happy engagement party.
2. Telling Sheila she'll have to get used to her husband putting work before his marriage.
3. Women appear to value faithfulness and romance more than men. Gerald has an affair and then dumps Eva, Arthur appears to think unfaithfulness is easy to forgive and Eric uses Eva for his sexual pleasure.
4. Eric says he didn't love her, she was just good fun.

Exam Practice

Answers might explore how men are shown to exploit women in the name of love, seeing them as trophies or sexual objects; there is a lack of actual love in the play, with the men often presenting economic power as a substitute for love.

Pages 60–61

Practice Questions

Use the mark scheme on page 80 to self-assess your strengths and weaknesses. The estimated grade boundaries are included so you can assess your progress towards your target grade.

Pages 62–63

Quick Test

1. Understanding of the whole text, use of textual evidence, awareness of the relevance of context, a well-structured essay and accurate writing.
2. Planning focuses your thoughts and allows you to produce a well-structured essay.
3. There are 8 marks available for using a range of accurate vocabulary, punctuation, and spelling in order to convey your ideas clearly and effectively.

Exam Practice

Answers might explore Eric's initial portrayal as a lazy and spoiled member of the middle class; his lack of morality undermines the family's idea that they are better than the working class; he is used to expose the hypocrisy of the middle class through the abuses of power and money that he reveals; Eric is also a symbol of hope that the younger generation can change; he begins to display socialist values, caring about the lives of others and wanting to stop inequality.

Pages 66–67 and pages 72–73

Exam Practice

Use the mark scheme on page 80 to self-assess your strengths and weaknesses. Work up from the bottom, putting a tick by things you have fully accomplished, a ½ by skills that are in place but need securing, and underlining areas that need particular development. The estimated grade boundaries are included so you can assess your progress towards your target grade.

Pages 68–69

Quick Test

1. Understanding of the whole text, use of textual evidence, awareness of the relevance of context, a well-structured essay and accurate writing.
2. Planning focuses your thoughts and allows you to produce a well-structured essay.
3. There are 8 marks available for using a range of accurate vocabulary, punctuation, and spelling in order to convey your ideas clearly and effectively.

Exam Practice

Answers might explore the Inspector's attempt to get the Birlings to acknowledge their guilt, linking to socialist ideas of responsibility and compassion; the older generation refuse to accept guilt, seeing the working class as below them and wanting to focus on capitalist values of profit and personal gain; the younger generation resist guilt, linking to their upbringing, but eventually accept their wrongs; this shows hope for the future and the importance of breaking away from tradition and selfishness.

Answers

Grade	AO1 (16 marks)	AO3 (16 marks)	AO4 (8 marks)
6–7+	A convincing, well-structured essay that answers the question fully. Clear interpretation of a range of different aspects of the play. Quotations and references are well-chosen and integrated into sentences. The response covers the whole play.	Exploration is linked to specific aspects of the play's contexts to show a detailed understanding. Context is integrated with interpretation.	Consistent high level of accuracy. Vocabulary and sentences are used to make ideas clear and precise.
4–5	A clear essay that always focuses on the exam question. Some interpretation of different aspects of the play. Quotations and references support ideas effectively. The response refers to different points in the play.	References to relevant aspects of context show a clear understanding.	Good level of accuracy. Vocabulary and sentences help to keep ideas clear.
2–3	The essay has some good ideas that are mostly relevant. There is an attempt to interpret a few aspects of the play. Some quotations and references are used to support the ideas.	Some awareness of how ideas in the play link to its context.	Reasonable level of accuracy. Errors do not get in the way of the essay making sense.